Sediment Transport in Coastal Waters

Sediment Transport in Coastal Waters

Special Issue Editor

Sylvain Ouillon

MDPI • Basel • Beijing • Wuhan • Barcelona • Belgrade

MDPI

Special Issue Editor
Sylvain Ouillon
Institut de Recherche pour le Développement
Université de Toulouse
France

Editorial Office
MDPI
St. Alban-Anlage 66
4052 Basel, Switzerland

This is a reprint of articles from the Special Issue published online in the open access journal *Water* (ISSN 2073-4441) from 2016 to 2018 (available at: https://www.mdpi.com/journal/water/special_issues/coastal_waters)

For citation purposes, cite each article independently as indicated on the article page online and as indicated below:

LastName, A.A.; LastName, B.B.; LastName, C.C. Article Title. *Journal Name* **Year**, *Article Number, Page Range.*

ISBN 978-3-03897-844-2 (Pbk)
ISBN 978-3-03897-845-9 (PDF)

Cover image courtesy of Saskia Keesstra.

Contents

About the Special Issue Editor

Sylvain Ouillon is physical coastal oceanographer, senior scientist at the Laboratoire d'Etudes Géophysiques et d'Océanographie Spatiale (LEGOS, University of Toulouse, France). He belongs to the Institute of Research for Development (IRD). Together with his research group, and through national and international cooperation, he analyses sediment transport from the integration of in situ data, remote sensing and numerical models at different time and space scales: large scales at the river basin or in river plumes to quantify sediment balance and its variability, local scales to study hydrosedimentary processes such as the dynamics of estuarine turbidity maxima. He also aims at developing new products on suspended particles from satellite data. He participated to more than twenty coastal sea cruises. Because of the strong interactions between living organisms and non-living matter in coastal and estuarine environments, he is convinced of the necessity to mix different scientific approaches to improve our knowledge of sediment transport mechanisms in these natural environments, and thus supported and participated to many pluridisciplinary studies with geochemists, microbiologists, biogeochemists. He has (co)supervised eight PhD students. Besides from research, he is leading the Master degree in Water-Environment-Oceanography of the University of Science and Technology of Hanoi (USTH, Vietnam) from 2011. Formation: Ph.D. in 1993, Institut National Polytechnique de Toulouse; Civil Engineer, ENSEEIHT, 1989.

Preface to "Sediment Transport in Coastal Waters"

Globally, while land erosion has increased in the last decades, sediment input to the ocean has decreased. Sediment transport and distribution at the land-ocean interface has huge impacts on morphodynamics of estuaries, deltas and coastal zones, on water quality (and related issues such as aquaculture), on navigation and harbor capability, on recreation areas, etc. The interface of 440,000 km long coastline in the world is subject to global change, with an increasing human pressure (land use, buildings, sand mining, dredging) and increasing population. Improving our knowledge on involved mechanisms and processes, monitoring the evolution of sedimentary stocks and anticipating changes in littoral and coastal zones is essential for this purpose.

Scientific objectives must be achieved to deepen our knowledge on processes, to refine sediment budgets (bedload and suspension), and to improve our observation and modelling capacities. Sediment transport which is mainly driven by varying dynamical forcings (currents, tides, wind, waves, turbulence, stratification, density currents) in the estuary, in the region of freshwater influence, on the shelf or in canyons may also be affected by chemical and biological processes.

Scientific research on sediment dynamics in the coastal zone and along the littoral zone has evolved considerably over the last decades. It benefits from a technological revolution that provides the community with cheaper or free tools for in situ study (e.g., sensors, gliders), remote sensing (satellite data, video cameras, drones) or modelling (open source models). These changes favour the transfer of developed methods to monitoring and management services, in tune with the needs of society.

The special issue of Water on "Sediment transport in coastal waters" aims at synthesizing and illustrating the current revolution in the scientific research related to coastal and littoral hydrosedimentary dynamics. The thirteen scientific papers published in this special issue reflect the diversity of concerns on which research in coastal sediment transport is based, and current trends—topics and preferred methods—to address them. I sincerely thank the authors and peer reviewers, as well as the journal's staff, for their contributions to this Special Issue.

Sylvain Ouillon
Special Issue Editor

water

MDPI

Editorial

Why and How Do We Study Sediment Transport? Focus on Coastal Zones and Ongoing Methods

Sylvain Ouillon

LEGOS, Université de Toulouse, IRD, CNES, CNRS, UPS, 14 Avenue Edouard Belin, 31400 Toulouse, France; sylvain.ouillon@ird.fr; Tel.: +33-56133-2935

Received: 22 February 2018; Accepted: 22 March 2018; Published: 27 March 2018

Abstract: Scientific research on sediment dynamics in the coastal zone and along the littoral zone has evolved considerably over the last four decades. It benefits from a technological revolution that provides the community with cheaper or free tools for in situ study (e.g., sensors, gliders), remote sensing (satellite data, video cameras, drones) or modelling (open source models). These changes favour the transfer of developed methods to monitoring and management services. On the other hand, scientific research is increasingly targeted by public authorities towards finalized studies in relation to societal issues. Shoreline vulnerability is an object of concern that grows after each marine submersion or intense erosion event. Thus, during the last four decades, the production of knowledge on coastal sediment dynamics has evolved considerably, and is in tune with the needs of society. This editorial aims at synthesizing the current revolution in the scientific research related to coastal and littoral hydrosedimentary dynamics, putting into perspective connections between coasts and other geomorphological entities concerned by sediment transport, showing the links between many fragmented approaches of the topic, and introducing the papers published in the special issue of *Water* on "Sediment transport in coastal waters".

Keywords: sediment transport; cohesive sediments; non cohesive sediments; sand; mud; coastal erosion; sedimentation; morphodynamics; suspended particulate matter; bedload

1. Introduction

On a global scale and over long periods, dissolved and particulate elemental fluxes are essential to improve our knowledge of geochemical cycles [1–4], in particular the carbon cycle, which plays a major role for the climate. Coastal areas bury 80% of continental organic carbon [5] and account for one quarter of global ocean primary production [6]. At a local or regional scale, the geomorphodynamics of river basins and coastal zones resulting from erosion, transport and deposition of particles, and the associated transport of nutrients and contaminants, require investigating sediment transport [7–11]. The interface of 440,000 km long coastline in the world is subject to global change, with an increasing human pressure (land use, buildings, sand mining, dredging) and increasing population. Improving our knowledge on involved mechanisms and processes, monitoring the evolution of sedimentary stocks and anticipating changes in littoral and coastal zones is essential for this purpose.

Particles are essentially grains of rocks detached from their matrix by the action of climate by air and water, or by tectonic movements, or particles of biogenic origin (calcareous tests, shells, corals, etc.) or from marine origin such as calcarenites [12–14]. Cohesive sediments (able to flocculate with each other, <20–30 μm in size, see [15,16]) have properties that coarser, non-cohesive sediments do not have (as reminder, mud is made of particles <62.5 μm or 1/16 mm in diameter and sand of particles between 62.5 μm and 2 mm in diameter). Cohesive sediments and biological material may stick and form aggregates, widely spread in estuaries and coastal environments. The spatial distribution of sediments, available for motion, strongly depends on the locally dominant energy source. Globally, wave energy delivered to the coast (2.5×10^9 kW) is in the order of tidal energy (2.2×10^9 kW) [17]. Fine sediments accumulate in tidal-dominating coasts, while wave-dominating coasts are mostly sandy or rocky. Coastal sediment dynamics that occur in estuaries and on continental shelves under multiple forcing (river, tide, wind, waves) may be distinguished from the littoral sediment dynamics, where the particles motion mainly results from the wave action, i.e., at depths less than their half-wavelength. Particle flows, which are irreversible in fluvial environments (from upstream to downstream), can be directed to the land or to the sea in estuaries and along the littoral.

The scientific questions associated with sediment transport are of three types: how much? how? what are the consequences? The first aims to improve the balances between geomorphological entities or compartments from the sources to the sinks. The second aims to better understand the mechanisms and processes involved in these movements and in successive transport phases, with some area-specific or sediment type-specific processes. The third crosses and extends the first two; its purpose is to quantify or even to anticipate the impacts of sediment dynamics on environments and ecosystems. These questions arise from upstream to downstream, with some variations, from continental sources (mostly bedrock) to the abyss, through the alluvial plains, deltas, coasts, continental shelves. Additionally, sediment transport in coastal waters must be studied with consideration of other domains for closure purpose.

The approach of these questions has recently evolved on two levels (which we will review): tools and methods, and societal emergency. Sediment transport, for a long time considered for civil engineering (polders, ports), has now very sensitive issues for society, by the growing needs for sand resources, for environmental issues (e.g., turbidity), or for safety, through extreme events that violently affect inhabited coastlines. Land less than 100 km from the coast represents 20% of the land area, home 60% of the major cities and 41% of the world population. Among the domains affected by erosion and transport, the societal concern about coastlines is therefore keen. Social issues lead to political awareness, funding, targeting of researcher profiles and thematic conversion of some researchers. As research funding is increasingly driven by policy (through project funding, distributed by state agencies or supranational as in Europe), sediment dynamics has become a field of finalized research, which can benefit from this new economy of knowledge.

Ultimately, scientific research in this area is constantly evolving in its objectives, methods, approaches, and relationship to society. In such context of very rapid evolution, this paper reviews recent changes to put into perspective connections between coasts and other geomorphological entities concerned by sediment transport, past and current research tracks, new methods and new tools. It aims to show the links between many fragmented approaches of the same scientific field. It finally introduces the 12 papers published in the special issue of *Water* devoted to this theme in 2016–2017.

2. Sediment Balances and Their Consequences

2.1. Sediment Routing and Forcing

Erosion and transport of unconsolidated sediments match energetic forcing. Rocks and soil breakdown results from physical weathering due to heat, water, frost and pressure, and to chemical weathering, then particles are eroded, i.e., removed, and transported. While rainfall and surface runoff are driving soil erosion, the fluvial stream (characterised by bottom shear and turbulence level) is driving stream erosion and is the main transport agent of all collected material. Along the sediment routing or "source-to-sink" systems, from mountains to abysses [18], fragments and grains of rocks undergo many episodes of resuspension, transport and deposition. From the basin head to the estuary, the decreasing slope of the watercourses induces a grain sorting which limits the transport to finer particles. Where tidal influence starts, particles enter the estuary or the delta. The "littoral" is classically considered as the coastal area where the wave action is sensitive. From the distributed and deposited particles on the continental shelf, the fraction brought back to the coast under the action of waves and currents design the shoreline (beaches, mudflats, delta, etc.), while another will be evacuated towards the abyss through canyons during intense energetic episodes. In the end, rocks deposited at the bottom become remineralised by diagenesis, are lithified and reintegrate the sediment rocks likely to undergo a new cycle (the rock cycle: transformation into metamorphic rocks under the action of temperature and pressure, or into igneous rocks, before being subjected again, after tectonic displacements, to weathering and erosion).

Six main domains of transport may be distinguished in this continuum by their dominant forcing (Figure 1): (I) watersheds where the flow is dominated by the balance between gravity and friction; (II) estuaries and deltas, where river flow is increasingly influenced by tidal propagation and/or intrusion of marine water (a typology of estuarine filters of river inputs is proposed by [19]); (III) the estuary-littoral transition on the shelf, which includes river plumes, bottom nepheloid layers and ROFIs (regions of freshwater influence, [20])—a mixing zone influenced by river flows, tides, waves, wind and the Coriolis force; (IV) the littoral zone, the earth-ocean interface zone governed locally by wave, wind and tidal forcing; (V) continental shelves beyond the littoral fringe, governed by dynamic processes such as the intrusion of mesoscale eddies, storms, cascading of dense water, etc.; and (VI) canyons that open out into the deep sea. Establishing sediment balances consists in estimating fluxes between these compartments or within these compartments and analysing their spatial and temporal variability at scales that vary over several orders of magnitude.

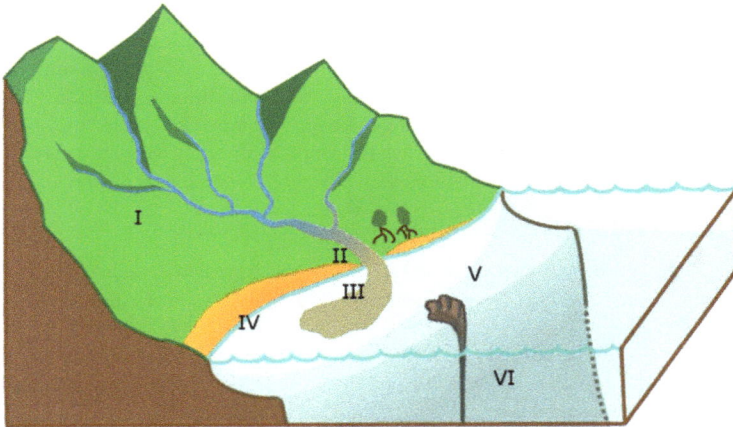

Figure 1. 6 main domains/geomorphological units of sediment transport studies. I the watershed; II estuary/delta; III the estuary-littoral transition, including the river plume, the bottom nepheloid layer and the ROFI zone; IV the littoral zone; V the continental shelf; VI canyon and deep sea.

Estuaries and deltas have been grouped in this general scheme but may be distinguished. In its downstream part affected by tides, a river with enough sediment load can develop a delta, a delta being defined by high morphologic dynamics, with lobes formation and rapid changes of mouths and morphology of the subaerial delta [21,22]. While a delta is qualified by river, wave or tide dominance [23,24], the river influence is less in an estuary which is dominated by wave or tide processes [25]. These classifications are based on long term evolution of the system. At shorter scale, the dominant processes on a same site can vary seasonally or interannually, under the double alternation dry season/rainy season and dry year/wet year, and the decline of sediment load of anthropogenic origin. While sediment are mainly deposited in the prodelta and coastal zone in the wet season, estuarine processes develop when the river influence decreases and favour estuarine deposition and consolidation in the dry season. In tropical areas, mangroves play an important role in filtering energy and matter between the river and the ocean, and buffer extreme events on these coastal areas.

It is clear that the major climatic variations—the alternation of ice ages and warmer, interglacial periods, with sea level variations of several tens of meters—impact all the land-ocean interface zones and control sedimentary plains, deltas and coastlines. With a 120-m sea level rise between −17,000 and −5000 BC at a rate of about one meter per century [26], the bed of old rivers became canyon (e.g., [27]), sedimentary plains and new deltas have formed, while melting glaciers tore up tremendous quantities of particles, transported downstream.

Geomorphological modelling is certainly a primary objective for the decades to come, to better understand and be able to consider the future of deltas and coastal zones under the effect of dynamic forcing. Models already exist (e.g., [28]), which take into account processes, in the state of our current knowledge. They are expected to be improved by integrating progressively more complex processes, expanded particle sizes, better represented forcing, etc. The main difficulty lies in building modelling capacity for multi-scale, integrative scenarios, to get realistic sedimentary balances. Over long periods of time, as wrote Hinderer [29] in his review, "the quantitative understanding of global sediment cycling over historic and geologic time and its response to allogenic forcing is still in its infancy and further research is needed towards a holistic view of sediment routing systems at various temporal and spatial scales and their coupling with global biogeochemical cycles".

The shortest time scales are also extremely important in sediment dynamics, since they determine turbulent fluctuations, and thus the small turbulent eddies without which there would be no transport in suspension—the action of turbulence acting against the gravity-driven particle settling.

Any sediment dynamics assessment therefore focuses on estimating particle fluxes between compartments or within a domain at scales and under forcing that must be specified because, in the Earth history, forcing endlessly change, and because human action profoundly impacts the situation [30–32]. Sediment transport also has other consequences than morphodynamics, which we review briefly.

2.2. Sediment Balances and Morphodynamic Evolution

On a global scale, the flux of particles from rivers to the sea (between domains I and II) is estimated at 20 billion tonnes of sediments per year [33–36], i.e., 630 tonnes per second, which correspond on average to a net denudation of a ~6.2 cm thick layer on all soils every 1000 years [17,37]. This flux is of the order of 13 to 40% of the particles eroded in watersheds (I: total soil loss estimated between 40 and 75×10^9 t yr^{-1}), the difference being explained by the massive downslope deposition of particles eroded within basins prior to their arrival at the marine interface, which is shaping landscapes [37–40].

Some of the eroded particles are deposited in the estuary or delta (within compartment II) while others form the river plume or the bottom nepheloid layer (III) which spread particles over the shelf (IV) and (V), a fraction of which are remineralised and another fraction reach the deep ocean (VI) by canyons. Export by canyons is triggered by sediment failure, river flooding, or dense shelf water cascading events [41–43]. On the example of the Ganges-Brahmaputra-Meghna—one of the two basins that bring the highest solid load to the ocean with the Amazon River, the sediment flow of about 1×10^9 t yr^{-1} is distributed between about 30% of deposit in the plains and deltaic wetlands, 40% to the subaqueous delta (continental shelf and littoral zone), and 30% that reach the Swatch of no Ground (canyon) where they are either stored or expelled to the deep-sea fan [44,45].

A very singular dynamic takes place between the estuary or the delta (II) and the shelf (III) with a bidirectional sediment transport regime. Sediment flow can be positive from the shelf to the estuary, especially in the dry season (e.g., [46–51]). In the Red River, for example, some of the sediments deposited in the subaqueous delta during the rainy season return in the dry season by tidal pumping to the estuary and deltaic wetlands, where deposition is three times greater than in the rainy season [52].

The littoral zone (IV) benefits from four major sources: fluvial allochthonous inputs, coastal erosion and cliff erosion along the shore, biogenic production (limestone tests, foraminifera, corals, etc.), and marine erosion. The shoreline change depends on the balance between inputs, exports, and the net subsidence (or uplift) attributed to tectonics, compaction, sedimentation and anthropogenic causes [53]. Apart from net subsidence and sand mining, the shoreline evolution reacts to the cross-shore redistribution of sands between the beach and the littoral zone (down-sizing beaches in winter, reconstitution in summer, see [54]), and longshore transport induced by oblique swells. Longshore net transport following an oblique wave can be very important; along the coast of Benin, it is estimated between 1.5×10^6 m^3 yr^{-1} [55] and 500,000 m^3 yr^{-1} [56] of sand per year; along the Senegal coast near Saint-Louis at 600–700,000 m^3 yr^{-1} [57]. Accretion and erosion on tidal flats are strongly dynamic as well. An example: while the intertidal zone of Haiphong Bay (one of the distributary of the Red-Thai Binh River system) has increased by 35×10^6 m^3 in 25 years (1989–2014), the area just north of the delta has lost 36×10^6 m^3 of sediment in the same period [58].

Issues associated with sediment transport may depend on the area of interest. The next section briefly reviews them.

2.3. Issues Associated with Environments and Ecosystems

Erosion, transfer and deposition of particles shape landscapes, alter the budget of geochemical cycles, move nutrients and affect the ecosystem health by their associated contaminants. Accumulation of particles (and associated nutrients and contaminants) in reservoirs reduces downstream export, limits reservoir storage capacity, and raises many challenges in management at medium to long term.

Particulate nutrients (particulate organic carbon, nitrogen, phosphorus) and other key elements for fertilization (like Fe) highly influence the ecology and biogeochemistry of aquatic environments by controlling the availability of dissolved nutrients, affecting light availability, influencing phytoplankton stocks, growth, grazing rates, and community structure, and affecting food webs [9,59–62]. Coastal or lagoon waters may become eutrophic. Amongst the consequences of eutrophication, harmful algae blooms may occur and poison ecosystems [63]. Finally, the nature of macro-aggregates, which include mineral particles, phytoplankton, transparent exopolymer particles (TEPs), among others, could contribute to the decrease in oxygen concentration and pH in the bottom estuarine layer by respiration/remineralisation processes [64]. Sediment dynamics should then be considered in an enlarged perspective aiming at ecosystem recovery and restoration [65].

The level of suspended particulate matter (SPM) concentration is not the only important factor: the duration of their exposure is important as well for the health of aquatic biota [66], and the control of light attenuation (turbidity induced) impact on primary production [60,61]. The residence time of water masses and associated SPM is also a key factor in the fate of the microbial compartment, and coupling a sedimentological study with a dynamic analysis (which provides residence time) is a way to estimate their impact on ecosystems.

Organic or metal contamination is a very sensitive concern in the hydrographic network, because water is withdrawn for consumption or irrigation of agriculture areas. This contamination also affects coastal waters, their ecosystems and their resources. Coastal contamination may be caused by mining activities (e.g., [67]) or results from industrial or urban influences (e.g., [68–70]). The physicochemical processes involved and their consequences on the contaminant deposition are themselves research topics, such as organic and inorganic matter chemistry in a salinity gradient (e.g., [71,72]). The final step in studying the fate of contaminants is to quantify their impact on living organisms (plants, animal tissues), which is the vast scientific field of ecotoxicology.

Coastal sediment dynamics also have major ecological consequences which are evident in the case of mangrove forests. Not only mangroves protect the coast from extreme events, but abundant flora and fauna develop there, and they serve as fish nurseries. The role of mangroves in the emission and pumping of greenhouse gases is also the subject of climate studies [73,74].

2.4. Global Issues Related to Climate and/or Anthropogenic Activities

The balance of chemical elements allows us to assess the recycling of continental crust and estimate the denudation rates of soils [4]. Continental erosion consumes CO_2 [75]. In particular, silicate weathering represents a critical step in long-term climate moderation [76]. However, at the geological time scale, oxidation of petrogenic organic carbon (derived from rocks weathering) releases atmospheric carbon [77]. The carbon budget associated with continental erosion thus depends on the balance between sinks (sequestration of POC from the terrestrial biosphere) and sources (release of atmospheric carbon) [18].

While changes in continental erosion affect the climate at geological scale, climate change has a rapid impact on erosion and particulate transport through changes in the hydrological regime and/or vegetation cover [78–80]. Example: the increase in rainfall variability and the earlier arrival of summer rains in northern Algeria have almost doubled the particulate flux every 10 years in the basin, during the last four decades [81]. The intensification of the water cycle as a result of global warming [82,83] should impact accordingly erosion and sediment transport.

At the coast, sea level rise increases risk of flooding [84,85]. Very recent work shows that the sea level rise may be higher at the coast than in the deep-sea ocean by a factor of 2 in the vicinity of Hong Kong [86]. The low frequency contribution of wave setups and runups is also a factor in sea level rise at the coast, so far largely underestimated, especially at decadal and centennial scales [87]. This process contributes to increase the shoreline vulnerability.

Land use also highly affect particle fluxes in all compartments. Half of the sediment supplies are trapped in reservoirs on basins with dams, on average. Vörösmarty, et al. [88] estimated the resulting decrease in particle flux at the coast to 25%, in 2003. Volumes of mud dredged in harbours and channels may be known from port authorities. On the other hand, sand mining, which is very high and more or less controlled, is hard to quantify. In 2000, Hooke [89] estimated that the volume of land displaced was on average 6 t yr^{-1} per capita, corresponding to $35 \times 10^9 \text{ t yr}^{-1}$ (e.g., around twice the sediment flux to the sea).

Changes in ecosystems fertilization is another stake related to sediment transport at global scale. The associated transport of nutrients is a key feature for fertilization or loss of fertile topsoil [59], and the global levels of biogeochemical flows (for Phosphorus and, above it, for Nitrogen) were identified amongst the most destabilizing effect of the Earth system [90].

3. Specific Hydrosedimentary Processes along the Sediment Pathways

3.1. Specific Processes per Study Area

The sediment flow from land to sea comes mainly from rivers (about $20 \times 10^9 \text{ t yr}^{-1}$), secondarily from lands eroded by the sea (about $0.4 \times 10^9 \text{ t yr}^{-1}$) [91]. Sediment routing encompasses particle fragmentation and sorting as the slope decreases, and particles are often deposited and stored for very long periods in valleys, flood plains, wetlands (e.g., [92]).

Basic mechanisms involved in transport are common to all compartments (start of motion, erosion, transport, flocculation, aggregation, settling in homogeneous or stratified waters, damping of turbulence by high concentrations, deposition). Some processes are specific to the study areas, such as tidal pumping and formation of the estuarine turbidity maximum (II), shelf distribution through the plume and bottom nepheloid layer dynamics (III), sediment transport induced by swash or uprush, or sandbar migrations (IV), turbidites (or underwater avalanches) in canyons (V > VI). Each study of one of these processes always closely associates hydrodynamics and sediment dynamics. Given the diversity of processes, researchers have focused on one type of sediment (cohesive or non-cohesive); however, more and more researchers focus on describing and modelling processes in the presence of mixed sediments.

3.2. Littoral and Coastal Hydrosedimentary Processes

Coastal sediment transport involves two main environments that largely depend on dominant energy forcing. Where swell and waves dominate, the available energy is such that the finest sediments are constantly maintained into suspension or resuspended, and only the coarse ones may deposit: these are sandy coasts, pebble beaches, cliffs or rocky areas. Wave power (which depends on their length and height) varies with their origin and their fetch, and is modulated by friction on the continental shelf. Their propagation angle is the driver of the littoral drift and thus determines if transport dominates along-shore or cross-shore. Where coasts are sheltered from waves and mainly forced by tides, as in semi-closed seas where the fetch is reduced, in wide continental shelf, or at the head of bays (e.g., Bay of Bengal), bottom shear stresses are weaker and allow fine particles to settle and consolidate, making them muddy environments.

Different dominant forcing in a cross-shore direction may induce reversed natures of soft bottoms from a region to another: in the microtidal Gulf of Lions (North-Western Mediterranean Sea) fed by the Rhône River, where waves dominate nearshore, coasts are sandy and mud particles are deposited at 30–40 m depth, while in the meso- to macro-tidal Gulf of Tonkin facing the Red River where tides dominate, mud is deposited and transported along the coast, while bottoms at 30–40 m depth are sandy.

75% of world coasts are neither beaches nor mudflats but hard rock cliffs [93]. Coastal cliff erosion is also a research topic related to long-term coastline evolution (e.g., [93,94]).

Nature of bottoms and dominant processes locally depend on sediment sources and forcing, in the short term (waves-current interactions, response to intense events such as cyclones, storms, floods, surges) and long-term (sea level rise, modification of the continental water cycle, subsidence and uplift [53,95]. Process studies must be able to explain and reconcile time and space scales associated with sediment transport.

We propose hereafter to review some important processes that set up as many research topics, to illustrate their large diversity. The reader is invited to complete it according to his field of interest with reviews as, for example, Nittrouer and Wright [96] and Brink [97] for transport on the continental shelves, Horner-Devine et al. [98] in coastal plumes, Syvitski et al. [99] for dynamics of coastal zone, or books by Dyer [100], Nielsen [101], Soulsby [102], Dean and Dalrymple [103], the Coastal Engineering Manual [104], Dronkers [105], Winterwerp and van Kesteren [106], van Rijn [107], Uncles and Mitchell [108], Mehta [109], amongst others.

3.2.1. Non Cohesive Sediments

The morphology of most sandy beaches is changing under wave conditions and is generally highly variable at the seasonal scale, with winter erosion and summer accretion. Wave conditions can be modulated by tides during spring-neap cycles. Except from mean seasonal forcing, the other key time scale for beach dynamics is the storm occurrence. Wave angles determine the main direction of transport (cross-shore or longshore), and the panel of their direction, strength, return periods, requests more in-depth analysis of basic mechanisms and processes (see [110–113]). Our knowledge has recently developed very fast on topics such as rip currents [114], beach cusps [115], sandbars and beachface evolution [116], double bars [117], swash [118–120], etc. A better understanding requires to study in parallel fundamental hydrodynamics processes, such as shoaling, wave runup, overwash, storm surges, rip curls, and fundamental sediment processes, such as bed load size distribution [121], armoring, and bed formation.

Sediment dynamics is also fast in sandy tidal inlets, which connect back-barrier lagoon to the open ocean. Specific hydrodynamical processes develop there, such as tidal transmission with phase lags and impact on flood-ebb asymmetry. Wave-current interactions and their effects on the wave setup in the inlet or at its mouth are also determinant in their dynamics [122,123].

Assessing the impact of extreme events is a hot topic in sediment dynamics [124,125]. Cutting-edge technologies enabled scientists to monitor the quick dynamics of sand beaches during the highly energetic winter 2013–2014 in Western Europe, study the involved processes and the beach recovery [117,126]. Storm-induced marine flooding is important as well (see a review in [125]). In winter 2013–2014, large overwash events were shown to be driven by infra gravity waves combined with high tides, and the sequence of steps was analysed in terms of sedimentation [127]. Washover deposits is amongst the main processes that need to be studied to interpret with better accuracy sediment records. Such analysis deserves different time scales involved, from event to paleoenvironment. As climate controls the rate of weathering and export of associated sediments over glacial-interglacial timescales (e.g., [128]), there is a climatic signature of sediment export to coasts and continental margins. Combined with extreme events such as tsunamis or storms, and affected by tectonic changes, they govern the sediment budget at long-term and the formation of beaches or lidos.

Processes are specific to some environments, such as sand spits, tombolos or barrier islands formation [129]. This is the case at river mouths, where the interaction between fluvial and coastal processes is responsible for ultimate sedimentary architecture such as tidal sand bars [130]. This is also the case in coral reef lagoons [131], where the transformation of waves over barrier reefs, wave setups and energy balances [132,133] govern sediment dynamics and sediment disturbances on corals [134,135].

3.2.2. Cohesive Sediments

Fine suspended particles—clay and fine silt—aggregate during their pathways under physical (collision, shear), chemical (electrochemical interaction) and microbiological effects (presence of sticking organic matter gluing particles together) [14,136].

In the water column, flocculation and deflocculation alter the particle size distribution, excess density, and aggregate settling velocity [137–142]. The influence of organic matter on aggregation is a specific topic [143–147]. Aggregation is one of the most complex processes in sediment transport since it requires deepening the interactions between physics, chemistry and biology, and developing the analysis of coupled mechanisms, such as the role of shear on the production of organic matter [148].

At the bottom, erodibility or erosion of cohesive sediments, which has been studied for a long time [149], remains difficult to assess and predict [150]. Bioturbation or biostabilisation due to the presence of macrophytes [151], seagrasses [152], biofilms or animals such as polychaete worms alter the seabed erodibility [153,154]. Finally, to better understand the hydrosedimentary processes in muddy environments, it is essential to improve our knowledge of sediment properties such as viscous and viscoelastic properties of mud (e.g., [155]), wave–mud interactions [156], and specific hydrodynamic processes, such as estuarine dynamics.

In meso- or macro-tidal estuaries, the tidal asymmetry in the estuary (with shorter and more energetic flows during flood tide than during ebb tide) may generate a tidal pumping which participates to the estuarine turbidity maximum zone formation. Sediments are then brought back from the sea to the estuary near the bottom, even in the absence of density circulation, and settle and consolidate under the control of spring-neap tide cycles. This process, described in the 1970s [46], is still under study (see [51,142]). This key mechanism of delta geomorphodynamics is very sensitive to changes in flows, and river water regulation induced by dams is enough to move the estuarine turbidity zone and cause the silting up of navigation channels and river ports, as on the Red River [157].

Wave action in estuaries may be important as well, since waves may fluidize subtidal seabeds and change bed erodibility. As stated by Green and Coco [158], "estuarine intertidal flats are excellent natural laboratories that offer opportunities for working on a number of fundamental problems in sediment transport". Fluidization of marine mud by waves [159] is a key factor for fluid mud dynamics and for the formation and migration of mud banks (e.g., [160–162]). Studies focus as well on hydrosedimentary processes in mangroves, which play a key role in wave attenuation and sediment trapping in muddy coasts [163]. Mangroves, natural filters of energy and matter, have many consequences on ecosystems and climate (by greenhouse gases balance).

Another series of hydrosedimentary processes relate to river plume dispersion and to the ROFI and continental shelf dynamics. However, if the river plume hydrodynamics has been studied in details for ~30–40 years (see a review in [98]), few studies were dedicated to the suspended matter characteristics (grain size distribution, excess of density, organic content, mixing processes at the interfaces) and their behaviour in the plume or at its boundaries, and in the bottom nepheloid layer (such as [164,165]).

3.2.3. Mixed Sediment

The integration of sediment heterogeneity in sediment transport modelling is still a challenge, due to the rapidly changing sediment characteristics in some coastal areas [166].

Natural environments often consist of mixed sediments—sand and mud—especially in the deposition areas at the land-ocean interface as estuary, delta, and mangrove forests. Studying specific processes with large particle size distributions are necessary to better understand and simulate sediment dynamics there, such as bed armouring or erodibility of mixed sediments [167–171]. Models of mixed sediment transport now enable us to study sorting dynamics of cohesive and non-cohesive sediments and to reproduce processes such as, for example, the landward fining of surficial sediment on intertidal mudflats, or the convexity of intertidal zones with respect to the mixing [171].

A fluff layer sometimes develops over sandy bottoms. Resuspension of the fluff layer is another kind of hydrosedimentary process over mixed sediment which needs to be better known and quantified [172,173].

4. A Targeted Research Too

The era seems far away when the credo of scientists was "science for science", as Henri Poincaré stated. In 1945, after the bombing of Hiroshima and the disruption it induced in outlooks, the societal factor entered the minds of researchers. It then appeared that "pure science is now in direct contact with material reality", in a word: "science proposes, and humanity disposes" [174]. National research efforts in developed countries during the 1950s and 1960s have been resolutely turned towards technologies (energy, transport, medicine). In Europe, research funding took a further step towards policy-oriented research when the Lisbon process was adopted in 2000, which in the following years would generalize project financing and thus targeted research.

Whether through public or private funding, it is now easier to find support for targeted research. Sediment transport is a topic in the mood, particularly well-suited as anthropogenic impacts predated those of the current climate change on sediment routing. Sediment budgets have been strongly constrained by human settlements and other activities since man settled in the Neolithic [89,175,176]. Human activities have both increased soil erosion [177,178], and decreased sediment supply to the oceans due to dam retention [176,179]. The decrease in river particulate discharges due to reservoirs exceeds 95% locally (like on the Nile and Ebro rivers), and is estimated at 25–30% of the total or 4–5 Gt yr^{-1} at global scale [88]. The Anthropocene has not only changed the sediment, carbon, nitrogen and phosphorus cycles over the past centuries but has introduced new man-made materials such as plastics or concrete which are now mixed with deposited sediments [180]. Locally, the deepening and narrowing of small and narrow estuaries has caused tidal amplification, enhancing risks towards hyper-turbid regimes, with many ecological and economic consequences [181]. Sand mining for building, glass, foundry and hydraulic fracturing or "fracking" is another growing pressure (196 million tons of sands in 2014 worldwide, according to the U.S. Geological Survey, with an estimated present rise of 5.5% per year according to a study by Freedonia Group, 2014 [182]). The production of sand for fracking wells was 19 times higher in 2013 than a decade earlier, according to the USGS. Sand is now the second most used natural resource in the world after water. The effects of sand mining in the mid/long term on geomorphology remain to be quantified. Trawling is another human activity which highly impact sediment dynamics and marine habitats. A related task is to assess the ecological impact of all these environmental changes, including the resilience capabilities of ecosystems [65]. Finally, coastal erosion, which is increasing due to the decrease in river sediment inputs and storms, tsunamis, surges whose effects are amplified over regions with increasing populations [88,183], and impacted by anthropogenic activities, make this scientific topic become social issues. Natural processes of sediment transport must be considered in connection with dam impoundments, climate change (on sea level, temperature, precipitation, hydrologic and meteorological regimes) [184,185], land subsidence or uplift, salt intrusion, to estimate resiliency of deltas and their ability to cope with future impacts [183,186].

Sediment dynamics are still involved in several fields of application. In civil engineering, dimensioning of coastal protection structures like groynes (submerged/emerged, at 90° or at an angle, etc.) or breakwaters must be optimized, and the impact of any other coastal development (sand bypassing, artificial river mouths, jetties, sand nourishment, etc.) assessed (e.g., [187,188]). New nature-based approaches, in this area, are reworking hydraulic engineering and coastal protection towards more sustainable and adaptable designs [189–193]. Sediment transport is also involved in areas where renewable energy production from tidal stream or wave power is foreseen [194,195].

Research in sediment transport, which both improves our knowledge and provides diagnoses to decision makers, is what our time has been asking for. Initially developed for civil engineering, this topic has been recognized as a full scientific field. Science to serve society must make it possible to better assess hazards and vulnerability of environments, to identify early warning signs of critical transitions, and the general society should perceive sediment dynamics as a critical matter requiring attention [185]. It is therefore necessary to continue efforts to better understand the involved processes, to monitor and analyse the evolution of sediment budgets in order to better adapt the information to be transferred to decision makers in suitable forms for planning purpose.

5. Old Topic, Emerging Questions and New Methods

5.1. A Science Field that Evolves with Technology

The progress of knowledge is strongly constrained by available means.

Following Philipp Forchheimer's pioneering hydraulic manual in 1914, renowned scientists such as Albert Shields, Hans Albert Einstein, Hunter Rouse, Ralph Alger Bagnold, Eugen Meyer-Peter, Ray Krone, Emmanuel Partheniades, Claude Migniot laid in the 1930s to 1960s the foundation on which engineers have dealt with practical problems. Mechanisms and hydrosedimentary processes were then mainly addressed on dimensional analyses, theoretical and/or statistical developments, and experiments in channels. Many concepts or basic formulations then issued entered into a common use as the Shields critical shear stress for transport (1936) [196], the Rouse's similarity of vertical profiles (1937) [197], bedload estimate from the excess of shear stress of Meyer-Peter and Müller (1948) [198] or on the basis of a probabilistic model of Einstein (1950) [199], the introduction of the current power concept by Bagnold (1966) [200], the deposition formula by Krone (1962) [201], the erosion formula of consolidated beds by Partheniades (1965) [202], the total transport models of Engelund and Hansen (1967) [203] or Ackers and White (1973) [204], etc. Resulting formulations for estimating or modelling bedload or suspension transport are still used or have been used as a basis for improved versions (e.g., [205–211]).

In the 1970s to 1990s, their successors broadened the field of study to coastal and littoral zones: (i) by undertaking large campaigns to study in situ coastal processes from physical, sedimentological and geological measurements using new sensors and technologies (current meters, CTD probes, turbidimeters or optical backscattering sensors, Pb-210 geochronology); (ii) developing large scale field or laboratory studies for littoral processes; (iii) taking advantage of the computer boom to develop process studies based on numerical models, and to generalize the use of numerical models in site studies; (iv) benefiting from first spatial images to monitor the Suspended Particulate Matter (SPM) distribution.

(i) Major technological and instrumental developments in the 1970s enabled researchers and engineers focused on sediment transport to introduce and adapt new approaches while contributing to improve them. It was the case in the measurement of suspended solids, particle size distribution (e.g., [212]), in sediment geochemical dating (e.g., [213]); flocculation/deflocculation was described and explained from observations [136]. And major campaigns were initiated (e.g., [214–216]).

(ii) In littoral zone, the knowledge of processes benefited from the conjunction of field analyses [110,207–209,217,218] and laboratory studies in oscillatory flow channels, in particular taking advantage of new methods such as the particle imaging velocimetry (PIV) [219–221]. A review of the progresses done from the 1980s to the 2000s was proposed by van Rijn et al. [113].

(iii) Numerical models both gained an acceptable level of representativeness to be involved in process studies [222], and their use became more widespread with cheaper computers with highly increasing capabilities. Numerical models became embedded in most projects. At that time, significant improvements were proposed in basic formulations (e.g., [102,223]), and new models for cohesive sediment transport (e.g., [210,224]), for sandy beach dynamics, either for longshore or cross-shore processes (e.g., [219,222,225,226]) and for large-scale geomorphological evolution analysis [220,227] became more and more realistic.

(iv) Spatial data of water colour started also to be considered in the 1970s and 1980s to monitor the dynamics of surface SPM in river plumes and coastal areas, or to determine shallow water bathymetry [228]. The contours and patterns of river plumes from remote sensing (mainly satellite observations) were initially compared to numerical simulations. SPM concentrations started to be quantified and mapped in the 80s–90s at the surface (e.g., [229–231], see a review in Acker et al. [232]), or even along a subsurface depth-profile under some conditions [233].

Since the 1990s–2000s and even more recently, tools available to researchers have evolved considerably: (i) an increasingly precise, miniaturized and cheaper instrumentation has appeared; (ii) the integration of numerical models into process studies including field or laboratory data has become widespread; (iii) free or cheap satellite data have become increasingly adapted, accurate and plentiful. These years have thus resulted in a convergence of tools. As many works have not ceased to emphasize for 40 years the high level of interactions between physics, chemistry, geochemistry and biology in the processes of suspended matter transformation, transport and fate, specialists of different disciplines have begun to strengthen their collaborations on this common topic, and multi-skilled researchers have become increasingly sought after by laboratories. Let us summarize the three major evolutionary aspects of these 1990s and 2000s:

(i) new sensors have revolutionised in situ measurement. A new generation of measurement instruments became available, with smaller and more accurate turbidimeters [234–237] that made it possible to design erodimeters and sea carousels [151,170], and multi-instrumented pilots for studying flocculation [137,138,143,238]; instruments to characterize cohesive sediments [239,240]; in situ grain size meters [241,242]; numerous acoustic instruments (ADCP to measure the suspended sediment flux; ADV, Altus, ABS) and optical sensors (including hyperspectral field radiometers to measure water colour at low cost) (e.g., [243–247]).

This instrumentation and researcher ingenuity have made leaps on flocculation studies [138,140,141] and on bed erodibility [170] in the 1990s–2000s. Aggregate characterization in lagoons was performed as well [248]. The phenomenal development of marine optics made it possible to determine SPM parameters from multi- or hyperspectral measurements such as their refractive index [249], the shape of their size distribution [250], their inherent optical properties [251] from which other parameters of nature or size are deduced.

Sandy littoral morphodynamics benefited from process studies in channels [252]. Field studies relied first on the use of echo sounders. The use of airborne images to map bathymetry and determine the beach slope (initiated during the Second World War) faced expensive image acquisition. Since the 90s and especially 2000s, the use of cheaper instrumentation such as video cameras has revolutionised the monitoring of high frequency evolution of beach dynamics with access to wave parameters, bathymetry and sandbar locations (see [253–255]). Temporal methods and spectral methods have been developed in parallel. X-band radar images can be used as well to infer shallow water bathymetry [256,257]. In the field, the accuracy of measurements was considerably improved by D-GPS and their acquisition time considerably reduced using quads in the intertidal zone and boats or scooters in the subtidal area.

(ii) In the 90s and 2000s, 2D and 3D models became sufficiently realistic—taking a free surface into account—to numerically study complex processes, such as the formation and dynamics of estuarine turbidity maximum [48,258], 3D transport of non-cohesive sediments in nonequilibrium situations [259], littoral morphodynamics [260,261], or particle dynamics in the presence of mixed sediments [262]. Models adapted to different classes of problems became available, some of them for free [263], and tools were proposed to evaluate their capacity (e.g., [264]). Many kinds of models are now available, for simulating the shoreline change such as GENESIS [225]; for simulating the 2D beach profile evolution such as XBeach [265]; for simulating the horizontal 2D or 3D coastal sediment transport at different scales such as SHORECIRC, ROMS, SELFE, Delft3D, MARS3D, SYMPHONIE, SWASH, amongst others. Most of local or regional hydrosedimentary models are now coupled to or forced by wave models such as SWAN or WAVEWATCH III, while coupling of coastline and fluvial dynamics models wears on (e.g., [266]). These available new methods make it possible to consider abandoning old simple methods such as the Brunn rule [267], a very simplified 2D model of shoreline response to sea level extensively used for over forty years [268,269].

(iii) Since the late 1990s and early 2000s, optical measurements over shallow waters are used to infer the water depth [270], and SPM maps derived from spatial data are used in calibration and validation of models, for analysis purpose, without data assimilation [271–274], or with assimilation [275–277].

Where are we now? How have the latest technological developments shifted our research?

In the 2010s, the novelty came from highly mobile vectors such as gliders (and other AUVs, Autonomous Underwater Vehicles), and drones or UAVs (Unmanned Aerial Vehicles). Gliders enable quick profiling along transects facing mouths, equipped with sensors such as CTD probes, turbidimeter, bio-optical sensors and, soon, in situ laser grain size meters (like the LISST 200-X). Many et al. [164] have recently documented sections of suspended matter in the surface plume and in the bottom nepheloid opposite the Rhône River mouth. Stereo restitution, which is adapted to build Digital Elevation Model (DEM) over sandy environments, may fail over mudflats which hold back residual tidal water. However, drones, equipped with a lidar or other, make it possible to map beaches or intertidal areas accurately, quickly and at a very low cost [278], and to observe the diachronic evolution from repeated surveys, using a GIS, at unprecedented temporal resolution. Coastal cliff erosion as well, which was used to be studied from stereographic aerial photographs, benefits now from lidar measurements by planes or drones, in addition to terrestrial laser scanning [93]. Most of these new techniques are costly effective and provide fast-scanning data at previously unobtainable precision. Furthermore, the possible high-frequency enables us to document strong spatial and temporal dynamics we could not measure one or a few decades ago (like the bathymetry changes during a storm from video).

Technological development even seems to be accelerating. Not only do satellite archives become long enough to distinguish between climatology and inter-annual variability of river plumes (e.g., [279,280]), but the number of satellites is increasing, with different characteristics in terms of spatial definition and revisit period: geostationary satellites (GOCI, with the ability to get eight scenes per day with a 1 h-temporal resolution, with 500 m spatial resolution); hyperspectral sensors with medium spatial resolution (such as MODIS with revisit every day and 200–1000 m resolution, Landsat8 with 30 m resolution and revisit every 16 days, or Sentinel-2 with 10–20 m resolution and revisit every 5 days); multispectral sensors at very high spatial resolution (such as Pleiades with 2 m resolution and daily revisit ability). Fusion methods are sometimes used to combine advantages of two sensors, one at high frequency and one with higher spatial definition (e.g., [281]). Time series of sufficiently resolved spatial images (such as Landsat) are also increasingly used to monitor shoreline evolution [282–287]. Other methods using a lidar or X-band radar may be used [257,288]. Drones are also equipped with multi- or hyperspectral sensors to map water colour and water quality parameters at high definition [289]. The potential of an instrument such as the LISST 100-X since its commercialization (about fifteen years) has not yet been encompassed that a second generation of instrument is available using holograms (LISST-HOLO, [290]), then a third generation (LISST-200X) to better measure larger particles and benefit from more robust optics. New sensors are regularly introduced, applied and compared to older ones: infrared laser turbidimeter adapted to very turbid water [291], comparison of acoustic and optical backscatter sensors in cohesive suspension [292,293], etc.

Most numerical models are now open-access. Old models are combining and transforming, like SCHISM from SELFE [294]. Numerical simulations are based on evolving models, and on refined preprocessing and postprocessing tools. New tools are emerging as well in the digitization of shorelines at high spatial definition on Google Earth. Global DEMs are available online (such as WorldDEM, at 12-m, from TandDEM-X and TerraSAR-X data). New methods are being developed and implemented to establish accurate DEMs of intertidal flats from images such as Landsat data. Their implementation at several years or decades intervals makes it possible to evaluate the erosion and accretion budgets (e.g., [58]).

Sediment transport models are steadily improving. While classical instrumentation is improving as well (see, e.g., [295] for bedload measurements), more accurate formulations are proposed to account for various mechanisms such as, for example, recently: the bedload grain size distribution [121]; the alternative approach, referred to as the "entrainment flux method" for quantifying the erosion properties of surficial sediments [173]; the integration of multiple classes in 3D models [296]; or new probabilistic formulations for bedload [297]. New types of models are developed, tested and evaluated on simple configurations to improve knowledge of basic mechanisms at a fine scale, e.g., direct numerical simulation of bedform evolution and/or sediment transport [298,299], emerging methods of smoothed particle hydrodynamics for fluid-flow interaction [194]. In-depth theoretical analysis is also ongoing on bedload (e.g., [300,301]), on bedforms [302], on the transition from bedload to suspended load [303,304], on the granular flow rheology in bedload transport [305], amongst others. These works, not yet adapted to the study of natural environments, precede the models evolution and move forward our knowledge of the involved mechanisms.

Finally, we note the increasing use of statistical methods of artificial intelligence or approaching them, adapted to process a growing amount of data: neural networks for the interpretation of concentration and turbulence profiles (e.g., to derive floc sizes and promote the use of artificial neural network to study flocculation, [306]) or in the analysis of sediment transport in water basins [307,308]; fuzzy logic; fusion methods and machine learning (e.g., in the concurrent analysis of forcing parameters, suspended sediment profiles and sediment concentration at the surface provided by remote sensing, [309]); Random Forest for classifications (e.g., to determine a mineralogical classification of suspended sediments in rivers after using scanning electron microscopy, [310,311]); support vector regression methods in satellite data analysis (e.g., [312], this Special Issue), etc.

All of these technological and scientific advances lead to very fast progress in monitoring sediment transport and in the knowledge of involved processes (e.g., [120]).

In the end, the information on sediment transport, which was based 30 or 40 years ago mainly on point measuring stations such as benthic stations, has become largely spatialised with technological development, notably through the use of GIS and the integration of spatial data (satellite, airborne, video or UAV) in studies, for model calibration or validation purposes. The increasingly frequent integration of spatial data in environmental or geophysical studies and their comparison with field data benefit from many recent tools such as data interpolation/spatialisation methods [313], bio-optical algorithms to accurately derive suspended sediment concentration from remote sensing [280,314–317], coupling of GIS and erosion model in water basins [318], and new metrics to analyse long time series of images (like centroids, north-south-east-westernmost points or skeleton of river plumes, see Gangloff et al. [319]).

Young researchers have always embraced cutting-edge technologies with ease and dexterity. Those who, among them, will adapt these new tools to our topic, with the support of senior colleagues and the vigour of their youth, will allow us to clear other aspects of sediment transport processes and monitoring. And there is not a month that goes without innovative works showing us changes in progress. The future upheaval could well be crowdfunding, with the emergence of easy-to-use applications on smartphones, such as HydroColor to measure a 3-band remote sensing reflectance and turbidity [320].

Finally, future improvements in our capabilities to understand, replicate and forecast sediment dynamics will arise not only from coupled hydrosedimentary process studies, but also from better monitoring of coastal oceanography. Among the future tools is the Surface Water Ocean Topography mission (SWOT) satellite, which will bring about a revolution in altimetry in 2021, since it will no longer provide nadir data along tracks, but altimetry maps, useful for oceanographic modelling. These data will enable us to calibrate/validate the friction coefficient (or roughness height) distribution in coastal zones, for example, and then improve the performance of hydrosedimentary models. These data should pave the way for an expanded use of data assimilation technique in coastal hydrosedimentary studies.

5.2. A Topic that Moves Forward through Interdisciplinary Approaches

Due to technological developments specific to each discipline, new knowledge in sediment transport have remained disciplinary for a long time, in hydraulics, fluid mechanics, geochemistry, microbiology, marine optics, remote sensing, modelling, etc. In parallel, while interdisciplinarity was encouraged during the 1970s and 1980s to account for complex phenomena, i.e., connections between objects of study [321], national or European funding has encouraged specialists from several disciplines to pool their respective know-how to address common scientific issues. Sediment dynamics is concerned, because of strong interactions between physics, chemistry, and biology in the particle transformation and fate in estuaries and coastal environments.

Synergy between disciplines multiplies opportunities for development and addresses complex issues. The practice of interdisciplinarity allows us to specify together what synergy can authorize beyond our respective results. Hydrologists, oceanographers, geomorphologists are concerned, and sedimentologists and geochemists as well, sharing their knowledge of sediment bottoms distribution, residence times and particle origins. The use of radionuclides, for example, enable to trace the variations in sediment dynamics and origin throughout an erosion/deposition episode such as a flood in a catchment, and determine the percentage of freshly formed particles from land erosion as compared to remobilized ones (e.g., [322]). Sediment source fingerprinting is an efficient tool in a source-to-sink perspective, of potential aid in catchment management [323–325]. At larger time scales, combining advanced provenance techniques with sediment budgets allows to reconstruct ancient systems [29].

Let us extend the spectrum of collaboration between earth sciences and life sciences, on the example of marine snow [212,326]. In coastal areas, as in the deep-sea ocean, particles fall as aggregates consisting of detritus, living organisms (notably microbial community) and inorganic matter [14]. In their pioneering paper, Alldredge and Silver [327] wrote "The greatest challenge to the study of marine snow at present is the development of appropriate technology to measure abundances and characteristics of aggregates in situ". These developments have come and allow for major steps in the aggregate characterization and in the determination of their settling velocity in estuaries and coastal areas. Transparent Exopolymer Particles (TEPs), a sticky organic matter resulting from the coagulation of colloids or secreted by living organisms (fish, algae, etc.), constitute a sticky matrix on which debris from rocks, silt, clays, and other suspended elements adhere [145–147]. The younger the organic matter, the more dense and sticky it is. Old organic matter gives less dense aggregates that can have a positive buoyancy, which may explain residual turbidity of bottom bays or harbours, with very high residence or renewal times (see, e.g., [328]). Microbial control on aggregate geometry, which affects particle settlement, can be quantified [329]. Bioturbation is another subject that calls for coupled physico-chemico-biological studies. Seabed erodibility is largely affected by algae, seagrasses, polychaete worms or cyanobacteria [152,330–333]. In addition, there are many examples of coastal ecosystem-based management (see [334]) where interdisciplinarity is requested.

Finally, integration of different scales into a systemic approach necessarily requires a multidisciplinary approach:

- integration of time scales to better understand the dynamics of extreme events at the coast [125] and anticipate the coastal impact of sea-level rise [84,87,335]. As Woodroffe and Murray-Wallace [95] explained, "coastal scientists presently have a relatively good understanding of coastal behaviour at millennial timescales, and process operation at contemporary timescale. However, there is a less certainty about how coasts change on decadal to century timescales";
- integration of spatial scales via the source-to-sink continuum (watershed-estuary-coastal zone, a version of the river continuum concept [336] enlarged to coastal zones, see, e.g., [18]). Such integration requires to develop monitoring, understanding and management tools adapted at this scale for, e.g., protection or restoration purposes [337]. For a long time, hydrologists and oceanographers have stayed at a distance around the estuary, this no-man's land, which is no longer the river but not yet the ocean. The needs of natural environment managers go beyond these divisions because it would be nonsense to fight coastal erosion without also acting upstream in the watershed. Integrated source-to-sink studies where models of hydrologists and oceanographers interact are still scarce but they are just starting (see, e.g., [338,339]);
- integration of time and space scales for long-term landscape changes [178,183,340], and filling the gap between stratigraphic models and process-based sediment transport models [25,341,342].

Now, interdisciplinary approach is expanding beyond nature and life sciences when their topic join societal concerns. In littoral and coastal zones, in the context of global change and growing population, it gives integrated coastal management (e.g., [343]). Risk and vulnerability analysis has become an important interdisciplinary domain linking Earth system and socio-system analysis from around 20 years [175]. The gap between Earth science and water resource management join as well in the composition of ecosphere and anthroposphere components and in the valuation of environmental services [175].

6. Highlights of Research Papers

Since "coasts are the nexus of the Anthropocene" [344], reviews [125,158,345] or special issues (e.g., [344,346,347]) are now frequently edited on their monitoring, processes, dynamics and vulnerability.

The papers published in this special issue reflect the diversity of concerns on which research in coastal sediment transport is based, and current trends—topics and preferred methods—to address them.

Two papers deal with two fundamental processes: erosion and aggregation. Their concerns already integrate the natural complexity since Mengual et al. [348] relate to erosion of sand–mud mixtures, and Fettweis and Lee [349] to the biological influence on aggregation in coastal waters.

Mengual et al. [348] propose a mathematical formulation of erosion suitable for all types of soft bottom, composed of sand, mud, or a mixture of sand and mud. This formulation is tested on 3D numerical simulations of sediment transport in the Bay of Biscay and compared to measurements made by acoustic profiler. Below a first critical fraction of mud (10–20%, corresponding to 3–6% of clay), pure sand erosion is prescribed. Above another critical fraction, a pure mud erosion law is applied. Several transition laws were tested and an abrupt exponential transition was shown to perform best, in agreement with experimental results from literature.

Fettweis and Lee [349] analyse the variability of aggregates measured in the Belgian coastal waters from 2004 to 2011. Here "aggregate" is preferred to "floc", following the recommendation of Milliman [350] who writes that "flocculation" suggests an electrochemical process and denies biological influences. The biological influence is there significant, since, as shown by the authors, SPM in the Turbidity Maximum Zone is sediment-enriched, dense, and made of settleable biomineral aggregates, while SPM in the offshore zone is a biomass-enriched, less dense and less settleable marine snow.

This special issue also contains works whose purpose is to quantify volumes of sediments transiting between morphological domains (estuary, beaches, spit, shelf) or which are redistributed between adjacent compartments, and which shape the coast at seasonal and interannual scales. Several illustrations concern sandy coasts in Senegal [57], Benin [351], Australia [352] and Vietnam [353], and one illustration concerns a muddy environment along the Mekong Delta [354]. Volumes of transported sediments serve as a basis for studying processes and for sensitivity analysis of shoreline changes under different forcing.

Shoreline changes on the wave-influenced Senegal River Delta in West Africa is highly dynamic. Abundant sand supply and strong wave-induced longshore drift have favoured the construction of a sand spit, which buffers wave energy and protect the back-barrier wetlands and lagoons, and is important in the regulation of the freshwater-saltwater balance and ecology of these areas. Sadio et al. [57] examined the spit dynamics from aerial photographs and satellite data between 1954 and 2015 and, using the longshore sediment transport rates they calculated from 1984 to 2015 via the re-analysis hindcast wave data, they analysed the mechanisms and processes behind these changes. Their study typically integrate field surveys, time-series of aerial and satellite data, and products of reanalysis, to enable us to better understand the impact of human action in this sand spit dynamics. Such scientific results are important in planning in future shoreline management and decision-making, between coastal protection and flooding of the lower delta plain of the Senegal River.

The morphological storm-event impact, seasonal cycles, trends of wave forcing, and beach's response along the Grand Popo beach in Benin by Abessolo Ondoa et al. [351] is based on three years and half of video data calibrated during a 10-day experiment, and on wave hindcast data. The alongshore shoreline position is affected seasonally, modulated by the wave height, and by winter storms, with 12 storms of averaged duration 1.6 days, mean erosion 3.1 m, and mean recovery duration of 15 days. Wave climate may amplify the impact of storms. This paper illustrates the use of video systems to monitor shoreline change and assess its dynamics at different scales, from events to seasons, at rather low cost, and prefigure long-term monitoring over sensitive beaches.

Mortlock et al. [352] analyse the impact of a storm along the Eastern Coast of Australia where beaches experienced one of their worst erosion in 40 years. They show that the obliquity of waves focused wave energy on coastal sections not equilibrated with such wave exposure under the prevailing south-easterly wave climate, and question the consequences of climate change on the regional wave climate, both for the mean state and extreme events.

Sandy beaches in Southeast Asia are affected both by paroxysmal but short storms such as cyclones (called typhoons there) and by atmospheric cold intrusions in winter (the "winter monsoon") responsible for 3-days to 3-weeks long strong persistent swell events. Almar et al. [353] investigated the shoreline response to a sequence of typhoon and monsoon events along the sandy beach of Nha Trang in Vietnam during a particularly active 2013–2014 season which encompassed the category 5 Haiyan typhoon. From continuous video monitoring, they show that long-lasting monsoon events have more persistent impact (longer beach recovery phase) than the typhoon. Using a shoreline equilibrium model, the seasonal shoreline behaviour is shown to be driven by the envelope of intra-seasonal events rather than monthly-averaged waves.

Vinh et al. [354] used the 3D numerical model Delft3D to study the fate of fine sediments in the estuaries and along the Mekong River delta. The model was calibrated and validated from 4 field campaigns. 50 scenarios corresponding to different wave and river discharge conditions enabled them to estimate the sediment dispersal and the longshore budget all along the delta, considering the occurrence of each condition. Such analysis of local vulnerability along a delta caused by typical forcing may serve as a scientific support to envisage protection measures in the most affected areas.

This special issue contains further illustrations of developments related to the spatialisation of data.

Quang et al. [355], on the example of the Cam Ranh Bay and Thuy Trieu lagoon, documented the spatio-temporal variations of turbidity from Landsat OLI data (at 30-m resolution) and analysed their dynamics from in situ data and from the distribution of bed shear stress obtained from a wave model. This work is an illustration of integrated studies, which combine satellite data, numerical models and field measurements to better assess sediment dynamics and its variability. Bottom reflectance was not an obstacle to the use of an empirical relationship to map turbidity from the remote sensing reflectance in the red band in their study.

Such a method was not applicable to the very shallow and oligotrophic waters of the lagoon of New Caledonia, where the environmental problem is not related to the amount of suspended sediments, but to their high toxicity. In their study, Wattelez et al. [312] introduced a Support Vector Regression (SVR) method and tested its capacity to map the turbidity distribution in a part of this coral reef lagoon. The model was trained with a large dataset of in situ turbidity, on coincident reflectance values from MODIS and on two other explanatory parameters: bathymetry and bottom colour. A comparison is done with a standard empirical inversion, which fail over such clear waters because of the bottom reflection of downwelling light. This paper illustrates the recent introduction of artificial intelligence and approaching methods in data processing, in the field of sediment dynamics, and the need to test and compare these tools with "classical" methods, to delineate their mutual advantages and drawbacks.

Zettam, et al. [356] use a coupled hydrological-erosion-transport model within a watershed (SWAT) to evaluate the contribution of different sub-basins to surface water and sediment flux, the dam's impact on water and sediment storage, and fluxes to the estuary, necessary for forcing estuarine and coastal studies. In their model, erosion is estimated using the modified universal soil loss equation (MUSLE) and sediment routing is based on a modified sediment transport equation from Bagnold. Since source-to-sink studies are expected to develop, it seemed important to us to include such a paper on sediment dynamics modelling at the watershed scale in this special issue.

Ohta, et al. [357] use a spatialized database on the size and geochemical composition of sediments in a watershed and adjacent coastal areas, and statistical analyses, to examine the transfer of particles from land to sea, and within coastal environment. The elemental concentrations of marine sediments are shown to vary with particle size. Geochemical features of silts and fine sands near the coasts reflect those of sediments in adjacent streams, while gravels and coarse sand do not, likely due to denudation of old basement rocks (Miocene-Pliocene siltstone) in a distant area of the Pacific Ocean (~100 km away) and to a strong bottom tidal current associated with the Kuroshio Current. This paper illustrates the information on particles origin provided by geochemistry, that are available to the oceanographers in the perspective of large integrated hydro-sedimentary studies.

Fernandez, et al. [358] offer another illustration of multidisciplinary approaches in sediment transport, combining geochemistry, dynamics and mineralogical analysis. They analysed vertical fluxes of particles and their composition (particle size distribution, geochemistry, mineralogy) in three bays impacted to varying degrees by nickel mining in the dry season. This enabled them to explain the main factors locally responsible for sediment transport (wind versus tide) and to show that particle aggregation led to a reduction in the metal concentration in the SPM, as identified by the decline in the metal solid-water distribution (or partition) coefficient K_d. This study illustrates that transport and associated processes may affect the composition of suspended matter, and underlines the benefit of integrating hydro-sedimentary dynamics, geochemistry and mineralogy in coastal environments.

7. Conclusions

The scientific research in coastal sediment dynamics is undergoing a profound current change. Two main factors, as we have seen, have encouraged the revolution of approaches: (1) a technological and instrumental development, so rapid that all the available tools renew almost completely in one or two decades, i.e., in a period shorter than the researcher's period of activity (with its corollary: the one who does not follow technological evolution is marginalized); (2) its applications concern growing societal concerns [359], which gives access to targeted founding. The interest related to its societal consequences adds to avowed or unacknowledged motivations of researchers (child dreams, will to understand, wish to venture on unknown territories of constantly evolving scientific questions). Let the scientific community take advantage of it and take the opportunity to show that this technological and mental boom is capable of improving knowledge, and consolidating the scientific foundation necessary for better management of coastal areas.

In terms of our practice, let us also remember the *challenge of comprehensiveness*, as stated by Edgard Morin [360]: "the increasingly wide, deep and serious imbalance between fragmented knowledge in the disciplines, on one hand, and multidimensional, comprehensive, transnational and global realities, and increasingly cross-cutting, polydisciplinary and even transdisciplinary problems, on the other hand". We are thus encouraged to develop research in sediment dynamics in two dimensions: personal work on selected specific issues, and a regular expansion of our knowledge and scientific culture beyond these issues, to understand the whole domain and possible interactions with other disciplines, to be able to always better situate our personal activity in the context and to follow the state of the art.

Everything moves in sciences and different approaches on a same topic develop in parallel, sometimes with differentiated speeds, where some dazzling acceleration move alongside more modest steps. "There are two kinds of science, applied and not yet applied" said George Porter, former President of the Royal Society of London. This encourages us, echoing Edgard Morin, to look after instrumental and analytical developments, the analysis of processes, modelling, and new approaches. Each of us is invited to balance its role as active participant, watch over its disciplinary field, and provide its practise in an expanded framework.

A last word because this new, multi-disciplinary and interdisciplinary scientific topic, previously abandoned to engineers, still bears no name other than that of its object of study: sediment transport and its consequences. How could we name it? *"Particulate transportology"*—from Latin roots: *trans* beyond, *portare* porter, *particula* small part, and Greek *logos* word, discourse, science—is neither practical, nor attractive, and would mix Latin and Greek roots. Would one dare to suggest *"metapheroclastology"*—from the Greek *metaphero*, I carry; *klastos*, broken, fragment, and *logos*? *Metapheroclastology*, what else? Don't hesitate to propose other options to our communities in the following years...

To conclude on another note, let us appreciate this special issue, very open to the scientific communities of the South and the North who have worked together. I thank our colleagues for sharing their results with the entire international community. Societal problems related to sediment dynamics, due to inland and coastal developments, concern the whole world. All countries with a maritime boundary are affected and have to deal with crisis situations, in industrialized, emerging, and developing countries. Scientific issues are common, and digital communication has allowed researchers to multiply worldwide collaborations, disrupting the production of knowledge. This special issue thus reflects the ongoing international cooperation between northern and southern countries on sensitive topics of common interest, thanks to the support of scientific research organizations such as the *Institute of Research for Development* in France, or through dedicated projects [57,351,353–356].

I thank the contributors, editors, reviewers and the staff of *Water*. I hope that the articles in this issue will give us food for thought. Happy reading, and good success in your respective projects!

Acknowledgments: For what they brought me in their approach to sediment dynamics, I warmly thank all the colleagues with whom I had the chance to collaborate since the end of the 1980s, starting with Jean Gruat and Benoît Le Guennec who were the first to teach me this topic and shared their passion with their students. This editorial is dedicated to them. I thank the *Institut de Recherche pour le Développement* for its continuous support. Finally, I thank two anonymous reviewers for their valuable comments and suggestions on the initial version of this paper.

Conflicts of Interest: The author declares no conflict of interest.

References

1. Martin, J.M.; Meybeck, M. Elemental mass-balance of material carried by major world rivers. *Mar. Chem.* **1979**, *7*, 173–206. [CrossRef]

2. Hedges, J.I.; Keil, R.G.; Benner, R. What happens to terrestrial organic matter in the ocean? *Org. Geochem.* **1997**, *27*, 195–212. [CrossRef]

3. Dupré, B.; Dessert, C.; Oliva, P.; Goddéris, Y.; Viers, J.; François, L.; Millot, R.; Gaillardet, J. Rivers, chemical weathering and Earth's climate. *C. R. Geosci.* **2003**, *335*, 1141–1160. [CrossRef]

4. Viers, J.; Dupré, B.; Gaillardet, J. Chemical composition of suspended sediments in World Rivers: New insights from a new database. *Sci. Total Environ.* **2009**, *407*, 853–868. [CrossRef] [PubMed]

5. Hedges, J.I.; Keil, R.G. Sedimentary organic matter preservation: An assessment and speculative synthesis. *Mar. Chem.* **1995**, *49*, 81–115. [CrossRef]

6. Smith, S.; Hollibaugh, J. Coastal metabolism and the oceanic organic carbon balance. *Rev. Geophys.* **1993**, *31*, 75–89. [CrossRef]

7. Foster, I.D.L.; Charlesworth, S.M. Heavy metals in the hydrological cycle: Trends and explanation. *Hydrol. Proc.* **1996**, *10*, 227–261. [CrossRef]

8. Ouillon, S. Erosion et transport solide: Ampleur et enjeux. *La Houille Blanche* **1998**, *2*, 52–58. [CrossRef]

9. Doney, S.C. The growing human footprint on coastal and open-ocean biogeochemistry. *Science* **2010**, *328*, 1512–1516. [CrossRef] [PubMed]

10. Vercruysse, K.; Grabowski, R.C.; Rickson, R.J. Suspended sediment transport dynamics in rivers: Multi-scale drivers of temporal variation. *Earth-Sci. Rev.* **2017**, *166*, 38–52. [CrossRef]

11. Zhou, Z.; Coco, G.; Townend, I.; Olabarrieta, M.; van der Wegen, M.; Gong, Z.; d'Alpaos, A.; Gao, S.; Jaffe, B.E.; Gelfenbaum, G.; et al. Is "Morphodynamic Equilibrium" an oxymoron? *Earth-Sci. Rev.* **2017**, *165*, 257–267. [CrossRef]

12. Chevillon, C. Skeletal composition of modern lagoon sediments in New Caledonia: Coral, a minor constituent. *Coral Reefs* **1996**, *15*, 199–207. [CrossRef]

13. Short, A.D. The distribution and impacts of carbonate sands on southern Australia beach-dune systems. In *Carbonate Beaches 2000*; Magoon, O.T., Robbins, L.L., Ewing, L., Eds.; ASCE: Reston, VA, USA, 2002; pp. 236–250. [CrossRef]

14. Droppo, I.G. Rethinking what constitutes suspended sediment. *Hydrol. Proc.* **2001**, *15*, 1551–1564. [CrossRef]

15. Migniot, C. Etude des propriétés physiques de différents sédiments très fins et de leur comportement sous des actions hydrodynamiques. *La Houille Blanche* **1968**, *7*, 591–620. [CrossRef]

16. Mehta, A.J. On estuarine cohesive sediment suspension behavior. *J. Geophys. Res.* **1989**, *94*, 14303–14314. [CrossRef]

17. Inman, D.L.; Jenkins, S.A. Energy and sediment budgets of the global coastal zone. In *Encyclopedia of Coastal Science*; Encyclopedia of Earth Science Series; Schwartz, M., Ed.; Springer: Dordrecht, The Netherland, 2005.

18. Leithold, E.L.; Blair, N.E.; Wegmann, K.W. Source-to-sink sedimentary systems and global carbon burial: A river runs through it. *Earth-Sci. Rev.* **2016**, *153*, 30–42. [CrossRef]

19. Dürr, H.H.; Laruelle, G.G.; van Kempen, C.M.; Slomp, C.P.; Meybeck, M.; Middelkoop, H. Worldwide typology of nearshore coastal systems: Defining the estuarine filter of river inputs to the Ocean. *Estuaries Coasts* **2011**, *34*, 441–458. [CrossRef]

20. Simpson, J.H. Physical processes in the ROFI regime. *J. Mar. Syst.* **1997**, *12*, 3–15. [CrossRef]

21. Coleman, J.M.; Gagliano, S.M. Cyclic sedimentation in the Mississippi river deltaic plain. *Gulf Coast Assoc. Geol. Soc. Trans.* **1964**, *14*, 67–80.

22. Allen, G.P.; Laurier, D.; Thouvenin, J.P. *Étude sédimentologique du delta de la Mahakam*; Notes Mem. 15, Total; Compagnies Française des Pétroles: Paris, France, 1979; p. 156.

23. Galloway, W.E. Process framework for describing the morphologic and stratigraphic evolution of deltaic depositional systems. In *Deltas, Models for Exploration*; Broussard, M.L., Ed.; Houston Geological Society: Houston, TX, USA, 1975; pp. 87–98.

24. Boyd, R.; Dalrymple, R.W.; Zaitlin, B.A. Classification of clastic coastal depositional environments. *Sediment. Geol.* **1992**, *80*, 139–150. [CrossRef]

25. Dalrymple, R.W.; Zaitlin, B.A.; Boyd, R. Estuarine facies models: Conceptual basis and stratigraphic implications: Perspective. *J. Sediment. Petrol.* **1992**, *62*, 1130–1146. [CrossRef]

26. Fleming, K.; Johnston, P.; Zwartz, D.; Yokoyama, Y.; Lambeck, K.; Chappell, J. Refining the eustatic sea-level curve since the Last Glacial Maximum using far- and intermediate-field sites. *Earth Planet. Sci. Lett.* **1998**, *163*, 327–342. [CrossRef]

27. Wetzel, A.; Szczygielski, A.; Unverricht, D.; Stattegger, K. Sedimentological and ichnological implications of rapid Holocene flooding of a gently sloping mud-dominated incised valley—An example from the Red River (Gulf of Tonkin). *Sedimentology* **2017**, *64*, 1173–1202. [CrossRef]

28. Syvitski, J.P.M.; Smith, J.N.; Calabrese, E.A.; Boudreau, B.P. Basin sedimentation and the growth of prograding deltas. *J. Geophys. Res.* **1988**, *93*, 6895–6908. [CrossRef]

29. Hinderer, M. From gullies to mountain belts: A review of sediment budgets at various scales. *Sediment. Geol.* **2012**, *280*, 21–59. [CrossRef]

30. Walling, D.E. Human impact on land-ocean sediment transfer by the world's rivers. *Geomorphology* **2006**, *79*, 192–216. [CrossRef]

31. Syvitski, J.P.M.; Milliman, J.D. Geology, geography, and humans battle for dominance over the delivery of fluvial sediment to the coastal ocean. *J. Geol.* **2007**, *115*, 1–19. [CrossRef]

32. Walling, D.E. *The Impact of Global Change on Erosion and Sediment Transport by Rivers: Current Progress and Future Challenges*; The United Nations World Water Development Report 3; UNESCO: Paris, France, 2009; p. 26.

33. Milliman, J.D.; Meade, R.H. World-wide delivery of river sediment to the oceans. *J. Geol.* **1983**, *91*, 1–2. [CrossRef]

34. Tamrazyan, G.P. Global peculiarities and tendencies in river discharge and wash-down of the suspended sediments—The Earth as a whole. *J. Hydrol.* **1989**, *107*, 113–131. [CrossRef]

35. Ludwig, W.; Probst, J.L. River sediment discharge to the oceans: Present-day controls and global budgets. *Am. J. Sci.* **1998**, *298*, 265–295. [CrossRef]

36. Walling, D.E.; Fang, D. Recent trends in the suspended sediment loads of the world's rivers. *Glob. Planet. Chang.* **2003**, *39*, 111–126. [CrossRef]

37. Wilkinson, B.H.; McElroy, B.J. The impact of humans on continental erosion and sedimentation. *Geol. Soc. Am. Bull.* **2007**, *119*, 140–156. [CrossRef]

38. Robinson, A.R. Relationships between soil erosion and sediment delivery. Erosion and Solid Matter Transport in Inland Waters. *IAHS Bull.* **1977**, *122*, 159–167.

39. Walling, D.E.; Webb, B.W. Erosion and sediment yield: A global overview. *IAHS Publ.* **1996**, *236*, 3–19.

40. Warrick, J.A.; Milliman, J.D.; Walling, D.E.; Wasson, R.J.; Syvitski, J.P.M.; Aalto, R.E. Earth is (mostly) flat: Apportionment of the flux of continental sediment over millennial time scales. *Geol. Forum* **2014**. [CrossRef]

41. Canals, M.; Puig, P.; Durrieu de Madron, X.; Heussner, S.; Palanques, A.; Fabres, J. Flushing submarine canyons. *Nature* **2006**, *444*, 354–357. [CrossRef] [PubMed]

42. Durrieu de Madron, X.; Wiberg, P.L.; Puig, P. Sediment dynamics in the Gulf of Lions: The impact of extreme events. *Cont. Shelf Res.* **2008**, *28*, 1867–1876. [CrossRef]

43. Palanques, A.; Puig, P.; Durrieu de Madron, X.; Sanchez-Vidal, A.; Pasqual, C.; Martin, J.; Calafat, A.; Heussner, S.; Canals, M. Sediment transport to the deep canyons and open-slope of the western Gulf of Lions during the 2006 intense cascading and open-sea convection period. *Prog. Ocean.* **2012**, *106*, 1–15. [CrossRef]

44. Goodbred, S.L., Jr.; Kuehl, S.A. Holocene and modern sediment budgets for the Ganges-Brahmaputra River System: Evidence for highstand dispersal to floodplain, shelf and deep-sea depocenters. *Geology* **1999**, *27*, 559–562. [CrossRef]

45. Wilson, C.A.; Goodbred, S.L., Jr. Construction and maintenance of the Ganges-Brahmaputra-Meghna delta: Linking process, morphology and stratigraphy. *Annu. Rev. Mar. Sci.* **2015**, *7*, 67–88. [CrossRef] [PubMed]

46. Allen, G.P.; Salomon, J.C.; Bassoullet, P.; du Penhoat, Y.; de Grandpré, C. Effects of tides on mixing and suspended sediment transport in macrotidal estuaries. *Sediment. Geol.* **1980**, *26*, 69–90. [CrossRef]

47. Jay, D.A.; Musiak, J.D. Particle trapping in estuarine tidal flows. *J. Geophys. Res.* **1994**, *99*, 20445–20461. [CrossRef]

48. Sottolichio, A.; Le Hir, P.; Castaing, P. Modeling mechanisms for the stability of the turbidity maximum in the Gironde estuary, France. *Proc. Mar. Sci.* **2000**, *3*, 373–386. [CrossRef]

49. Mitchell, S.B.; Uncles, R.J. Estuarine sediments in macrotidal estuaries: Future research requirements and management challenges. *Ocean Coast. Manag.* **2013**, *79*, 97–100. [CrossRef]

50. Toublanc, F.; Brenon, I.; Coulombier, T. Formation and structure of the turbidity maximum in the macrotidal Charente estuary (France): Influence of fluvial and tidal forcing. *Estuar Coast. Shelf Sci.* **2016**, *169*, 1–14. [CrossRef]

51. Burchard, H.; Schuttelaars, H.M.; Ralston, D.K. Sediment trapping in estuaries. *Annu. Rev. Mar. Sci.* **2018**, *10*, 371–395. [CrossRef] [PubMed]

52. Lefebvre, J.-P.; Ouillon, S.; Vinh, V.D.; Arfi, R.; Panche, J.-Y.; Mari, X.; Thuoc, C.V.; Torreton, J.P. Seasonal variability of cohesive sediment aggregation in the Bach Dang-Cam Estuary, Haiphong (Vietnam). *Geo-Mar. Lett.* **2012**, *32*, 103–121. [CrossRef]

53. Brown, S.; Nicholls, R.J. Subsidence and human influences in mega deltas: The case of the Ganges-Brahmaputra-Meghna. *Sci. Total Environ.* **2015**, *527–528*, 362–374. [CrossRef] [PubMed]

54. Winant, C.D.; Inman, D.L.; Nordstrom, C.E. Description of seasonal changes using empirical eigenfunctions. *J. Geophys. Res.* **1975**, *80*, 1979–1986. [CrossRef]

55. Migniot, C. Action des courants, de la houle et du vent sur les sediments. *La Houille Blanche* **1977**, *1*, 9–47. [CrossRef]

56. Almar, R.; Kestenare, E.; Reyns, J.; Jouanno, J.; Anthony, E.J.; Laibi, R.; Hemer, M.; du Penhoat, Y.; Ranasinghe, R. Response of the Bight of Benin (Gulf of Guinea, West Africa) coastline to anthropogenic and natural forcing, Part1: Wave climate variability and impacts on the longshore sediment transport. *Cont. Shelf Res.* **2015**, *110*, 48–59. [CrossRef]

57. Sadio, M.; Anthony, E.J.; Diaw, A.T.; Dussouillez, P.; Fleury, J.T.; Kane, A.; Almar, R.; Kestenare, E. Shoreline Changes on the Wave-Influenced Senegal River Delta, West Africa: The Roles of Natural Processes and Human Interventions. *Water* **2017**, *9*, 357. [CrossRef]

58. Tong, S.S. Cartographie par télédétection des espaces intertidaux du Vietnam. Ph.D. Thesis, University de Reims Champagne-Ardenne, Reims, France, 2016.

59. Beusen, A.H.W.; Dekkers, A.L.M.; Bouwman, A.F.; Ludwig, W.; Harrison, J. Estimation of global river transport of sediments and associated particulate C, N and P. *Glob. Biogeochem. Cycles* **2005**, *19*, GB4S05. [CrossRef]

60. Turner, A.; Millward, G.E. Suspended particles: Their role in estuarine biogeochemical cycles. *Estuar. Coast. Shelf Sci.* **2002**, *55*, 857–883. [CrossRef]

61. Bilotta, G.S.; Brazier, R.E. Understanding the influence of suspended solids on water quality and aquatic biota. *Water. Res.* **2008**, *42*, 2849–2861. [CrossRef] [PubMed]

62. Hickey, B.M.; Kudela, R.M.; Nash, J.D.; Bruland, K.W.; Peterson, W.T.; MacCready, P.; Lessard, E.J.; Jay, D.A.; Banas, N.S.; Baptista, A.M.; et al. River Influences on Shelf Ecosystems: Introduction and synthesis. *J. Geophys. Res.-Oceans* **2010**, *115*, C00B17. [CrossRef]

63. Anderson, D.M.; Burkholder, J.M.; Cochlan, W.P.; Glibert, P.M.; Gobler, C.J.; Heil, C.A.; Kudela, R.M.; Parsons, M.L.; Rensel, J.E.J.; Townsend, D.W.; et al. Harmful algal blooms and eutrophication: Examining linkages from selected coastal regions of the United States. *Harmful Algae* **2008**, *8*, 39–53. [CrossRef] [PubMed]

64. Annane, S.; St-Amand, L.; Starr, M.; Pelletier, E.; Ferreyra, G.A. Contribution of transparent exopolymeric particles (TEP) to estuarine particulate organic carbon pool. *Mar. Ecol. Prog. Ser.* **2015**, *529*, 17–34. [CrossRef]

65. Duarte, C.M.; Borja, A.; Carstensen, J.; Elliott, M.; Krause-Jensen, D.; Marba, N. Paradigms in the recovery of estuarine and coastal ecosystems. *Estuaries Coasts* **2015**, *38*, 1202–1212. [CrossRef]

66. Newcombe, C.P.; MacDonald, D.D. Effects of suspended sediments on aquatic ecosystems. *N. Am. J. Fish. Manag.* **1991**, *11*, 72–82. [CrossRef]

67. Hédouin, L.; Bustamante, P.; Churlaud, C.; Pringault, O.; Fichez, R.; Warnau, M. Trends in concentrations of selected metalloid and metals in two bivalves from the SW lagoon of New Caledonia. *Ecotoxicol. Environ. Saf.* **2009**, *72*, 372–381. [CrossRef] [PubMed]

68. Stoichev, T.; Amouroux, D.; Wasserman, J.C.; Point, D.; De Diego, A.; Bareille, G.; Donard, O.F.X. Dynamics of mercury species in surface sediments of a macrotidal estuarine-coastal system (Adour River, Bay of Biscay). *Estuar. Coast. Shelf Sci.* **2004**, *59*, 511–521. [CrossRef]

69. Navarro, P.; Amouroux, D.; Nghi, D.T.; Rochelle-Newall, E.; Ouillon, S.; Arfi, R.; Thuoc, C.V.; Mari, X.; Torréton, J.P. Fate and tidal transport of butyltin and mercury compounds in the waters of the tropical Bach Dang estuary (Haiphong, Vietnam). *Mar. Poll. Bull.* **2012**, *64*, 1789–1798. [CrossRef] [PubMed]

70. Pang, H.J.; Lou, Z.H.; Jin, A.M.; Yan, K.K.; Jiang, Y.; Yang, X.H.; Chen, C.T.A.; Chen, X.G. Contamination, distribution, and sources of heavy metals in the sediments of Andong tidal flat, Hangzhou bay, China. *Cont. Shelf Res.* **2015**, *110*, 72–84. [CrossRef]

71. Sholkovitz, E.R. Flocculation of dissolved organic and inorganic matter during the mixing of river water and seawater. *Geochim. Cosmochim. Acta* **1976**, *40*, 831–845. [CrossRef]

72. Sholkovitz, E.R. The flocculation of dissolved Fe, Mn, Cu, Ni, Co and Cd during estuarine mixing. *Earth Planet. Sci. Lett.* **1978**, *41*, 77–86. [CrossRef]

73. Bouillon, S.; Borges, A.V.; Castaned-Moya, E.; Diele, K.; Dittmar, T.; Duke, N.C.; Kristensen, E.; Lee, S.Y.; Marchand, C.; Middelburg, J.J.; et al. Mangrove production and carbon sinks: A revision of global budget estimates. *Glob. Biogeochem. Cycles* **2008**, *22*, GB2013. [CrossRef]

74. Kristensen, E.; Bouillon, S.; Dittmar, T.; Marchand, C. Organic carbon dynamics in mangrove ecosystems: A review. *Aquat. Bot.* **2008**, *89*, 201–219. [CrossRef]

75. Ludwig, W.; Amiotte-Suchet, P.; Munhoven, G.; Probst, J.L. Atmospheric CO_2 consumption by continental erosion: Present-day control and implications for the last glacial maximum. *Glob. Planet. Chang.* **1998**, *16–17*, 107–120. [CrossRef]

76. Gaillardet, J.; Dupré, B.; Louvat, P.; Allègre, C.J. Global silicate weathering and CO_2 consumption rates deduced from the chemistry of large rivers. *Chem. Geol.* **1999**, *159*, 3–30. [CrossRef]

77. Galy, V.; Peucker-Ehrenbrink, B.; Eglinton, T. Global carbon export from the terrestrial biosphere controlled by erosion. *Nature* **2015**, *521*, 204–207. [CrossRef] [PubMed]

78. Goudie, A.S. Global warming and fluvial geomorphology. *Geomorphology* **2006**, *79*, 384–394. [CrossRef]

79. Jerolmack, D.J.; Paola, C. Shredding of environmental signals by sediment transport. *Geophys. Res. Lett.* **2010**, *37*, L19401. [CrossRef]

80. Knight, J.; Harrison, S. The impacts of climate change on terrestrial Earth surface systems. *Nat. Clim. Chang.* **2013**, *3*, 24–29. [CrossRef]

81. Achite, M.; Ouillon, S. Recent changes in climate, hydrology and sediment load in the Wadi Abd, Algeria (1970–2010). *Hydrol. Earth Syst. Sci.* **2016**, *20*, 1355–1372. [CrossRef]

82. Labat, D.; Goddéris, Y.; Probst, J.L.; Guyot, J.L. Evidence for global runoff increase related to climate warming. *Adv. Water Resour.* **2004**, *27*, 631–642. [CrossRef]

83. Bates, B.C.; Kundzewicz, Z.W.; Wu, S.; Palutikof, J.P. (Eds.) *Climate Change and Water*; Technical Paper of the Intergovernmental Panel on Climate Change; IPCC Secretariat: Geneva, Switzerland, 2008; 210p.

84. Cazenave, A.; Le Cozannet, G. Sea level rise and its coastal impacts. *Earth Future* **2013**, *2*, 15–34. [CrossRef]

85. Dieng, H.B.; Cazenave, A.; Meyssignac, B.; Ablain, M. New estimate of the current rate of sea level rise from a sea level budget approach. *Geophys. Res. Lett.* **2017**, *44*. [CrossRef]

86. Xu, X.-Y.; Birol, F.; Cazenave, A. Evaluation of Coastal Sea Level Offshore Hong Kong from Jason-2 Altimetry. *Remote Sens.* **2018**, *10*, 282. [CrossRef]

87. Melet, A.; Meyssignac, B.; Almar, R.; Le Cozannet, G. Underestimated wave contribution to sea level change and rise at the coast. *Nat. Clim. Chang.* **2018**, *8*, 234–239. [CrossRef]

88. Vörösmarty, C.J.; Meybeck, M.; Fekete, B.; Sharma, K.; Green, P.; Syvitski, J.P.M. Anthropogenic sediment retention: Major global impact from registered river impoundments. *Glob. Planet. Chang.* **2003**, *39*, 169–190. [CrossRef]

89. Hooke, R.L.B. On the history of humans as geomorphic agents. *Geology* **2000**, *28*, 843–846. [CrossRef]

90. Steffen, W.; Richardson, K.; Rockström, J.; Cornell, S.E.; Fetzer, I.; Bennett, E.M.; Biggs, R.; Carpenter, S.R.; de Vries, W.; de Wit, C.A.; et al. Planetary boundaries: Guiding human development on a changing planet. *Science* **2015**, *347*, 1259855. [CrossRef] [PubMed]

91. Syvitski, J.P.M.; Peckham, S.D.; Hilberman, R.; Mulder, T. Predicting the terrestrial flux of sediment to the global ocean: A planetary perspective. *Sediment. Geol.* **2003**, *162*, 5–24. [CrossRef]

92. Bhattacharya, J.P.; Copeland, P.; Lawton, T.F.; Holbrook, J. Estimation of source area, river paleo-discharge, paleoslope, and sediment budgets of linked deep-time depositional systems and implications for hydrocarbon potential. *Earth-Sci. Rev.* **2016**, *153*, 77–110. [CrossRef]

93. Rosser, N.J.; Petley, D.; Lim, M.; Dunning, S.; Allison, R.J. Terrestrial laser scanning for monitoring the process of hard rock coastal cliff erosion. *Q. J. Eng. Geol. Hydrogeol.* **2005**, *38*. [CrossRef]

94. Lim, M.; Rosser, N.J.; Petley, D.N.; Keen, M. Quantifying the Controls and Influence of Tide and Wave Impacts on Coastal Rock Cliff Erosion. *J. Coast. Res.* **2011**, *27*, 46–56. [CrossRef]

95. Woodroffe, C.D.; Murray-Wallace, C.V. Sea-level and coastal change: The past as a guide to the future. *Quat. Sci. Rev.* **2012**, *54*, 4–11. [CrossRef]

96. Nittrouer, C.A.; Wright, L.D. Transport of particles across continental shelves. *Rev. Geophys.* **1994**, *32*, 85–113. [CrossRef]

97. Brink, K.H. Cross-shelf exchange. *Annu. Rev. Mar. Sci.* **2016**, *8*, 59–78. [CrossRef] [PubMed]

98. Horner-Devine, A.R.; Hetland, R.D.; MacDonald, D.G. Mixing and Transport in Coastal River Plumes. *Ann. Rev. Fluid Mech.* **2015**, *47*, 569–594. [CrossRef]
99. Syvitski, J.P.M.; Harvey, N.; Wolanski, E.; Burnett, W.C.; Perillo, G.M.E.; Gornitz, V. Dynamics *of the* coastal zone. In *Coastal Fluxes in the Anthropocene: The Land-Ocean Interactions in the Coastal Zone Project of the International Geosphere-Biosphere Programme*; Crossland, C.J., Kremer, H.H., Lindeboom, H.J., Crossland, J.I.M., Le Tissier, M.D.A., Eds.; Springer: Berlin, Germany, 2005; pp. 39–94.
100. Dyer, K.R. *Coastal and Estuarine Sediment Dynamics*; John Wiley & Sons: Chichester, UK, 1986; 342p.
101. Nielsen, P. *Coastal Bottom Boundary Layers and Sediment Transport*; World Scientific: Singapore, 1992; 324p.
102. Soulsby, R. *Dynamics of Marine Sands*; Thomas Telford: London, UK, 1998; 250p.
103. Dean, R.G.; Dalrymple, R.A. *Coastal Processes with Engineering Applications*; Cambridge University Press: Cambridge, UK, 2002; 488p.
104. U.S. Army Corps of Engineers. *Coastal Engineering Manual (CEM)*; Engineer Manual 1110-2-1100; U.S. Army Corps of Engineers: Washington, DC, USA, 2002; Volume 6.
105. Dronkers, J. *Dynamics of Coastal Systems*; Advanced Series on Ocean Engineering; World Scientific: Singapore, 2005; Volume 25.
106. Winterwerp, J.C.; van Kesteren, W. *Introduction to the Physics of Cohesive Sediment in the Marine Environment*; Elsevier Developments in Sedimentology 56; Elsevier: Cambridge, UK, 2004.
107. Van Rijn, L.C. *Principles of Sediment Transport in Rivers, Estuaries and Coastal Seas*; Aqua Publications: Amsterdam, The Netherlands, 2005.
108. Uncles, R.J.; Mitchell, S.B. (Eds.) *Estuarine and Coastal Hydrography and Sediment Transport*; Cambridge University Press: Cambridge, UK, 2017. [CrossRef]
109. Mehta, A.J. *An Introduction to Hydraulics of Fine Sediment Transport*; Advanced Series on Ocean Engineering; World Scientific: Singapore, 2013; Volume 38.
110. Wright, L.D.; Short, A.D. Morphodynamic variability of surf zones and beaches: A synthesis. *Mar. Geol.* **1984**, *56*, 93–118. [CrossRef]
111. Lippmann, T.C.; Holman, R.A. The spatial and temporal variability of sand bar. *J. Geophys. Res.* **1990**, *95*, 11575–11590. [CrossRef]
112. Coco, G.; Murray, A.B. Patterns in the sand: From forcing templates to self-organization. *Geomorphology* **2007**, *91*, 271–290. [CrossRef]
113. Van Rijn, L.C.; Ribberink, J.S.; van der Werf, J.; Walstra, D.J.R. Coastal sediment dynamics: Recent advances and future research needs. *J. Hydraul. Res.* **2013**, *51*, 475–493. [CrossRef]
114. Castelle, B.; Scott, T.; Brander, R.; McCarroll, R.J. Rip current types, Circulation and hazards. *Earth-Sci. Rev.* **2016**, *163*. [CrossRef]
115. Almar, R.; Coco, G.; Bryan, K.R.; Huntley, D.A.; Short, A.D.; Senechal, N. Video observations of beach cusp morphodynamics. *Mar. Geol.* **2008**, *254*, 216–223. [CrossRef]
116. Ruessink, G.; Blenkinsopp, C.; Brinkkemper, J.; Castelle, B.; Dubarbier, B.; Grasso, F.; Puleo, J.A.; Lanckriet, T. Sandbar and beach-face evolution on a prototype coarse sandy barrier. *J. Coast. Eng.* **2015**. [CrossRef]
117. Castelle, B.; Marieu, V.; Bujan, S.; Splinter, K.; Robinet, A.; Sénéchal, N.; Ferreira, S. Impact of the winter 2013–2014 series of severe Western Europe storms on a double-barred sandy coast: Beach and dune erosion and megacusp embayments. *Geomorphology* **2015**, *238*. [CrossRef]
118. Horn, D. Measurements and modeling of beach groundwater flow in the swash-zone: A review. *Cont. Shelf Res.* **2006**, *26*, 622–652. [CrossRef]
119. Puleo, J.A.; Torres-Freyermuth, A. The second international workshop on swash-zone processes. *Coast. Eng.* **2016**, *115*, 1–7. [CrossRef]
120. Chardon-Maldonado, P.; Pintado-Patiño, J.C.; Puleo, J.A. Advances in swash-zone research: Small-scale hydrodynamic and sediment transport processes. *Coast. Eng.* **2016**, *115*, 8–25. [CrossRef]
121. Recking, A. A generalized threshold model for computing bed load grain size distribution. *Water Resour. Res.* **2016**, *52*. [CrossRef]
122. De Swart, H.E.; Zimmerman, J.T.F. Morphodynamics of tidal inlets. *Annu. Rev. Fluid Mech.* **2009**, *41*, 203–229. [CrossRef]
123. Dodet, G.; Bertin, X.; Bruneau, N.; Fortunato, A.B.; Nahon, A.; Roland, A. Wave-current interactions in a wave-dominated tidal inlets. *J. Geophys. Res. Oceans* **2013**, *118*, 1587–1605. [CrossRef]

124. Huang, W.P. Modelling the effects of typhoons on morphological changes in the estuary of Beinan, Taiwan. *Cont. Shelf Res.* **2017**, *135*, 1–13. [CrossRef]

125. Chaumillon, E.; Bertin, X.; Fortunato, A.B.; Bajo, M.; Schneider, J.-L.; Dezileau, L.; Walsh, J.P.; Michelot, A.; Chauveau, E.; Créach, A.; et al. Storm-induced marine flooding: Lessons from a multidisciplinary approach. *Earth-Sci. Rev.* **2017**, *165*, 151–184. [CrossRef]

126. Masselink, G.; Castelle, B.; Scott, T.; Dodet, G.; Suanez, S.; Jackson, D.W.T.; Floc'h, F. Extreme wave activity during 2013/2014 winter and morphological impacts along the Atlantic coast of Europe. *Geophys. Res. Lett.* **2016**, *45*. [CrossRef]

127. Baumann, J.; Chaumillon, E.; Bertin, X.; Schneider, J.L.; Guillot, B.; Schmutz, M. Importance of infragravity waves for the generation of washover deposits. *Mar. Geol.* **2017**, *391*, 20–35. [CrossRef]

128. Hein, C.J.; Galy, V.; Galy, A.; France-Lanord, C.; Kudrass, H.; Schwenk, T. Post-glacial climate forcing of surface processes in the Ganges-Brahmaputra river basin and implications for carbon sequestration. *Earth Planet. Sci. Lett.* **2017**, *478*, 89–101. [CrossRef]

129. Otvos, E.G. Coastal barriers—Nomenclature, processes, and classification issues. *Geomorphology* **2012**, *139–140*, 39–52. [CrossRef]

130. Leuven, J.R.F.W.; Kleinhans, M.G.; Weisscher, S.A.H.; van der Vegt, M. Tidal sand bar dimensions and shapes in estuaries. *Earth-Sci. Rev.* **2016**, *161*, 204–223. [CrossRef]

131. Larcombe, P.; Ridd, P.V.; Prytz, A.; Wilson, B. Factors controlling suspended sediment on inner-shelf coral reefs, Townsville, Australia. *Coral Reefs* **1995**, *14*, 163–171. [CrossRef]

132. Bonneton, P.; Lefebvre, J.P.; Bretel, P.; Ouillon, S.; Douillet, P. Tidal modulation of wave-setup and wave-induced currents on the Aboré coral reef, New Caledonia. *J. Coast. Res.* **2007**, *50*, 762–766.

133. Lowe, R.J.; Falter, J.L. Oceanic forcing of coral reefs. *Ann. Rev. Mar. Sci.* **2015**, *7*, 43–66. [CrossRef] [PubMed]

134. Ouillon, S.; Douillet, P.; Lefebvre, J.P.; Le Gendre, R.; Jouon, A.; Bonneton, P.; Fernandez, J.M.; Chevillon, C.; Magand, O.; Lefèvre, J.; et al. Circulation and suspended sediment transport in a coral reef lagoon: The southwest lagoon of New Caledonia. *Mar. Poll. Bull.* **2010**, *61*, 269–296. [CrossRef] [PubMed]

135. Erftemeijer, P.L.A.; Riegl, B.; Hoeksema, B.W.; Todd, P.A. Environmental impacts of dredging and other sediment disturbances on corals: A review. *Mar. Poll. Bull.* **2012**, *64*, 1737–1765. [CrossRef] [PubMed]

136. Eisma, D. Flocculation and de-flocculation of suspended matter in estuaries. *Neth. J. Sea Res.* **1986**, *20*, 183–199. [CrossRef]

137. Winterwerp, J.C. A simple model for turbulence induced flocculation of cohesive sediment. *J. Hydraul. Res.* **1998**, *36*, 309–326. [CrossRef]

138. Winterwerp, J.C. A heuristic formula for turbulence-induced flocculation of cohesive sediment. *Estuar. Coast. Shelf Sci.* **2006**, *68*, 195–207. [CrossRef]

139. Khelifa, A.; Hill, P.S. Models for effective density and settling velocity of flocs. *J. Hydraul. Res.* **2006**, *44*, 390–401. [CrossRef]

140. Fettweis, M. Uncertainty of excess density and settling velocity of mud flocs derived from in situ measurements. *Estuar. Coast. Shelf Sci.* **2008**, *78*, 426–436. [CrossRef]

141. Verney, R.; Lafite, R.; Brun-Cottan, J.C.; Le Hir, P. Behaviour of a floc population during a tidal cycle: Laboratory experiments and numerical modelling. *Cont. Shelf Res.* **2011**, *31*, S64–S83. [CrossRef]

142. Vinh, V.D.; Ouillon, S.; Uu, D.V. Estuarine Turbidity Maxima and variations of aggregate parameters in the Cam-Nam Trieu estuary, North Vietnam, in early wet season. *Water* **2018**, *10*, 68. [CrossRef]

143. Maggi, F. Biological flocculation of suspended particles in nutrient-rich aqueous ecosystems. *J. Hydrol.* **2009**, *376*, 116–125. [CrossRef]

144. Maggi, F.; Tang, F.H.M. Analysis of the effect of organic matter content on the architecture and sinking of sediment aggregates. *Mar. Geol.* **2015**, *363*, 102–111. [CrossRef]

145. Mari, X.; Torréton, J.P.; Chu, V.T.; Lefebvre, J.P.; Ouillon, S. Seasonal aggregation dynamics along a salinity gradient in the Bach Dang estuary, North Vietnam. *Estuar. Coast. Shelf Sci.* **2012**, *96*, 151–158. [CrossRef]

146. Yamada, Y.; Fukuda, H.; Uchimiya, M.; Motegi, C.; Nishino, S.; Kikuchi, T.; Nagata, T. Localized accumulation and a shelf-basin gradient of particles in the Chukchi Sea and Canada Basin, western Arctic. *J. Geophys. Res. Oceans* **2015**, *120*, 4638–4653. [CrossRef]

147. Mari, X.; Passow, U.; Migon, C.; Burd, A.B.; Legendre, L. Transparent exopolymer particles: Effects on carbon cycling in the ocean. *Prog. Oceanogr.* **2017**, *151*, 13–37. [CrossRef]

148. Chen, T.Y.; Skoog, A. Aggregation of organic matter in coastal waters: A dilemma of using a Couette flocculator. *Cont. Shelf Res.* **2017**, *139*, 62–70. [CrossRef]

149. Parchure, T.M.; Mehta, A.J. Erosion of soft cohesive sediment deposits. *J. Hydraul. Eng.* **1985**, *111*. [CrossRef]

150. Grabowski, R.C.; Droppo, I.G.; Wharton, G. Erodibility of cohesive sediment: The importance of sediment properties. *Earth-Sci. Rev.* **2011**, *105*, 101–120. [CrossRef]

151. Amos, C.L.; Bergamasco, A.; Umgiesser, G.; Cappucci, S.; Cloutier, D.; DeNat, L.; Flindt, M.; Bonardi, M.; Cristante, S. The stability of tidal flats in Venice Lagoon—The results of in-situ measurements using two benthic, annular flumes. *J. Mar. Syst.* **2004**, *51*, 211–241. [CrossRef]

152. Ganthy, F.; Soissons, L.; Sauriau, P.G.; Verney, R.; Sottolochio, A. Effects of short flexible seagrass Zostera noltei on flow, erosion and deposition processes determined using flume sediments. *Sedimentology* **2015**, *62*, 997–1023. [CrossRef]

153. Heinzelmann, C.; Wallisch, S. Benthic settlement and bed erosion. A review. *J. Hydraul. Res.* **1991**, *29*, 355–371. [CrossRef]

154. Le Hir, P.; Monbet, Y.; Orvain, F. Sediment erodibility in sediment transport modelling: Can we account for biota effects? *Cont. Shelf Res.* **2007**, *27*, 1116–1142. [CrossRef]

155. Jiang, F.; Mehta, A.J. Mudbanks of the Southwest Coast of India IV: Mud viscoelastic properties. *J. Coast. Res.* **1995**, *11*, 918–926.

156. Beyramzade, M.; Siadatmousavi, S.M. Implementation of viscoelastic mud-induced energy attenuation in the third-generation wave model, SWAN. *Ocean Dyn.* **2018**, *68*, 47–63. [CrossRef]

157. Vinh, V.D.; Ouillon, S.; Thanh, T.D.; Chu, L.V. Impact of the Hoa Binh dam (Vietnam) on water and sediment budgets in the Red River basin and delta. *Hydrol. Earth Syst. Sci.* **2014**, *18*, 3987–4005. [CrossRef]

158. Green, M.O.; Coco, G. Review of wave-driven sediment resuspension and transport in estuaries. *Rev. Geophys.* **2014**, *52*, 77–117. [CrossRef]

159. Foda, A.M.; Hunt, J.R.; Chou, H.-T. A nonlinear model for the fluidization of marine mud by waves. *J. Geophys. Res.* **1993**, *98*, 7039–7047. [CrossRef]

160. Allison, M.A.; Lee, M.T.; Ogston, A.S.; Aller, R.C. Origin of Amazon mudbanks along the northeast of South America. *Mar. Geol.* **2000**, *163*. [CrossRef]

161. Anthony, E.; Gardel, A.; Gratiot, N.; Proisy, C.; Allison, M.A.; Dolique, F.; Fromard, F. The Amazon-influenced muddy coast of South America: A review of mud-bank-shoreline interactions. *Earth-Sci. Rev.* **2010**, *103*. [CrossRef]

162. Muraleedharan, K.R.; Kumar, P.K.D.; Kumar, S.P.; Srijith, B.; John, S.; Kumar, K.R.N. Observed salinity changes in the Alappuzha mud bank, southwest coast of India and its implication to hypothesis of mudbank formation. *Cont. Shelf Res.* **2017**, *137*, 39–45. [CrossRef]

163. Furukawa, K.; Wolanski, E.; Mueller, H. Currents and sediment transport in mangrove forests. *Estuar. Coast. Shelf Sci.* **1997**, *44*, 301–310. [CrossRef]

164. Many, G.; Bourrin, F.; de Madron, X.D.; Pairaud, I.; Gangloff, A.; Doxaran, D.; Ody, A.; Verney, R.; Menniti, C.; le Berre, D.; et al. Particle assemblage characterization in the Rhone River ROFI. *J. Mar. Syst.* **2016**, *157*, 39–51. [CrossRef]

165. Friedrichs, C.T.; Wright, L.D.; Hepworth, D.A.; Kim, S.C. Bottom-boundary-layer processes associated with fine sediment accumulation in coastal seas and bays. *Cont. Shelf Res.* **2000**, *20*, 807–841. [CrossRef]

166. Holland, K.T.; Elmore, P.A. A review of heterogeneous sediments in coastal environments. *Earth-Sci. Rev.* **2008**, *89*, 116–134. [CrossRef]

167. Mitchener, H.; Torfs, H. Erosion of mud/sand mixtures. *Coast. Eng.* **1996**, *29*, 1–25. [CrossRef]

168. Panagiotopoulos, I.; Voulgaris, G.; Collins, M.B. The influence of clay on the threshold of movement of fine sandy beds. *Coast. Eng.* **1997**, *32*, 19–43. [CrossRef]

169. Van Ledden, M.; van Kesteren, W.G.M.; Winterwerp, J.C. A conceptual framework for the erosion behavior of sand-mud mixtures. *Cont. Shelf Res.* **2004**, *24*, 1–11. [CrossRef]

170. Jacobs, W.; Le Hir, P.; van Kesteren, W.; Cann, P. Erosion threshold of sand-mud mixtures. *Cont. Shelf Res.* **2011**, *31*, S14–S25. [CrossRef]

171. Zhou, Z.; Coco, G.; van der Wegen, M.; Gong, Z.; Zhang, C.; Townend, I. Modeling sorting dynamics of cohesive and non-cohesive sediments on intertidal flats under the effect of tides and wind waves. *Cont. Shelf Res.* **2015**, *104*, 76–91. [CrossRef]

172. Jago, C.F.; Jones, S.E. Observation and modelling of the dynamics of benthic fluff resuspended from a sandy bed in the southern North Sea. *Cont. Shelf Res.* **1998**, *18*, 1255–1282. [CrossRef]

173. Rooni, M.; Winterwerp, J.C. Surficial sediment erodibility from time-series measurements of suspended sediment concentrations: Development and validation. *Ocean Dyn.* **2017**, *67*, 691–712. [CrossRef]

174. George, A. L'humanisme scientifique. In *Les Grands Appels de L'homme Contemporain*; Editions du Temps Présent: Paris, France, 1946; pp. 5–33.

175. Meybeck, M. Global analysis of river systems: From Earth system controls to Anthopocene syndromes. *Phil. Trans. R. Soc. Lond. B* **2003**, *358*, 1935–1955. [CrossRef] [PubMed]

176. Syvitski, J.P.M.; Vörösmarty, C.J.; Kettner, A.J.; Green, P. Impact of Humans on the flux of terrestrial sediment to the global coastal ocean. *Science* **2005**, *308*, 376. [CrossRef] [PubMed]

177. Milliman, J.D.; Qin, Y.S.; Ren, M.E.; Saito, Y. Man's influence on the erosion and transport of sediment by Asian Rivers: The Yellow River (Huanghe) example. *J. Geol.* **1987**, *95*, 751–762. [CrossRef]

178. Wilkinson, B.H. Humans as geologic agents: A deep-time perspective. *Geology* **2005**, *33*, 161–164. [CrossRef]

179. Syvitski, J.P.M.; Kettner, A. Sediment flux and the Anthropocene. *Phil. Trans. R. Soc. A* **2011**, *369*, 957–975. [CrossRef] [PubMed]

180. Waters, C.N.; Zalasiewicz, J.; Summerhayes, C.; Barnosky, A.D.; Poirier, C.; Galuszka, A.; Cearreta, A.; Edgeworth, M.; Ellis, E.C.; Ellis, M.; et al. The Anthropocene is functionally and stratigraphically distinct from the Holocene. *Science* **2016**, *351*, 137. [CrossRef] [PubMed]

181. Winterwerp, J.C. On the Response of Tidal Rives to Deepening and Narrowing—Risks for a Regime towards Hyper-Turbid Conditions. Final Report of the Research Program 'LTV Safety and Accessibility', WWF, 2013. Available online: http://www.wwf.de/fileadmin/fm-wwf/Publikationen-PDF/Report_Risks_for_a_regime_shift_towards_hyper-turbid_conditions.pdf (accessed on 14 February 2018).

182. World Industrial Silica Sand. Available online: https://www.freedoniagroup.com/industry-study/world-industrial-silica-sand-3237.htm (accessed on 14 February 2018).

183. Syvitski, J.P.M.; Saito, Y. Morphodynamics of deltas under the influence of humans. *Glob. Planet. Chang.* **2007**, *57*, 261–282. [CrossRef]

184. Nicholls, R.J.; Cazenave, A. Sea-level rise and its impact on coastal zones. *Science* **2010**, *328*, 1517. [CrossRef] [PubMed]

185. Garcia-Ruiz, J.M.; Begueria, S.; Lana-Renault, N.; Nadal-Romero, E.; Cerda, A. Ongoing and emerging questions in water erosion studies. *Land Degrad. Dev.* **2017**, *28*, 5–21. [CrossRef]

186. Nittrouer, C.A.; Mullarney, J.C.; Allison, M.A.; Ogston, A.S. Introduction to the special issue on sedimentary processes building a tropical delta yesterday, today, and tomorrow: The Mekong System. *Oceanography* **2017**, *30*, 10–21. [CrossRef]

187. Garel, E.; Sousa, C.; Ferreira, O. Sand bypass and updrift beach evolution after jetty construction at an ebb-tidal delta. *Estuar Coast. Shelf Sci.* **2015**, *167*, 4–13. [CrossRef]

188. Nienhuis, J.H.; Ashton, A.D.; Nardin, W.; Fagherazzi, S.; Giosan, L. Alongshore sediment bypassing as a control on river mouth morphodynamics. *J. Geophys. Res. Earth Surf.* **2016**, *121*, 664–683. [CrossRef]

189. Gedan, K.B.; Kirwan, M.L.; Wolanski, E.; Barbier, E.B.; Silliman, B.R. The present and future role of coastal wetland vegetation in protecting shorelines: Answering recent challenges to the paradigm. *Clim. Chang.* **2011**, *106*, 7–29. [CrossRef]

190. Van Slobbe, E.; de Vriend, H.J.; Aarninkhof, S.; Lulofs, K.; de Vries, M.; Dircke, P. Building with Nature: In search of resilient storm surge protection strategies. *Nat. Hazards* **2013**, *66*, 1461–1480. [CrossRef]

191. Cheong, S.M.; Silliman, B.; Wong, P.P.; van Wesenbeeck, B.; Kim, C.K.; Guannel, G. Coastal adaptation with ecological engineering. *Nat. Clim. Chang.* **2013**, *3*, 787–791. [CrossRef]

192. De Schipper, M.A.; de Vries, S.; Ruessink, G.; de Zeeuw, R.C.; Rutten, J.; van Gelder-Maas, C.; Stive, M.J.F. Initial spreading of a mega feeder nourishment: Observations of the Sand Engine pilot project. *Coast. Eng.* **2016**, *111*, 23–38. [CrossRef]

193. Van der Nat, A.; Vellinga, P.; Leemans, R.; van Slobbe, E. Ranking coastal flood protection designs from engineering to nature-based. *Ecol. Eng.* **2016**, *87*, 80–90. [CrossRef]

194. Stansby, P.K. Coastal hydrodynamics—Present and future. *J. Hydraul. Res.* **2013**, *51*, 341–350. [CrossRef]

195. Thiébot, J.; Bailly du Bois, P.; Guillou, S. Numerical modeling of the effect of tidal stream turbines on the hydrodynamics and the sediment transport—Application to the Alderney Race (Raz Blanchard), France. *Renew. Energy* **2015**, *75*, 356–365. [CrossRef]

196. Shields, A. Anwendung der Ähnlichkeitsmechanik und der Turbulenzforschung auf die Geschiebebewegung. Available online: http://repository.tudelft.nl/islandora/search/author%3AShields?collection=research (accessed on 5 May 2017).

197. Rouse, H. Modern conceptions of the mechanics of fluid turbulence. *Trans. Am. Soc. Civ. Eng.* **1937**, *102*, 463–554.

198. Meyer-Peter, E.; Müller, R. Formulas for bed-load transport. In Proceedings of the 2nd Meeting of the International Association for Hydraulic Structures Research, Delft, The Netherlands, 7 June 1948; pp. 39–64.

199. Einstein, H.A. *Bed-Load Function for Sediment Transportation in Open Channel Flows*; US Department of Agriculture: Washington, DC, USA, 1950.

200. Bagnold, R.A. *An Approach to the Sediment Transport Problem form General Physics*; U.S. Geological Survey Professional Paper, 422–J; US Government Printing Office: Washington, DC, USA, 1966.

201. Krone, R.B. *Flume Studies of the Transport of Sediment in Estuarial Shoaling Processes*; Tech Rep; Hydraulic Eng Lab and Sanitary Eng Res Lab, University California: Berkeley, CA, USA, 1962.

202. Partheniades, E. Erosion and Deposition of Cohesive Soils. *J. Hydraul. Div.* **1965**, *91*, 105–139.

203. Engelund, F.; Hansen, E. *A Monograph on Sediment Transport in Alluvial Streams*; Teknich Forlag, Technical Press: Copenhagen, Denmark, 1967. Available online: https://repository.tudelft.nl/islandora/object/uuid:81101b08-04b5-4082-9121-861949c336c9 (accessed on 3 January 2018).

204. Ackers, P.; White, W.R. Sediment transport: New approach and analysis. *J. Hydraul. Div. ASCE* **1973**, *99*, 2041–2060.

205. Ariathurai, C.R. A Finite Element Model for Sediment Transport in Estuaries. Ph.D. Thesis, Davis, University of California, Berkeley, CA, USA, 1974.

206. Miller, M.C.; McCave, I.N.; Komar, P.D. Threshold of sediment motion under unidirectional currents. *Sedimentology* **1977**, *24*, 507–527. [CrossRef]

207. Bailard, J.A. An energetics total load sediment transport model for a plane sloping beach. *J. Geophys. Res. Oceans* **1981**, *86*, 938–954. [CrossRef]

208. Bailard, J.A.; Inman, D.L. An energetics bedload model for a plane sloping beach: Local transport. *J. Geophys. Res. Oceans* **1981**, *86*, 2035–2043. [CrossRef]

209. Holman, R.A.; Bowen, A.J. Bars, bumps, and holes—Models for the generation of complex beach topography. *J. Geophys. Res. Oceans* **1982**, *87*, 457–468. [CrossRef]

210. Mehta, A.J.; Hayter, E.J.; Parker, W.R.; Krone, R.B.; Teeter, A.M. Cohesive sediment transport I: Process description. *J. Hydraul. Eng.* **1989**, *115*, 1076–1093. [CrossRef]

211. Camenen, B.; Larson, M. A general formula for non-cohesive bed load sediment transport. *Estuar. Coast. Shelf Sci.* **2005**, *63*, 249–260. [CrossRef]

212. McCave, I.N. Vertical flux of particles in the ocean. *Deep-Sea Res.* **1975**, *22*, 491–502. [CrossRef]

213. Nittrouer, C.A.; Sternberg, R.W.; Carpenter, R.; Bennet, J.T. The use of Pb-210 geochronology as a sedimentological tool: Application to the Washington continental shelf. *Mar. Geol.* **1979**, *31*, 297–316. [CrossRef]

214. Teisson, C.; Ockenden, M.; Le Hir, P.; Kranenburg, C.; Hamm, L. Cohesive sediment transport processes. *Coast. Eng.* **1993**, *21*, 129–162. [CrossRef]

215. Kuehl, S.A.; Nittrouer, C.A.; Allison, M.A.; Faria, L.E.C.; Dukat, D.A.; Jaeger, J.M.; Pacioni, T.D.; Figueiredo, A.G.; Underkoffler, E.C. Sediment deposition, accumulation, and seabed dynamics in an energetic fine-grained coastal environment. *Cont. Shelf Res.* **1996**, *16*, 787–815. [CrossRef]

216. Allison, M.A.; Kineke, G.C.; Gordon, E.S.; Goni, M.A. Development and reworking of a seasonal flood deposit on the inner continental shelf off the Atchafalaya River. *Cont. Shelf Res.* **2000**, *20*, 2267–2294. [CrossRef]

217. Kraus, N.C. Application of portable traps for obtaining point measurements of sediment transport rates in the surf zone. *J. Coast. Res.* **1987**, *3*, 139–152.

218. Miller, H.C. Field measurements of longshore sediment transport during storms. *Coast. Eng.* **1999**, *36*, 301–321. [CrossRef]

219. Roelvink, J.A.; Stive, M.F. Bar-generating cross-shore flow mechanisms on a beach. *J. Geophys. Res.* **1989**, *94*, 4785–4800. [CrossRef]

220. De Vriend, H.J.; Capobianco, M.; Chesher, T.; de Swart, H.E.; Latteux, B.; Stive, M.J.F. Approaches to long-term modelling of coastal morphology: A review. *Coast. Eng.* **1993**, *21*, 225–269. [CrossRef]

221. Dibajnia, M.; Watanabe, A. Transport rate under irregular sheet flow conditions. *Coast. Eng.* **1998**, *35*, 167–183. [CrossRef]

222. De Vriend, H.J.; Stive, M.J.F. Quasi-3D modeling of nearshore currents. *Coast. Eng.* **1987**, *11*, 565–601. [CrossRef]

223. Van Rijn, L.C. Mathematical modelling of suspended sediment in non-uniform flows. *J. Hydraul. Eng.* **1986**, *112*, 433–455. [CrossRef]

224. Teisson, C. Cohesive suspended sediment transport: Feasibility and limitations of numerical modelling. *J. Hydraul. Res.* **1991**, *29*, 755–769. [CrossRef]

225. Hanson, H. GENESIS-A Generalized Shoreline Change Numerical Model. *J. Coast. Res.* **1989**, *5*, 1–27.

226. Kamphuis, J.W. Alongshore sediment transport rate. *J. Waterw. Port Coast. Ocean Eng.* **1991**, *117*, 624–640. [CrossRef]

227. Stive, M.; Roelvink, D.; de Vriend, H. Large-scale coastal evolution concept. In Proceedings of the 22th Coastal Engineering Conference, Delft, The Netherlands, 2–6 July 1990; ASCE: New York, NY, USA, 1991; pp. 1017–1027.

228. Lyzenga, D.R. Passive remote-sensing techniques for mapping water depth and bottom features. *Appl. Opt.* **1978**, *17*, 379–383. [CrossRef] [PubMed]

229. Tassan, S.; Sturm, B. An algorithm for the retrieval of sediment content in turbid coastal waters data. *Int. J Remote Sens.* **1986**, *7*, 643–655. [CrossRef]

230. Stumpf, R.P.; Pennock, J.R. Calibration of a general optical equation for remote sensing of suspended sediments in a moderately turbid estuary. *J. Geophys. Res.* **1989**, *94*, 14363–14371. [CrossRef]

231. Ouillon, S.; Forget, P.; Froidefond, J.M.; Naudin, J.J. Estimating suspended matter concentrations from SPOT data and from field measurements in the Rhône river plume. *Mar. Technol. Soc. J.* **1997**, *31*, 15–20.

232. Acker, J.; Ouillon, S.; Gould, R.W., Jr.; Arnone, R.A. Measuring Marine Suspended Sediment Concentrations from Space: History and Potential. In *Proceedings of the 8th International Conference Remote Sensing for Marine and Coastal Environments, Halifax, NS, Canada, 17–19 May 2005*; Altarum/AMRS: Ann Arbor, MI, USA, 2005.

233. Ouillon, S. An inversion method for reflectance in stratified turbid waters. *Int. J. Remote Sens.* **2003**, *24*, 535–548. [CrossRef]

234. Bunt, J.A.C.; Larcombe, P.; Jago, C.F. Quantifying the response of optical backscatter devices and transmissometers to variations in suspended particulate matter. *Cont. Shelf Res.* **1999**, *19*, 1199–1220. [CrossRef]

235. Sutherland, T.F.; Lane, P.M.; Amos, C.L.; Downing, J. The calibration of optical backscatter sensors for suspended sediment of varying darkness levels. *Mar. Geol.* **2000**, *162*, 587–597. [CrossRef]

236. Guillen, J.; Palanques, A.; Puig, P.; Durrieu de Madron, X.; Nyffeler, F. Field calibration of optical sensors for measuring suspended sediment concentration in the western Mediterranean. *Scient. Mar.* **2000**, *64*, 427–435. [CrossRef]

237. Davies-Colley, R.J.; Smith, D.G. Turbidity, suspended sediment, and water clarity: A review. *J. Am. Water Res. Assoc.* **2001**, *37*, 1085–1101. [CrossRef]

238. Mietta, F.; Chassagne, C.; Manning, A.J.; Winterwerp, J.C. Influence of shear rate, organic matter content, pH and salinity on mud flocculation. *Ocean Dyn.* **2009**, *59*, 751–763. [CrossRef]

239. Berlamont, J.; Ockenden, M.; Toorman, E.; Winterwerp, J. The characterization of cohesive sediment properties. *Coast. Eng.* **1993**, *21*, 105–128. [CrossRef]

240. Neukermans, G.; Loisel, H.; Meriaux, X.; Astoreca, R.; Mckee, D. In situ variability of mass-specific beam attenuation and backscattering of marine particles with respect to particle size, density, and composition. *Limnol. Oceanogr.* **2012**, *57*, 124–144. [CrossRef]

241. Agrawal, Y.C.; Pottsmith, H.C. Instrument for particle size and settling velocity observations in sediment transport. *Mar. Geol.* **2000**, *16*, 89–114. [CrossRef]

242. Mikkelsen, O.A.; Pejrup, M. The use of LISST-100 in-situ laser particle sizer for estimates of floc size, density and settling velocity. *Geo Mar. Lett.* **2001**, *20*, 187–195. [CrossRef]

243. IOCCG. *Remote Sensing of Ocean Colour in Coastal, and Other Optically-Complex, Waters*; Report of the International Ocean-Colour Coordinating Group, n°3; Sathyendranath, S., Ed.; IOCCG: Darmouth, NS, Canada, 2000.

244. Fugate, C.D.; Friedrichs, C.T. Determining concentration and fall velocity of estuarine particle populations using ADV, OBS and LISST. *Cont. Shelf Res.* **2002**, *22*, 1867–1886. [CrossRef]

245. Gartner, J.W. Estimating suspended solids concentrations from backscatter intensity measured by acoustic Doppler current profiler in San Francisco Bay, California. *Mar. Geol.* **2004**, *211*, 169–187. [CrossRef]

246. Tessier, C.; Le Hir, P.; Lurton, X.; Castaing, P. Estimation of suspended sediment concentration from backscatter intensity of Acoustic Doppler Current Profiler. *C. R. Geosci.* **2008**, *340*, 57–67. [CrossRef]

247. Jourdin, F.; Tessier, C.; Le Hir, P.; Verney, R.; Lunven, M.; Loyer, S.; Lusven, A.; Filipot, J.-F.; Lepesqueur, J. Dual-frequency ADCPs measuring turbidity. *Geo-Mar. Lett.* **2014**, *34*, 381–397. [CrossRef]

248. Jouon, A.; Ouillon, S.; Douillet, P.; Lefebvre, J.P.; Fernandez, J.-M.; Mari, X.; Froidefond, J.M. Spatio-temporal variability in suspended particulate matter concentration and the role of aggregation on size distribution in a coral reef lagoon. *Mar. Geol.* **2008**, *256*, 36–48. [CrossRef]

249. Twardowski, M.S.; Boss, E.; McDonald, J.B.; Pegau, W.S.; Barnard, A.H.; Zaneveld, J.R.V. A model for estimating bulk refractive index from the optical backscattering ratio and the implications for understanding particle composition in case I and case II waters. *J. Geophys. Res. Oceans* **2001**, *106*, 14129–14142. [CrossRef]

250. Boss, E.; Twardowski, M.S.; Herring, S. Shape of the particulate beam attenuation spectrum and its inversion to obtain the shape of the particulate size distribution. *Appl. Opt.* **2001**, *40*, 4885–4893. [CrossRef] [PubMed]

251. Werdell, P.J.; Frantz, B.A.; Bailey, S.W.; Feldman, G.C.; Boss, E.; Brando, V.E.; Dowell, M.; Hirata, T.; Lavender, S.J.; Lee, Z.P.; et al. Generalized ocean color inversion model for retrieving marine inherent optical properties. *Appl. Opt.* **2013**, *52*, 2019–2037. [CrossRef] [PubMed]

252. O'Donogue, T.; Wright, S. Flow tunnel measurements of velocities and sand flux in oscillatory sheet flow for well sorted and graded sands. *Coast. Eng.* **2004**, *51*, 1163–1184. [CrossRef]

253. Stockdon, H.F.; Holman, R.A. Estimation of wave phase speed and nearshore bathymetry from video imagery. *J. Geophys. Res. Oceans* **2000**, *105*, 22015–22033. [CrossRef]

254. Aarninkhof, S.; Turner, I.L.; Dronkers, T.; Caljouw, M.; Nipius, L. A video-technique for mapping intertidal beach bathymetry. *Coast. Eng.* **2003**, *49*, 275–289. [CrossRef]

255. Bergsma, E.; Conley, D.C.; Davidson, M.A.; O'Hare, T.J. Video-based nearshore bathymetry estimation in macro-tidal environments. *Mar. Geol.* **2016**, *374*. [CrossRef]

256. Bell, P.S. Shallow water bathymetry derived from an analysis of X-band marine radar images. *Coast. Eng.* **1999**, *37*. [CrossRef]

257. Bell, P.S.; Bird, C.O.; Plater, A.J. A temporal waterline approach to mapping intertidal areas using X-band marine radar. *Coast. Eng.* **2016**, *107*, 84–101. [CrossRef]

258. Brenon, I.; Le Hir, P. Modelling the turbidity maximum in the Seine estuary (France): Identification of formation processes. *Estuar. Coast. Shelf Sci.* **1999**, *49*, 525–544. [CrossRef]

259. Wu, W.M.; Rodi, W.; Wenka, T. 3D numerical modelling of flow and sediment transport in open channels. *J. Hydraul. Eng. ASCE* **2000**, *126*, 4–15. [CrossRef]

260. Lesser, G.R.; Roelvink, J.A.; van Kester, J.A.T.M.; Stelling, G.S. Development and validation of a three-dimensional morphological model. *Coast. Eng.* **2004**, *51*, 883–915. [CrossRef]

261. Roelvink, J.A. Coastal morphodynamic evolution techniques. *Coast. Eng.* **2006**, *53*, 277–287. [CrossRef]

262. Le Hir, P.; Cayocca, F.; Waeles, B. Dynamics of sand and mud mixtures: A multiprocess-based modelling strategy. *Cont. Shelf Res.* **2011**, *31*, S135–S149. [CrossRef]

263. Coastal Models. Available online: http://csdms.colorado.edu/wiki/Coastal_models (accessed on 14 February 2018).

264. Sutherland, J.; Peet, A.H.; Soulsby, R.L. Evaluating the performance of morphological models. *Coast. Eng.* **2004**, *51*, 917–939. [CrossRef]

265. Roelvink, D.; Reniers, A.; van Dongeren, A.; van Thiel de Vries, J.; McCall, R.; Lescinski, J. Modelling storm impacts on beaches, dunes and barrier islands. *Coast. Eng.* **2009**, *56*, 1133–1152. [CrossRef]

266. Ashton, A.D.; Hutton, E.W.H.; Kettner, A.J.; Xing, F.; Kallumadikal, J.; Nienhuis, J.; Giosan, L. Progress in coupling models of coastline and fluvial dynamics. *Comput. Geosci.* **2013**, *53*, 21–29. [CrossRef]

267. Brunn, P. Sea level rise as a cause of shore erosion. *ASCE J. Waterw. Harb. Div.* **1962**, *88*, 117–130.

268. Cooper, J.A.G.; Pilkey, O.H. Sea-level rise and shoreline retreat: Time to abandon the Bruun Rule. *Glob. Planet. Chang.* **2004**, *43*, 157–171. [CrossRef]

269. Le Cozannet, G.; Garcin, M.; Yates, M.; Idier, D.; Meyssignac, B. Approaches to evaluate the recent impacts of sea-level rise on shoreline changes. *Earth-Sci. Rev.* **2014**, *138*, 47–60. [CrossRef]

270. Lee, Z.P.; Carder, K.L.; Mobley, C.D.; Steward, R.G.; Patch, J.S. Hyperspectral remote sensing for shallow waters: 2. Deriving bottom depths and water properties by optimization. *Appl. Opt.* **1999**, *38*, 3831–3843. [CrossRef] [PubMed]

271. Ouillon, S.; Douillet, P.; Andréfouet, S. Coupling satellite data with in situ measurements and numerical modeling to study fine suspended sediment transport: A study for the lagoon of New Caledonia. *Coral Reefs* **2004**, *23*, 109–122. [CrossRef]

272. Zhao, H.H.; Chen, Q.; Walker, N.D.; Zheng, Q.A.; MacIntyre, H.L. A study of sediment transport in a shallow estuary using MODIS imagery and particle tracking simulation. *Int. J. Remote Sens.* **2011**, *32*, 6653–6671. [CrossRef]

273. Carniello, L.; Silvestri, S.; Marani, M.; D'Alpaos, A.; Volpe, V.; Defina, A. Sediment dynamics in shallow tidal basins: In situ observations, satellite retrievals, and numerical modeling in the Venice Lagoon. *J. Geophys. Res. Earth Surf.* **2014**, *119*, 802–815. [CrossRef]

274. Li, Y.D.; Li, X.F. Remote sensing observations and numerical studies of a super typhoon-induced suspended sediment concentration variation in the East China Sea. *Ocean Model.* **2016**, *104*, 187–202. [CrossRef]

275. Stroud, J.R.; Lesht, B.M.; Schwab, D.J.; Beletsky, D.; Stein, M.L. Assimilation of satellite images into a sediment transport model of Lake Michigan. *Water Resour. Res.* **2009**, *45*. [CrossRef]

276. El Serafy, G.Y.; Eleveld, M.A.; Blaas, M.; van Kessel, T.; Gaytan Aguilar, S.; van der Woerd, H.J. Improving the description of the suspended particulate matter concentrations in the southern North Sea through assimilating remotely sensed data. *Ocean Sci. J.* **2011**, *46*, 179–204. [CrossRef]

277. Margvelashvili, N.; Andrewartha, J.; Herzfeld, M.; Robson, B.J.; Brando, V.E. Satellite data assimilation and estimation of a 3D coastal sediment transport model using error-subspace emulators. *Environ. Model. Softw.* **2013**, *40*, 191–201. [CrossRef]

278. Jaud, M.; Grasso, F.; Le Dantec, N.; Verney, R.; Delacourt, C.; Ammann, J.; Deloffre, J.; Grandjean, P. Potential of UAVs for monitoring mudflat morphodynamics (Application to the Seine estuary, France). *ISPRS Int. J. Geo-Inf.* **2016**, *5*, 50. [CrossRef]

279. Loisel, H.; Mangin, A.; Vantrepotte, V.; Dessailly, D.; Dat, D.N.; Garnesson, P.; Ouillon, S.; Lefebvre, J.P.; Mériaux, X.; Thu, P.M. Variability of suspended particulate matter concentration in coastal waters under the Mekong's influence from ocean color (MERIS) remote sensing over the last decade. *Remote Sens. Environ.* **2014**, *150*, 218–230. [CrossRef]

280. Dogliotti, A.I.; Ruddick, K.; Guerrero, R. Seasonal and inter-annual turbidity variability in the Rio de la Plata from 15 years of MODIS: El Niño dilution effect. *Estuar. Coast. Shelf Sci.* **2016**, *182*, 27–39. [CrossRef]

281. Pan, Y.; Shen, F.; Wei, X. Fusion of Landsat-8/OLI and GOCI Data for hourly mapping of Suspended Particulate Matter at high spatial resolution: A case study in the Yangtze (Changjiang) Estuary. *Remote Sens.* **2018**, *10*, 158. [CrossRef]

282. McFeeters, S.K. The use of the Normalized Difference Water Index (NDWI) in the delineation of open water features. *Int. J. Remote Sens.* **1996**, *17*, 1425–1432. [CrossRef]

283. Gens, R. Remote sensing of coastlines: Detection, extraction and monitoring. *Int. J. Remote Sens.* **2010**, *31*, 1819–1836. [CrossRef]

284. Liu, Y.; Huang, H.; Qiu, Z.; Fan, J. Detecting coastline change from satellite images based on beach slope estimation in a tidal flat. *Int. J. Appl. Earth Obs. Geoinf.* **2013**, *23*, 165–176. [CrossRef]

285. Feyisa, G.L.; Meilby, H.; Fensholt, R.; Proud, S.R. Automated water extraction index: A new technique for surface water mapping using Landsat imagery. *Remote Sens. Environ.* **2014**, *140*, 23–35. [CrossRef]

286. Garcia-Rubio, G.; Huntley, D.; Russell, P. Evaluating shoreline identification using optical satellite images. *Mar. Geol.* **2015**, *359*, 96–105. [CrossRef]

287. Sagar, S.; Roberts, D.; Bala, B.; Lymburner, L. Extracting the intertidal extent and topography of the Australian coastline from a 28 year time series of Landsat observations. *Remote Sens. Environ.* **2017**, *195*, 153–169. [CrossRef]

288. Cracknell, A.P. Remote sensing techniques in estuaries and coastal zones—An update. *Int. J. Remote Sens.* **1999**, *20*, 485–496. [CrossRef]

289. Su, T.C. A study of a matching pixel by pixel (MPP) algorithm to establish an empirical model of water quality mapping, as based on unmanned aerial vehicle (UAV) images. *Int. J. Appl. Earth Obs. Geoinf.* **2017**, *58*, 213–224. [CrossRef]

290. Graham, G.W.; Davies, E.J.; Nimmo-Smith, W.A.M.; Bowers, D.G.; Braithwaite, K.M. Interpreting LISST-100X measurements of particles with complex shape using digital in-line holography. *J. Geophys. Res.* **2012**, *117*, C05034. [CrossRef]
291. Shao, Y.; Maa, J. Comparisons of different instruments for measuring suspended cohesive sediment concentrations. *Water* **2017**, *9*, 968. [CrossRef]
292. Sahin, C.; Verney, R.; Sheremet, A.; Voulgaris, G. Acoustic backscatter by suspended cohesive sediments: Field observations, Seine Estuary, France. *Cont. Shelf Res.* **2017**, *134*, 39–51. [CrossRef]
293. Felix, D.; Albayrak, I.; Boes, R.M. In-situ investigation on real-time suspended sediment measurement techniques: Turbidimetry, acoustic attenuation, laser diffraction (LISST) and vibrating tube densimetry. *Int. J. Sediment Res.* **2018**, in press. [CrossRef]
294. Zhang, Y.; Ye, F.; Stanev, E.V.; Grashorn, S. Seamless cross-scale modeling with SCHISM. *Ocean Model.* **2016**, *102*, 64–81. [CrossRef]
295. Frings, R.M.; Vollmer, S. Guidelines for sampling bedload transport with minimum uncertainty. *Sedimentology* **2017**, *64*, 1630–1645. [CrossRef]
296. Franz, G.; Leitão, P.; Pinto, L.; Jauch, E.; Fernandes, L.; Neves, R. Development and validation of a morphological model for multiple sediment classes. *Int. J. Sediment Res.* **2017**, *32*, 585–596. [CrossRef]
297. Li, J.D.; Sun, J.; Lin, B. Bed-load transport rate based on the entrainment probabilities of sediment grains by rolling and lifting. *Int. J. Sediment Res.* **2018**, in press. [CrossRef]
298. Kidanemariam, A.G.; Uhlmann, M. Direct numerical simulation of pattern formation in subaqueous sediment. *J. Fluid Mech.* **2014**, *750*, R2. [CrossRef]
299. Elghannay, H.; Tafti, D. LES-DEM simulations of sediment transport. *Int. J. Sediment Res.* **2018**, in press. [CrossRef]
300. Ancey, C.; Heyman, J. A microstructural approach to bed load transport: Mean behavior and fluctuations of particle transport rates. *J. Fluid Mech.* **2014**, *744*, 129–168. [CrossRef]
301. Ancey, C.; Bohorquez, P.; Heyman, J. Stochastic interpretation of the advection-diffusion equation and its relevance to bed load transport. *J. Geophys. Res. Earth Surf.* **2015**, *120*, 2529–2551. [CrossRef]
302. Charru, F.; Andreotti, B.; Claudin, P. Sand ripples and dunes. *Annu. Rev. Fluid Mech.* **2013**, *45*, 469. [CrossRef]
303. Chiodi, F.; Claudin, P.; Andreotti, B. A two-phase flow model of sediment transport: Transition from bedload to suspended load. *J. Fluid Mech.* **2014**, *755*, 561–581. [CrossRef]
304. Houssais, M.; Ortiz, C.P.; Durian, D.J.; Jerolmack, D.J. Onset of sediment transport is a continuous transition driven by fluid shear and granular creep. *Nat. Commun.* **2015**, *6*, 6527. [CrossRef] [PubMed]
305. Maurin, R.; Chauchat, J.; Frey, P. Dense granular flow rheology in turbulent bedload transport. *J. Fluid Mech.* **2016**, *804*, 490–512. [CrossRef]
306. Sahin, C.; Gruner, H.A.A.; Ozturk, M.; Sheremet, A. Floc size variability under strong turbulence: Observations and artificial neural network modeling. *Appl. Ocean Res.* **2017**, *68*, 130–141. [CrossRef]
307. Boukhrissa, Z.A.; Khanchoul, K.; Le Bissonnais, Y.; Tourki, M. Prediction of sediment load by sediment rating curve and neural network (ANN) in El Kebir catchment, Algeria. *J. Earth Syst. Sci.* **2013**, *122*, 1303–1312. [CrossRef]
308. Malik, A.; Kumar, A.; Piri, J. Daily suspended sediment concentration simulation using hydrological data of Pranhita River Basin, India. *Comput. Elect. Agric.* **2017**, *138*, 20–28. [CrossRef]
309. Pannimpullath, R.R.; Jourdin, F.; Charantonis, A.A.; Yala, K.; Rivier, A.; Badran, F.; Thiria, S.; Guillou, S.; Lecker, F.; Gohin, F.; et al. Construction of multiyear time-series profiles of suspended particulate inorganic matter concentrations using machine learning approach. *Remote Sens.* **2017**, *9*, 1320. [CrossRef]
310. Pinet, S.; Lartiges, B.; Martinez, J.-M.; Ouillon, S. A SEM-based method to determine the mineralogical composition and the particle size distribution of suspended sediments applied to the Amazon River basin. *Int. J. Sediment Res.* **2018**, in press.
311. Pinet, S.; Martinez, J.-M.; Ouillon, S.; Lartiges, B.; Villar, R.E. Variability of apparent and inherent optical properties of sediment-laden waters in large river basins—Lessons from in situ measurements and bio-optical modeling. *Opt. Express* **2017**, *25*, A283–A310. [CrossRef] [PubMed]
312. Wattelez, G.; Dupouy, C.; Lefèvre, J.; Ouillon, S.; Fernandez, J.M.; Juillot, F. Application of the Support Vector Regression method for turbidity assessment with MODIS on a shallow coral reef lagoon (Voh-Koné-Pouembout, New Caledonia). *Water* **2017**, *9*, 737. [CrossRef]

313. Li, J.; Heap, A.D. Spatial interpolation methods applied in the environmental sciences: A review. *Environ. Model. Softw.* **2014**, *53*, 173–189. [CrossRef]

314. Kong, J.L.; Sun, X.M.; Wong, D.W.; Chen, Y.; Yang, J.; Yan, Y.; Wang, L.X. A semi-analytical model for remote sensing retrieval of suspended sediment concentration in the Gulf of Bohai, China. *Remote Sens.* **2015**, *7*, 5373–5397. [CrossRef]

315. Constantin, S.; Doxaran, D.; Constantinescu, S. Estimation of water turbidity and analysis of its spatio-temporal variability in the Danube River plume (Black Sea) using MODIS satellite data. *Cont. Shelf Res.* **2016**, *112*, 14–30. [CrossRef]

316. Han, B.; Loisel, H.; Vantrepotte, V.; Mériaux, X.; Bryère, P.; Ouillon, S.; Dessailly, D.; Xing, Q.; Zhu, J. Development of a semi-analytical algorithm for the retrieval of Suspended Particulate Matter from remote sensing over clear to very turbid waters. *Remote Sens.* **2016**, *8*, 211. [CrossRef]

317. Yang, X.P.; Sokoletsky, L.; Wei, X.D.; Shen, F. Suspended sediment concentration mapping based on the MODIS satellite imagery in the East China inland, estuarine, and coastal waters. *Chin. J. Ocean. Limnol.* **2017**, *35*, 39–60. [CrossRef]

318. Toubal, A.K.; Achite, M.; Ouillon, S.; Dehni, A. Soil erodibility mapping using the RUSLE model to prioritize erosion control in the Wadi Sahouat basin, North-West of Algeria. *Environ. Monit. Assess.* **2018**, *190*, 210. [CrossRef] [PubMed]

319. Gangloff, A.; Verney, R.; Doxaran, D.; Ody, A.; Estournel, C. Investigating Rhône River plume (Gulf of Lions, France) dynamics using metrics analysis from the MERIS 300m Ocean Color archive (2002-2012). *Cont. Shelf Res.* **2017**, *144*, 98–111. [CrossRef]

320. Leeuw, T.; Boss, E. The HydroColor App: Above water measurements of remote sensing reflectance and turbidity using a smartphone camera. *Sensors* **2018**, *18*, 256. [CrossRef] [PubMed]

321. Morin, E. *Science Avec Conscience*; Fayard: Paris, France, 1982.

322. Gourdin, E.; Evrard, O.; Huon, S.; Lefevre, I.; Ribolzi, O.; Reyss, J.L.; Sengtaheuanghoung, O. Suspended sediment dynamics in a southeast Asian mountainous catchment: Combining river monitoring and fallout radionuclides tracers. *J. Hydrol.* **2014**, *519*, 1811–1823. [CrossRef]

323. Zebracki, M.; Eyrolle-Boyer, F.; Evrard, O.; Claval, D.; Mourier, B.; Gairoard, S.; Cagnat, X.; Antonelli, C. Tracing the origin of suspended sediment in a large Mediterranean river by combining continuous river monitoring and measurement of artificial and natural radionuclides. *Sci. Total Environ.* **2015**, *502*, 122–132. [CrossRef] [PubMed]

324. Collins, A.L.; Pulley, S.; Foster, I.D.L.; Gellis, A.; Porto, P.; Horowitz, A.J. Sediment source fingerprinting as an aid to catchment management: A review of the current state of knowledge and a methodological decision-tree for end-users. *J. Environ. Manag.* **2017**, *194*, 86–108. [CrossRef] [PubMed]

325. Laceby, J.P.; Evrard, O.; Smith, H.G.; Blake, W.H.; Olley, J.M.; Minella, J.P.G.; Owens, P.N. The challenges and opportunities of addressing particle size effects in sediment source fingerprinting: A review. *Earth-Sci. Rev.* **2017**, *169*, 85–103. [CrossRef]

326. Fowler, S.W.; Knauer, G.A. Role of large particles in the transport of elements and organic compounds through the oceanic water column. *Prog. Oceanogr.* **1986**, *16*, 147–194. [CrossRef]

327. Alldredge, A.L.; Silver, M.W. Characteristics, Dynamics and Significance of Marine Snow. *Prog. Oceanogr.* **1988**, *20*, 41–82. [CrossRef]

328. Mari, X.; Rochelle-Newall, E.; Torréton, J.P.; Pringault, O.; Jouon, A.; Migon, C. Water residence time: A regulatory factor of the DOM to POM transfer efficiency. *Limnol. Oceanogr.* **2007**, *52*, 808–819. [CrossRef]

329. Nguyen, T.H.; Tang, F.H.M.; Maggi, F. Optical measurement of cell colonization patterns on individual suspended sediment aggregates. *J. Geophys. Res. Earth Surf.* **2017**, *122*, 1794–1807. [CrossRef]

330. Black, K.S.; Tolhurst, T.J.; Paterson, D.M.; Hagerthey, S.E. Working with natural cohesive sediments. *J. Hydraul. Eng.* **2002**, *128*, 2–8. [CrossRef]

331. Sanford, L.P. Modeling a dynamically varying mixed sediment bed with erosion, deposition, bioturbation, consolidation, and armoring. *Comput. Geosci.* **2008**, *34*, 1263–1283. [CrossRef]

332. Hagadorn, J.W.; Mcdowell, C. Microbial influence on erosion, grain transport and bedform genesis in sandy substrates under unidirectional flow. *Sedimentology* **2012**, *59*, 795–808. [CrossRef]

333. Droppo, I.G.; D'Andrea, L.; Krishnappan, B.G.; Jaskot, C.; Trapp, B.; Basuvaraj, M.; Liss, S.N. Fine-sediment dynamics: Towards an improved understanding of erosion and sediment transport. *J. Soil Sediment.* **2015**, *15*, 467–479. [CrossRef]

334. Barbier, E.B.; Koch, E.W.; Silliman, B.R.; Hacker, S.D.; Wolanski, E.; Primavera, J.; Granek, E.F.; Polasky, S.; Aswani, S.; Cramer, L.A.; et al. Coastal ecosystem-based management with non linear ecological functions and values. *Science* **2008**, *319*, 321–323. [CrossRef] [PubMed]

335. Ranasinghe, R. Assessing climate change impacts on open sandy coasts: A review. *Earth-Sci. Rev.* **2016**, *160*, 320–332. [CrossRef]

336. Vannote, R.R.; Minshall, G.W.; Cummins, K.W.; Sedell, J.R.; Cushing, C.E. The river continuum concept. *Can. J. Fish. Aquat. Sci.* **1980**, *37*, 130–137. [CrossRef]

337. Kemp, G.P.; Day, J.W.; Rogers, J.D.; Giosan, L.; Peyronnin, N. Enhancing mud supply from the Lower Missouri River to the Mississippi River Delta USA: Dam bypassing and coastal restoration. *Estuar. Coast. Shelf Sci.* **2016**, *183*, 304–313. [CrossRef]

338. Alexandridis, T.K.; Monachou, S.; Skoulikaris, C.; Kalopesa, E.; Zalidis, G.C. Investigation of the temporal relation of remotely sensed coastal water quality with GIS modeled upstream soil erosion. *Hydrol. Proc.* **2015**, *29*, 2373–2384. [CrossRef]

339. Passy, P.; Le Gendre, R.; Garnier, J.; Cugier, P.; Callens, J.; Paris, F.; Billen, G.; Riou, P.; Romero, E. Eutrophication modelling chain for improved management strategies to prevent algal blooms in the Seine Bight. *Mar. Ecol. Prog. Ser.* **2016**, *543*, 107–125. [CrossRef]

340. Hooke, R.L.; Martin-Duque, J.F.; Pedraza, J. Land transformation by humans: A review. *GSA Today* **2012**, *22*, 4–10. [CrossRef]

341. Nittrouer, C.A.; Austin, J.A.; Field, M.E.; Kravitz, J.H.; Syvitski, J.P.M.; Wiberg, P.L. (Eds.) *Continental Margin Sedimentation: From Sediment Transport to Sequence Stratigraphy*; Special Publication Number 27 of the International Association of Sedimentologists; Wiley: New York, NY, USA, 2009.

342. Armitage, J.J.; Allen, P.A.; Burgess, P.M.; Hampson, G.J.; Whittaker, A.C.; Duller, R.A.; Michael, N.A. Sediment transport model for the Eocene Escanilla sediment-routing system: Implications for the uniqueness of sequence stratigraphic architectures. *J. Sediment. Res.* **2015**, *85*, 1510–1524. [CrossRef]

343. Mazé, C.; Dahou, T.; Ragueneau, O.; Danto, A.; Mariat-Roy, E.; Raimonet, M.; Weibstein, J. Knowledge and power in integrated coastal management. For a political anthropology of the sea combined with the sciences of the marine environment. *C. R. Geosci.* **2017**, *349*, 359–368. [CrossRef]

344. Newton, A.; Harff, J.; You, Z.J.; Zhang, H.; Wolanski, E. Sustainability of Future Coasts and Estuaries: A synthesis. *Estuar. Coast. Shelf Sci.* **2016**, *183*, 271–274. [CrossRef]

345. Ramesh, R.; Chen, Z.; Cummins, V.; Day, J.; d'Elia, C.D.; Dennison, B.; Forbes, D.L.; Glaser, B.; Glavovic, B.; Kremer, H.; et al. Land–ocean interactions in the coastal zone: Past, present & future. *Anthropocene* **2015**, *12*, 85–98. [CrossRef]

346. Mishra, D.R.; Gould, R.W., Jr. Preface: Remote sensing in coastal environments. *Remote Sens.* **2016**, *8*, 665. [CrossRef]

347. Manighetti, I.; de Wit, R.; Duval, S.; Seyler, P. Vulnerability of intertropical littoral areas. *C. R. Geosci.* **2017**, *349*, 235–237. [CrossRef]

348. Mengual, B.; Hir, P.L.; Cayocca, F.; Garlan, T. Modelling Fine Sediment Dynamics: Towards a Common Erosion Law for Fine Sand, Mud and Mixtures. *Water* **2017**, *9*, 564. [CrossRef]

349. Fettweis, M.; Lee, B.J. Spatial and Seasonal Variation of Biomineral Suspended Particulate Matter Properties in High-Turbid Nearshore and Low-Turbid Offshore Zones. *Water* **2017**, *9*, 694. [CrossRef]

350. Milliman, J.D. Les matières solides dans les mers littorales: Flux et devenir. *Nature et Ressources* **1990**, *26*, 12–22. (In French) Adapted with authorization from: Milliman, J.D. Flux and fate of fluvial sediment and water in coastal sea. In *Ocean Margin Processes in Global Change*; Mantoura, R.F.C., Martin, J.M., Wollast, R., Eds.; Dahlem Workshop Report; Wiley: Chichester, UK, 1990; pp. 69–90.

351. Abessolo Ondoa, G.; Bonou, F.; Tomety, F.S.; du Penhoat, Y.; Perret, C.; Degbe, C.G.E.; Almar, R. Beach Response to Wave Forcing from Event to Inter-Annual Time Scales at Grand Popo, Benin (Gulf of Guinea). *Water* **2017**, *9*, 447. [CrossRef]

352. Mortlock, T.R.; Goodwin, I.D.; McAneney, J.K.; Roche, K. The June 2016 Australian East Coast Low: Importance of Wave Direction for Coastal Erosion Assessment. *Water* **2017**, *9*, 121. [CrossRef]

353. Almar, R.; Marchesiello, P.; Almeida, L.P.; Thuan, D.H.; Tanaka, H.; Viet, N.T. Shoreline Response to a Sequence of Typhoon and Monsoon Events. *Water* **2017**, *9*, 364. [CrossRef]

354. Vinh, V.D.; Ouillon, S.; Thao, N.V.; Tien, N.N. Numerical Simulations of Suspended Sediment Dynamics Due to Seasonal Forcings in the Mekong Coastal Area. *Water* **2016**, *8*, 255. [CrossRef]

355. Quang, N.H.; Sasaki, J.; Higa, H.; Huan, N.H. Spatiotemporal Variation of Turbidity Based on Landsat 8 OLI in Cam Ranh Bay and Thuy Trieu Lagoon, Vietnam. *Water* **2017**, *9*, 570. [CrossRef]

356. Zettam, A.; Taleb, A.; Sauvage, S.; Boithias, L.; Belaidi, N.; Sánchez-Pérez, J.M. Modelling Hydrology and Sediment Transport in a Semi-Arid and Anthropized Catchment Using the SWAT Model: The Case of the Tafna River (Northwest Algeria). *Water* **2017**, *9*, 216. [CrossRef]

357. Ohta, A.; Imai, N.; Tachibana, Y.; Ikehara, K. Statistical Analysis of the Spatial Distribution of Multi-Elements in an Island Arc Region: Complicating Factors and Transfer by Water Currents. *Water* **2017**, *9*, 37. [CrossRef]

358. Fernandez, J.M.; Meunier, J.D.; Ouillon, S.; Moreton, B.; Douillet, P.; Grauby, O. Dynamics of suspended sediments during a dry season and their consequences on metal transportation in a coral reef lagoon impacted by mining activities, New Caledonia. *Water* **2017**, *9*, 338. [CrossRef]

359. Rockström, J.; Steffen, W.; Noone, J.; Persson, Å.; Chapin, F.S., III; Lambin, E.F.; Lenton, T.M.; Scheffer, M.; Folke, C.; Schellnhuber, H.J.; et al. A safe operating space for humanity. *Nature* **2009**, *461*, 472–475. [CrossRef]

360. Morin, E. Introduction aux journées thématiques. In *Relier Les Connaissances, le Défi du XXIe Siècle*; Morin, E., Ed.; Seuil: Paris, France, 1999; 478p.

water

MDPI

Article

Modelling Fine Sediment Dynamics: Towards a Common Erosion Law for Fine Sand, Mud and Mixtures

Baptiste Mengual [1],*, Pierre Le Hir [1], Florence Cayocca [2] and Thierry Garlan [3]

[1] IFREMER/DYNECO/DHYSED, centre de Bretagne, ZI de la pointe du Diable CS 10070, 29280 Plouzané, France; Pierre.Le.Hir@ifremer.fr

[2] AAMP (Agence des Aires Marines Protégées), 16 quai de la Douane, 29200 Brest, France; florence.cayocca@aires-marines.fr

[3] SHOM/DOPS/HOM/Sédimentologie, 13 rue du Châtellier CS 92803, 29228 Brest, France; thierry.garlan@shom.fr

* Correspondence: bapt.mengual@hotmail.fr; Tel.: +33-(0)298-224-491

Received: 7 June 2017; Accepted: 18 July 2017; Published: 27 July 2017

Abstract: This study describes the building of a common erosion law for fine sand and mud, mixed or not, in the case of a typical continental shelf environment, the Bay of Biscay shelf, characterized by slightly energetic conditions and a seabed mainly composed of fine sand and muddy sediments. A 3D realistic hydro-sedimentary model was used to assess the influence of the erosion law setting on sediment dynamics (turbidity, seabed evolution). A pure sand erosion law was applied when the mud fraction in the surficial sediment was lower than a first critical value, and a pure mud erosion law above a second critical value. Both sand and mud erosion laws are formulated similarly, with different parameters (erodibility parameter, critical shear stress and power of the excess shear stress). Several transition trends (linear or exponential) describing variations in these erosion-related parameters between the two critical mud fractions were tested. Suspended sediment concentrations obtained from simulations were compared to measurements taken on the Bay of Biscay shelf with an acoustic profiler over the entire water column. On the one hand, results show that defining an abrupt exponential transition improves model results regarding measurements. On the other hand, they underline the need to define a first critical mud fraction of 10 to 20%, corresponding to a critical clay content of 3–6%, below which pure sand erosion should be prescribed. Both conclusions agree with results of experimental studies reported in the literature mentioning a drastic change in erosion mode above a critical clay content of 2–10% in the mixture. Results also provide evidence for the importance of considering advection in this kind of validation with in situ observations, which is likely to considerably influence both water column and seabed sediment dynamics.

Keywords: sand-mud mixture erosion; numerical modelling; non-cohesive to cohesive transition

1. Introduction

The transport of fine sediments can be assumed to mainly occur in suspension. Suspended transport is generally simulated by solving an advection/diffusion equation, assuming that sediment particles have the same velocity as water masses, except the vertical settling component (e.g., [1]). Such an equation involves sink and source terms at the bed boundary, which are deposition and erosion fluxes under conditions defined by the hydraulic forcing and the behaviour and composition of both the suspended and deposited sediments. This means that dealing with fine sediment dynamics, whether or not cohesive, requires the formulation of an erosion law. Such an erosion law should be applicable for fine sands (and even medium sands under strong shear stresses) as well as for mud, and naturally for mixtures of sand and mud. Despite these similarities between the suspended transport of fine sand and mud, their erosion processes have generally been investigated separately due to their contrasting behaviours.

An abundant literature is available on cohesive sediments, based on the existence of a critical shear stress for erosion (τ_e in N·m^{-2}) and several empirical relationships between the erosion flux E (in kg·m^{-2}·s^{-1}) and excess shear stress (i.e., the difference between the actual shear stress (τ in N·m^{-2}) and the critical value τ_e, either normalized by the latter or not). The parameters of such an erosion law involve bed characteristics, which may concern electrochemical forces, mineral composition, and organic matter content [2], or pore water characteristics [3]. They also depend on the consolidation state [4], and may be altered by biota effects [5]. Bulk density has often been proposed as a proxy for characterizing the bed, but the plasticity index has also been suggested [6], as well as the undrained cohesion and the sodium adsorption ratio [3].

Literature on sand erosion laws is scarce, partly because of experimental difficulties linked to simultaneous settling and resuspension processes, and partly because sand transport has mainly been considered through the formulation of transport capacity, even in the case of suspension. The need to simultaneously simulate transport of mud and (fine) sand, and their mixture, updates the need for an erosion law for fine sand, often named *pick up* functions [7]. *Pick up* functions link the erosion rate with the particle characteristics (size and density) and the shear stress (or, equivalently to the Shields parameter) [8,9]. As for cohesive sediment erosion laws, the dependence on the forcing stress is either absolute (~τ) or relative (~τ/τ_e), involving (or not) a threshold value τ_e.

The analogy between erosion formulations for sand and mud can be highlighted, especially considering the Partheniades-Ariathurai law [10,11] for cohesive sediments, the more usual in numerical models. The erosion law can then be written as

$$E = E_0 \cdot (\tau/\tau_e - 1)^n, if\ \tau \geq \tau_e \tag{1}$$

where E_0 is an erodibility parameter in kg·m^{-2}·s^{-1} and n a power function of the sediment composition. In Equation (1), E_0 and τ_e are functions of the sediment composition and its consolidation state in the case of cohesive sediments, or functions of the particle diameter and density in the case of non-cohesive sediments. Assuming a similar erosion law for the whole range of mixed sediments, the problem becomes the assessment of the critical shear stress for erosion on the one hand, and the erosion factor E_0 on the other hand, in the full transition range between cohesive and non-cohesive materials. Following Van Ledden et al. [12], the proxy for such a transition could be clay content. An alternative proxy could be mud content (hereafter referred to as f_m), considering that the clay to silt ratio is often uniform in a given study area [13].

Experiments have provided evidence that the resistance to erosion (i.e., critical shear stress τ_e) increases when mud is added to sand, either because of electrochemical bonds which take effect in binding the sand grains or because a cohesive matrix takes place between and around sand grains (e.g., [6,7,14–17]). Literature on erosion rate for mixed sediments is much less abundant, but a significant decrease (several orders of magnitude) in the erosion rate with mud content is most often reported (e.g., [14,17,18]). For instance, Smith et al. [17] presented laboratory measurements showing a decrease of about two orders of magnitude when the clay fraction in the mixture increased from 0% to 5–10%, and up to one order of magnitude more when the clay fraction increased from 10% to 30%.

From the point of view of modelling, Le Hir et al. [1] reviewed the different approaches developed in the last 20 years to manage sand/mud erosion in numerical process-based models. A non-cohesive and a cohesive regime separated by one or two critical mud fractions were commonly introduced and simulated by specific erosion laws. Le Hir et al. [1] and Waeles et al. [19], partly followed by Bi and Toorman [20], used a three-stage erosion law. Below a first critical mud fraction f_{mcr1}, they considered a non-cohesive regime where the erosion flux of any class of sandy and muddy sediments remains proportional to its respective concentration in the mixture, but is computed according to a pure sand erosion law (with a potential modulation of the power applied to the excess shear stress). Starting from a value characteristic of a pure sand bed, critical shear stress in this first regime is either kept constant [1], or linearly increases with the mud fraction f_m [19,20]. Above a second critical mud fraction f_{mcr2}, these authors defined a cohesive regime: Waeles et al. [19] and Le Hir et al. [1] formulate the erosion law using the relative mud concentration (the concentration of mud in the space between sand grains), considered as more relevant than the mud density in the case of sand/mud mixtures, which was in agreement with observations from Migniot [21] or Dickhudt et al. [22]. Between f_{mcr1} and f_{mcr2}, they ensured the continuity by prescribing a linear variation of E_0 and τ_e between non-cohesive and cohesive erosion settings. Carniello et al. [23] used a two-stage erosion law built by Van Ledden [24]: below a critical mud fraction f_{mcr}, the erosion factor of the sand fraction is steady and the one for the mud fraction varies slightly according to the factor $1/(1 - f_m)$, while above f_{mcr} the erosion rate is the same for sand and mud fractions and logarithmically decreases according to a power law. Regarding the critical shear stress for erosion, it first slightly increases with f_m and then varies linearly to reach the mud shear strength. Dealing only with the critical shear stress, Ahmad et al. [25] proposed an alternative to the Van Ledden [24] expression: without any critical mud fraction, τ_e varies linearly with f_m for low values of f_m and more strongly for high f_m values, using a parameter representing the packing of the sand sediment in the mixture. Generally speaking, the transitional erosion rate between the two regimes is poorly documented.

The aim of this paper is to fit an erosion law for mixed sediments to be applied in environments dominated by fine sediment, such as continental shelves, including our area of interest, the Bay of Biscay continental shelf (hereafter referred to as BoBCS). In these environments that develop outside the coastal fringe where wave impact is higher and where stronger tidal currents may take place, a common sedimentological feature is a mixture of fine sands and mud. In addition, bathymetric gradients are often gentle so that the shear stress gradients are likely to be small: as a consequence, the critical shear stress for erosion of surficial material in equilibrium with such environments is also likely to be low contrasted. Last, suspended sediment concentrations (hereafter referred to as SSC) are rather low in such "deep" coastal waters (at least they are much lower than in shallow waters), and sediment exchanges remain very small: as a consequence, freshly deposited sediment does not consolidate under its own weight, and erosion is so small that the surficial sediment can hardly become over-consolidated. This means that in this context, consolidation appears as a second order process, and that when the surficial sediment is muddy, its shear strength never reaches high values. On the contrary, the surface shear strength is likely to remain close to typical values for fine sand in the order of 0.1–0.2 Pa (for sands, this range corresponds to the decreasing trend of the Shields parameter with diameter, when the latter is below 0.6 mm).

Due to difficulties in measuring erosion fluxes (and not only shear strengths) for sand and mud mixtures, the assessment of the erosion law for fine sediment of the BoBCS is achieved by means of continuous measurements in the field, completed by a realistic hydro-sedimentary model of the specific area for testing the erosion law and its parameterization. In this way, the paper also proposes a methodology for fitting the erosion law in a poorly known natural environment. The paper is organized as follows. Section 2 gives the strategy for assessing the erosion law and describes measurements acquisitions on the one hand, and the modelling framework on the other. The description of the model includes the main features of hydrodynamic and sediment models, and the validation of hydrodynamics computation. Section 3 details the successive steps which led to the building of a new erosion law for sand/mud mixtures. Section 4 presents model results obtained in different erosion settings along with an assessment of their relevance with respect to observations. Section 5 discusses the main results in light of previous studies, followed by a short conclusion in Section 6.

2. Strategy and Modelling Background

2.1. Strategy for Assessing an Erosion Law, and Its Application to the BoBCS

Most often erosion laws have been investigated by comparing the critical erosion shear stress and erosion fluxes deduced from the tested law with measurements in a laboratory flume. However, this process has some difficulties. When the sediment is placed or settled in a flume, its behaviour is likely to differ from natural sediment one, especially when cohesive properties begin to develop. In addition, while critical shear stresses can be determined straightforwardly, the erosion flux of the sandy fraction is questionable, mainly because computations generally apply to rough erosion, while measurements often concern a net erosion flux, resulting from this rough erosion and possible simultaneous deposition, which are not easy to characterize. This concern does not apply to the mud fraction.

Here, an alternative methodology is proposed. Selecting an environment where local sediment is representative of the study area, a continuous measurement of suspended sediment concentration and local forcing is used to test the erosion law by simulating observed features by means of modelling. In the likely case of non-negligible horizontal gradients, a full 3D modelling is preferred, as it accounts for all processes, including erosion, deposition and horizontal advection. Deploying field measurements over several months increases the probability of investigating very different forcing conditions, especially if a winter period is selected. The use of an acoustic current profiler enables a simultaneous measurement of local forcing (both current patterns and waves) and resulting resuspension over the whole water column, by analysing the acoustic backscatter. The former observation is used for validating hydrodynamics simulations, while the latter can be compared to predicted *SSC* according to tested erosion parameters.

2.2. Measurements Used for Erosion Law Assessment and Model Validation

As in many continental shelves (e.g., [26,27]), the surficial sediment of the BoBCS is mainly composed of fine sand (~200 μm) and mud, since water depths exceed the depth of wave-induced frequent reworking, typically about 20–30 m, for the wave regime of the Bay of Biscay. The erosion law for fine sediment mixtures has been fitted by selecting measurements in a rather representative environment of the shelf, both in terms of sediment cover, hydrodynamic forcing and near bottom turbidity. A station located near the coast of southern Brittany, close to *Le Croisic* (hereafter referred to as *LC* station, Figure 1) meets these criteria: the local water depth is 23 m on average and the exposure to waves is attenuated by a rocky bank localised in the southwest nearby. Despite a tidal range of 4–5 m, tidal currents remain low, and flow is likely to be controlled by wind-induced currents [28]. The seabed is constituted by muddy sands (d_{10} of 7.5 μm, d_{50} of 163 μm, 5.1% clay (% < 4 μm) and 25% mud (% < 63 μm) contents), and exhibits some gradients around the station, with muddy facies to the north, and more sandy ones to the south (Figure 1).

For measurements, a mooring line was deployed over the whole water column for two months between 25 November 2007 and 31 January 2008. This period corresponded to typical winter hydrodynamic and meteorological conditions: mainly south-westerly winds, high rates of river discharge (e.g., Loire river: average flow rate of 1161 $m^3 \cdot s^{-1}$, and maximum flow rate of 2240 $m^3 \cdot s^{-1}$), and rather large swells (H_s (significant wave height) peaks > 3 m, T_p (peak period) 10–18 s).

Several instruments were placed along the mooring line. A *MS5-Hydrolab* sensor (OTT Hydromet, Kempten, Germany) providing temperature and salinity measurements (at 60 min intervals) was fixed 1.50 m below the sea surface. An upward looking acoustic Doppler profiler (Acoustic Wave And Current profiler, 1000 KHz, Nortek AS, Vangkroken 2, Norway; hereafter referred to as *AWAC*) was fixed on a structure placed near the seabed to record the backscattered acoustic intensity (at 30 min intervals), the intensity and direction of the currents (1 min of time-integrated data at 30 min intervals), water elevation (at 30 min intervals), as well as wave height, period, and direction (at 60 min intervals). The return echo was sampled over the entire water column with 25 cm-thick cells. In this configuration, the first sampling cell of the *AWAC* profiler was located 1.67 m above the seabed (including the frame height and the *AWAC* blank distance). At the same elevation, a turbidity sensor (*WETlabs*; Sea-Bird Scientific, Bellevue, Washington, DC, USA) provided hourly measurements for calibration. This turbidity signal was further transformed to *SSC* in mg·L^{-1} using water samples. The backscatter index (*BI*) from the *AWAC* profiler was evaluated from the sonar equation [29], following the procedure described by Tessier et al. [30], in particular by considering the geometrical attenuation for spherical spreading, the signal attenuation induced by the water, and the geometric correction linked to the expansion of the backscattering volume with increasing distance from the source. Given that *SSC* derived from the turbidity sensor did not exceed 100 mg·L^{-1}, the signal attenuation caused by the particles was disregarded when estimating backscatter [30]. Then, an empirical relationship was established between the *BI* of first *AWAC* cell and *SSC* measurements of turbidity sensor, following Tessier et al. [30]:

$$10log_{10}(SSC) = c_1 \cdot BI + c_2 \tag{2}$$

We finally obtained a determination coefficient R^2 of 0.78 with $c_1 = 0.42794$ and $c_2 = 32.8907$. Changes in *SSC* concentrations in the water column could thus be quantified (Section 4).

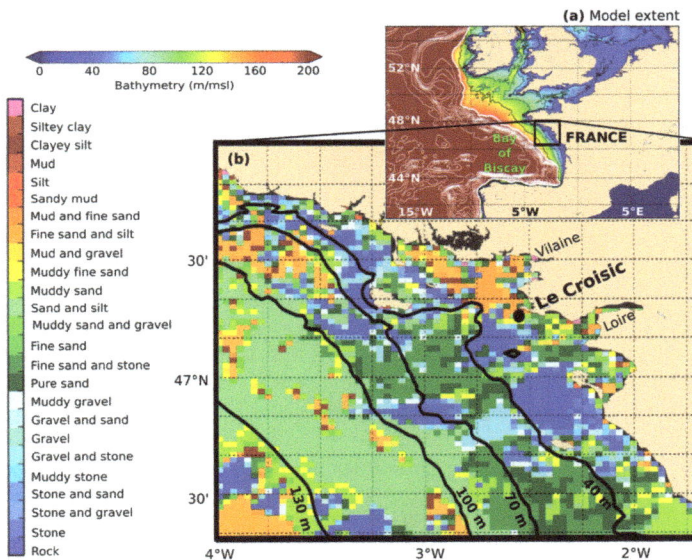

Figure 1. Geographic extent of the 3D model configuration with its bathymetry (in meters with respect to mean sea level) (**a**). Initial condition for the seabed compartment (at the resolution of the model) (**b**), over the zone surrounding *Le Croisic* station (indicated by a black dot). In (**a**), the thickest white line represents the 180 m isobath, which can be considered as the external boundary of the continental shelf. In both subplots, black lines refer to the 40-m, 70-m, 100-m, and 130-m isobaths over the shelf.

2.3. Hydrodynamics Models (Waves, Currents)

2.3.1. Brief Description

Current patterns and advection of suspended sediments were computed with the three-dimensional MARS3D model (3D hydrodynamic Model for Applications at Regional Scale, IFREMER), described in detail by Lazure and Dumas [31]. This code solves the primitive equations under classical Boussinesq and hydrostatic pressure assumptions. The model configuration spreads from 40°58′39″ N to 55° N in latitude, and from 18°2′1″ W to 9°30′ E in longitude (Figure 1) with a uniform horizontal resolution of 2500 m (822 × 624 horizontal grid points). The vertical discretization is based on 40 generalized sigma layers. For water depths lower than 15 m, sigma levels are uniformly distributed through the water column. Above this depth, a resolution increase is prescribed in the lower and upper parts of the water column, following the formulation of Song and Haidvogel [32]. For instance, this setting leads to a bottom cell of 40 cm for a total water depth of 25 m. The numerical scheme uses a mode splitting technique with an iterative and semi-implicit method which allows simultaneous calculation of internal and external modes at the same time step. In our case, the time step is fixed at 150 s. Vertical viscosity and diffusivity for temperature, salinity and momentum were obtained using the k-ε turbulence closure of Rodi [33]. Simulations were performed with the realistic forcings detailed in Table 1.

Table 1. Realistic forcings used in 3D simulations.

Forcing	Source
Initial & boundary conditions (3D velocities, temperature, salinity)	GLORYS global ocean reanalysis [34]
Wave (Significant height, peak period, bottom excursion and orbital velocities)	WaveWatch III hindcast [35]
Meteorological conditions (Atmospheric pressure, wind, temperature, relative humidity, cloud cover)	ARPEGE model [36]
Tide (14 components)	FES2004 solution [37]
River discharge (flow and SSC)	Daily runoff data (French freshwater office)

Regarding wave forcing, a wave hindcast database built with the WaveWatchIII (WWIII) model (realistic and validated configuration of Boudière et al. [35]) enabled the computation of bottom wave-induced shear stresses (τ_w in N·m^{-2}). The wave-induced shear stress was computed according to the formulation of Jonsson [38], with a wave-induced friction factor determined following Soulsby et al. [39]. Then, the total bottom shear stress τ was computed from the estimated τ_w and from the current-induced shear stress (τ_c in N·m^{-2}) provided by the hydrodynamic model, according to the formulation of Soulsby [40], i.e., accounting for a non-linear interaction between waves and currents. Both wave and current shear stresses were computed by considering a skin roughness length z_0 linked to a 200 μm sand, representative of the sandy facies widely encountered on the BoBCS ($z_0 = k_s/30 = 2 \times 10^{-5}$ m with k_s the Nikuradse roughness coefficient).

2.3.2. Hydrodynamic Validation of the Model

Figure 2 shows model validation regarding hydrodynamics at the *LC* station from 1 December 2007 to 18 January 2008 in terms of significant wave height (H_S), surface temperature and salinity (T_S and S_S, respectively), and current intensity and direction (Vel_{INT} and Vel_{DIR}, respectively).

Measurements highlight typical late autumn/winter energetic conditions. Average H_s were around 1.5 m and occasionally exceeded 4 m in stormy conditions. Peak periods (not illustrated here) ranged from 8 to 18 s (around 10 s on average). Even during "calm" periods, H_S values are generally no lower than 0.8 m. Figure 2a illustrates the ability of the model of Boudière et al. [35] to describe H_S over the period, with a root mean square error (*RMSE*) of 0.24 cm and a R^2 of 0.95. Measured and modelled T_S and S_S are illustrated in Figure 2b,c respectively, and demonstrate the correct response of the model with respect to observations with $RMSE/R^2$ values of 0.5 °C/0.86 for T_S and 2 PSU/0.7 for S_S. The weaker correlation obtained for S_S is mainly due to model underestimation around 2 January 2008 (i.e., about 5 PSU). It should be underlined that the model accurately reproduces the abrupt change (decrease) in surface temperature and salinity on 11 December 2007, linked to the veering of the Loire river plume caused by easterly winds. This plume advection led to stratification which in turn influenced the vertical profiles of currents in terms of direction and intensity (Figure 2e,g, respectively), which are well represented by the model. More generally, the model provides an appropriate response regarding the direction and intensity of the current (Figure 2d,f) over the entire water column throughout the period. For instance, measured bottom current velocities (0.11 m·s^{-1} on average, max of 0.44 m·s^{-1}) are correctly reproduced by the model with a *RMSE* of 0.05 m·s^{-1}.

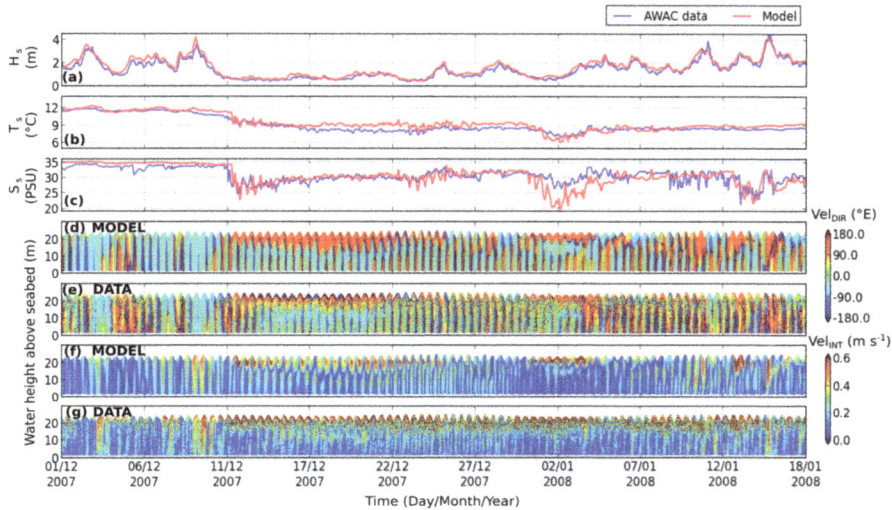

Figure 2. Validation of the hydrodynamic model over most of the simulated period at *LC* station. Model results are compared with *AWAC* measurements in terms of (**a**) significant wave height, (**b**) surface temperature, (**c**) surface salinity, current direction ((**d,e**), respectively), and current intensity ((**f,g**), respectively).

2.4. Sediment Transport Model

The sediment model used in the present 3D modelling system is described in Le Hir et al. [1]. The model makes it possible to simulate the transport and changes in all kinds of sediment mixtures under the action of hydrodynamic forcing, including that of waves and currents. According to the model concept, the consolidation state of the bottom sediment is linked to the relative mud concentration, i.e., the concentration of mud in the space between the sand grains ($C_{rel\,mud}$). Given that the surficial sediment in our study area is weakly consolidated (erosion actually occurs in surficial layers only, typically a few mm or cm thick) and mainly composed of fine sand, consolidation was disregarded and $C_{rel\,mud}$ was set at a constant value of 550 kg·m^{-3} (representative of pre-consolidated sediment according to Grasso et al. [41]). In addition, bed load was not taken into account in the present application, assuming that surficial sediment in our study area is mainly composed of mixtures of mud and fine sand. In our case, three sediment classes (for which the mass concentrations are the model state variables) are considered: a fine sand (*S1*), a non-cohesive material with a representative size of 200 μm, and two muddy classes (*M1* and *M2*), which can be distinguished by their settling velocity, in order to be able to schematically represent the vertical dispersion of cohesive material over the shelf. The sediment dynamics was computed with an advection/dispersion equation for each sediment class, representing transport in the water column, as well as exchanges at the water/sediment interface linked to erosion and deposition processes. Consequently, the concentrations of suspended sediment, the related horizontal and vertical fluxes, and the corresponding changes in the seabed (composition and thickness) are simulated. This section details the way of managing sediment deposition, seabed initialization, and technical aspects linked to vertical discretization within the seabed compartment.

The erosion law establishment and the numerical modelling experiment aiming to fit an optimal setting will be addressed independently in Section 3.

2.4.1. Managing Sediment Deposition

The deposition flux D_i for each sediment class i is computed according to the Krone law:

$$D_i = W_{s,i} \cdot SSC_i \cdot (1 - \frac{\tau}{\tau_{d,i}}) \tag{3}$$

In Equation (3), τ is the total bottom shear stress (waves and currents; see Section 2.3.1), $W_{s,i}$ is the settling velocity, SSC_i is the suspended sediment concentration, and $\tau_{d,i}$ is the critical shear stress for deposition. In the present study, the latter was set to a very high value (1000 N·m^{-2}) so it is ineffective: considering that consolidation processes are negligible near the interface and that, as a result, the critical shear stress for erosion remains low (Section 3), deposited sediments can be quickly resuspended in the water column if the hydrodynamic conditions are sufficiently intense, which replaces the role played by the term between parentheses in Equation (3) [1].

The settling velocity of non-cohesive sediment ($W_{s,S1}$) is computed according to the formulation of Soulsby [40]. The one related to the mud M1 ($W_{s,M1}$) is assumed to vary as a function of its concentration in the water column (SSC_{M1}) and the ambient turbulence according to the formulation of Van Leussen [42]:

$$W_{s,M1} = max\left\{ W_{s,min} ; min\left\{ W_{s,max} ; k \cdot SSC_{M1}{}^m \cdot \left(\frac{1+a \cdot G}{1+b \cdot G^2} \right) \right\} \right\}, with\ G = \sqrt{\frac{\varepsilon}{\nu}} \tag{4}$$

where k, m, a and b are empirical constants, G is the absolute velocity gradient, ε is the turbulent dissipation rate, and ν is the water viscosity. For a and b, respective values of 0.3 and 0.09, set by Van Leussen [42] from experiments, were used. In agreement with the settling velocity setting used by Tessier et al. [43] in a modelling study of turbidity over the southern Brittany continental shelf and recent experiments by Verney et al. [44] in similar environments in the Gulf of Lions, k and m were respectively set to 0.005 and 0.7. Following Tessier et al. [43], $W_{s,M1}$ is limited by minimum ($W_{s,min}$) and maximum ($W_{s,max}$) values, respectively set to 0.1 mm·s^{-1} and 4 mm·s^{-1}, the latter being reached for $SSC \geq 700$ mg·L^{-1} (thus ignoring the hindered settling process which actually does not occur in the range of SSC over the shelf). Lastly, the mud M2 linked to ambient turbidity in the water column is assumed to have a constant very low settling velocity set to 2.5×10^{-6} m·s^{-1}.

After resolving the sediment transport equation, the actual deposition on the bottom can be computed. In agreement with Le Hir et al. [1], sand and mud particles are deposited successively, by filling the pores or by creating new layers. A volume concentration of well-sorted sediment ($C_{vol\ sort} = 0.58$) was attributed to sediment when only one class of non-cohesive sediment is present. However, the bed concentration can reach a higher value if several classes are mixed ($C_{vol\ mix} = 0.67$). These typical volume concentrations [45] led to mass concentrations assuming a fixed sediment density ρs (2600 kg·m^{-3}) for all classes. In the case of simultaneous deposition of sand and mud, the sand is first deposited at a concentration which depends on surficial sediment composition: in the case of mixed sediment, the sand is first mixed with the initial mixture until C_{mix} (= $C_{vol\ mix} \cdot \rho_s = 1742$ kg·m^{-3}) is reached. The remaining sand is deposited with the concentration C_{sort} (= $C_{vol\ sort} \cdot \rho_s = 1508$ kg·m^{-3}), from which an increase in the thickness of the layer can be deduced. The same kind of deposition occurs when the surficial sediment is only comprised of sand. In addition, the thickness of any layer is limited to $dz_{sed,\ max}$, a numerical parameter of the model. Any deposition of excess sand leads to the creation of a new layer (see Section 2.4.2). Next, mud is deposited: it progressively fills up pores between the sand grains until either C_{mix} or $C_{rel\ mud}$ is reached. Considering these criteria, mud is mixed within the initial and new deposits starting from the water/sediment interface. If any excess mud remains after the mixing step, it is added to the upper sediment layer, contributing to its thickening.

2.4.2. Sediment Discretization within the Seabed

In the model, an initial sediment height ($h_{sed, ini}$) is introduced, and the seabed is vertically discretized in a given number of layers of equivalent thicknesses ($dz_{sed, ini}$). An optimal vertical discretization of the sediment was assessed by Mengual [46]. By means of sensitivity analyses, it was shown that beyond a 1/3 mm resolution within the seabed compartment (i.e., thickness of each layer $dz_{sed, ini}$), the SSC response of the model in the water column did not change anymore. According to the conclusions drawn from this sensitivity analysis, the initial sediment thickness $h_{sed, ini}$ was set at 0.03 m, corresponding to 90 sediment layers of thickness $dz_{sed, ini}$ (1/3 mm).

As previously mentioned in Section 2.4.1, a new layer is created when the actual deposition leads to a thickness of the surficial sediment layer higher than a maximum value $dz_{sed, max}$, corresponding to a parameter of the model. Nevertheless, a maximum number of layers in the seabed compartment (N_{max}) needs to be defined in order to make computational costs acceptable. While N_{max} is reached, a fusion of the two sediment layers located at the base of the sedimentary column occurs. By this way, the creation of new deposited layers becomes once again possible. The parameters $dz_{sed, max}$ and N_{max} control changes of the "sediment vertical discretization" during the simulation, and are likely to influence the sediment dynamics. To prevent any variations in the SSC response of the model linked to changes in the seabed vertical resolution, the maximum thickness of sediment layers, $dz_{sed, max}$, was set at the same value than the initial one $dz_{sed, ini}$.

This seabed management constitutes a compromise between the likely maximum erosion depth in most of the shelf and the possibility of new layers being deposited in other places.

2.4.3. Sediment Facies Initialization for the Application to the BoBCS

The seabed was initialized using the distribution of the three sediment classes according to existing surficial sediment maps (e.g., [47]) and for the sand and mud fractions, new sediment samples taken in the BoBCS (Figure 1). No sediment initialization was undertaken in areas indicated as rocky on sediment maps. The mud $M2$, characterized by a very low settling velocity, is linked to river inputs and was also uniformly initialized (20 kg·m^{-3}) on the seabed over the entire shelf. This particle class enables the representation of a non-negligible part of the ambient turbidity of a few mg·L^{-1} near the coast.

Given the sand fraction f_s of the mixed sediment, a relationship could be derived between $C_{rel\ mud}$ (relative mud concentration set at a constant value of 550 kg·m^{-3}) and the bulk sediment density (C_{bulk}). The latter can be written as:

$$C_{bulk} = \frac{C_{rel\ mud}}{1 + f_s \cdot \left(\frac{C_{rel\ mud}}{\rho_s} - 1 \right)} \tag{5}$$

For large sand fraction (above 90%), the given relative concentration of 550 kg·m^{-3} becomes unlikely, so that C_{bulk} is maximised by C_{max}, which corresponds to a dense sediment (C_{sort} or C_{mix} depending on the properties of the sediment mixture, i.e., well-sorted or mixed sediments). This formulation could be validated by comparing with recent data in two sectors of the BoBCS [48]: bulk densities of 1540 kg·m^{-3} and 1380 kg·m^{-3} were obtained for 85% and 75% of sand, respectively, while the application of Equation (5) with sand fractions f_s of 0.85 and 0.75 led to C_{bulk} values of 1668 kg·m^{-3} and 1346 kg·m^{-3} in fair agreement. Sediment concentrations linked to each particle class were then deduced according to their respective fraction.

Lastly, the sediment distribution has been prescribed uniformly along the vertical dimension (i.e., in each layer of the seabed compartment) in the absence of 3D data on grain size distributions.

3. Erosion Law Setting: Building and Numerical Experiment

3.1. General Formulation

As previously evoked in the introduction, the Partheniades-Ariathurai law [10,11], the most commonly used for cohesive sediments, can also represent the *pick up* process of fine sands. Such an erosion law is then assumed whatever the sediment composition. In the present study, the erosion flux E is thus expressed following Equation (1). A simultaneous erosion of sandy and muddy fractions is assumed, according to their respective concentration in the mixture.

This erosion law involves three parameters, E_0 (erodibility parameter, kg·m^{-2}·s^{-1}), τ_e (critical shear stress, N·m^{-2}), and n (hereafter referred to as "erosion-related parameters"), which are set at different values depending on the mud content f_m (<63 µm) of the surficial sediment. The concept of critical mud fraction is retained, with the definition of a first critical fraction f_{mcr1} below which a non-cohesive behaviour is prescribed, and of a second one, f_{mcr2}, above which the sediment is assumed to behave as pure mud. Two sets of erosion-related parameters will be defined below and above these mud fractions (see Sections 3.2 and 3.3 for sand and mud, respectively). As already mentioned in Section 1, a transition in erosion-related parameters has to be prescribed between f_{mcr1} and f_{mcr2} (between pure sand and pure mud parameters) to manage the erosion of "transitional" sand/mud mixtures. Such a transition is investigated in Section 3.4.

3.2. Pure Sand Erosion

Below the first critical mud fraction f_{mcr1}, the non-cohesive erosion regime is prescribed by defining a first set of erosion-related parameters linked to pure sand erosion in Equation (1): $E_{0,sand}$, $\tau_{e,sand}$, and n_{sand}. Le Hir et al. [1] suggested to compute f_{mcr1} as a function of the sand mean diameter D ($f_{mcr1} = \alpha_0 \times D$ with $\alpha_0 = 10^3$ m^{-1}), leading to a value of 20% considering the fine sand $S1$ of 200 µm (considered as a reference value hereafter). However, the model erosion dynamics is likely to significantly vary while the surficial mud fraction f_m is close to f_{mcr1}. Such a sensitivity is addressed in Section 4.2.

In Equation (1), parameters linked to pure sand erosion $E_{0,sand}$, $\tau_{e,sand}$, and n_{sand} were deduced from numerical simulations and empirical formulations. According to Van Rijn [49] and many other numerical models (e.g., [23,24]), the best fit for n_{sand} is 1.5. The critical shear stress $\tau_{e,sand}$ was determined from the Shields critical mobility parameter computed according to the formulation of Soulsby [40], leading to a value of 0.15 N·m^{-2} for a sand of 200 µm.

Considering that equilibrium conditions are usually met for non-cohesive sediment transport, and that such an equilibrium requires compensation between the erosion rate and the deposition flux, many authors formulate the erosion rate (or *pick up* function) as:

$$E = W_s \cdot C_{ref} \qquad (6)$$

where W_s is the settling velocity and C_{ref} is a reference concentration which characterizes the equilibrium. The description of C_{ref} is generally associated with the reference height, h_{ref}, the distance from the bed where the concentration is considered. In point of fact, this location has to be the one where the equilibrium between deposition and erosion is considered, with respective values that largely depend on the reference height, because of large concentration gradients near the bed. From the point of view of sediment modelling, this means that the deposition flux at the base of the water column has to be expressed at the exact location where the erosion flux is considered, that is, at the reference height where the equilibrium concentration is given. Van Rijn [49] used the concept of equilibrium concentration as a boundary condition of the computation of suspended sediment profile and fitted the expression:

$$C_{ref} = 0.015 \rho_s D / \left(h_{ref} \cdot D_*^{0.3} \right) \cdot (\tau / \tau_e - 1)^{1.5} \qquad (7)$$

where D_* is the non-dimensional median sand diameter ($D_* = D(\sigma g/\nu^2)^{1/3}$ with $\sigma = (\rho_s/\rho - 1)$, ρ being the water density, g the gravitational acceleration, and ν the water viscosity). Using the C_{ref} formulation of Van Rijn [49] in Equation (6) enables us to express the $E_{0,sand}$ constant in the "Partheniades" form of the erosion law as:

$$E_{0,sand} = \frac{0.015 \, \rho_s \, D \, W_{s,S1}}{h_{ref} D_*^{0.3}} \tag{8}$$

For D = 200 μm (S1), h_{ref} = 0.02 m, and $W_{s,S1}$ = 2.5 cm·s^{-1}, Equation (8) leads to $E_{0,sand}$ = 5.94 × 10^{-3} kg·m^{-2}·s^{-1}. The relevance of this $E_{0,sand}$ value was assessed by simulating an equilibrium state under a steady current and by comparing the depth-integrated horizontal sediment flux with some standard transport capacity formulations. Using a 1DV version of the code, several computations of fine sand resuspension were performed under different flow intensities (Vel_{INT}), and once the equilibrium was reached (deposition = resuspension) the total transport (Q_{sand}) was computed as:

$$Q_{sand} = \frac{1}{\rho_s} \int_{k=1}^{n} Vel_{INT}(k) \cdot SSC_{S1}(k) \cdot dz(k) \tag{9}$$

where $dz(k)$ refers to the thickness of the cell k (in m) in the water column (discretized in n layers along the vertical dimension) and $SSC_{S1}(k)$ to the suspended sediment concentration of sand (S1) in the cell k. Sand transport rates Q_{sand} were then compared to the rates deduced from the formulations of Van Rijn [50], Engelund and Hansen [51], and Yang [52] for similar flow velocities (hereafter $Q_{sand,VR1984}$, $Q_{sand,EH1967}$, and $Q_{sand,Y1973}$ respectively). Results are illustrated in Figure 3. The results obtained by Dufois and Le Hir [53], who also used an advection/diffusion model to predict sand transport rates for a wide range of current conditions and numerous sand diameters, have been added in Figure 3 ($Q_{sand,DLH2015}$). Figure 3 shows that the sand transport rates obtained from our computations are in a consistent range regarding those obtained with other formulations or studies cited in the literature, demonstrating the suitability of our $E_{0,sand}$ parameter.

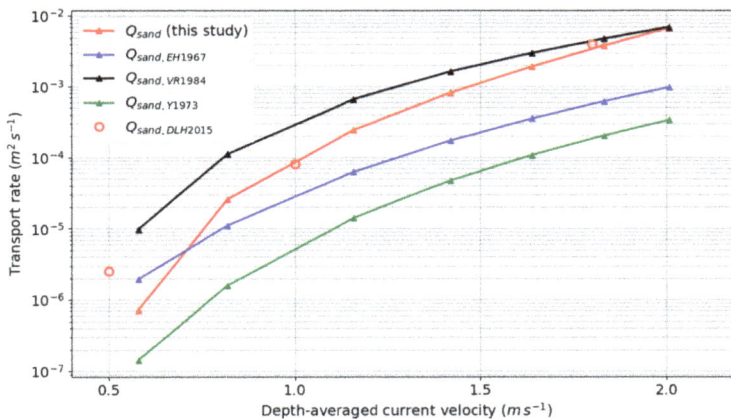

Figure 3. Sand (200 μm diameter) transport rates computed with a 1DV model using the pure sand erosion law, and obtained for different flow intensities (solid red curve). For identical flow intensities and sand diameter, sand transport rates deduced from empirical formulations of Van Rijn [50] (black curve), Engelund and Hansen [51] (blue curve), and Yang [52] (green curve) are illustrated. Empty red circles refer to the modelling results of Dufois and Le Hir [53] representing transport rates of a 200 μm sand under different flow intensities from an advection/diffusion model.

The erosion-related parameters $E_{0,sand}$, $\tau_{e,sand}$, and n_{sand} for fine sand are summarized in Table 2 and constitute reference parameters characterizing pure sand erosion.

Table 2. Erosion parameters for pure sand and pure mud sediment.

Erosion Regime	E_0 (kg·m^{-2}·s^{-1})	τ_e (N·m^{-2})	n
Non-cohesive (pure sand)	$E_{0,sand} = 5.94 \times 10^{-3}$	$\tau_{e,sand} = 0.15$	$n_{sand} = 1.5$
Cohesive (pure mud)	$E_{0,mud} = 10^{-5}$	$\tau_{e,mud} = 0.1$	$n_{mud} = 1$

3.3. Pure Mud Erosion

Above the second critical mud fraction f_{mcr2} (reference value of 70% according to the default value used by Le Hir et al. [1]), the cohesive erosion regime is prescribed by defining a second set of erosion-related parameters linked to pure mud erosion in Equation (1): $E_{0,mud}$, $\tau_{e,mud}$, and n_{mud}.

As frequently specified in the Partheniades-Ariathurai formulation, the n_{mud} exponent was set to 1. Given the lack of established formulation of the erosion factor for pure mud, experimental approaches are often used to calibrate it for specific materials, preferably in situ when possible. For this purpose, a specific device had been designed: the "erodimeter" is described in Le Hir et al. [7]. It consists of a small recirculating flume where a unidirectional flow is increased step by step and interacts with a sediment sample carefully placed at the bottom after transfer from a cylindrical core. When measurements are made on board an oceanographic vessel, the test can be considered as *quasi* in situ. On the BoBCS, erosion tests had been conducted on board the Thalassa N/O: a few of them were performed on muddy sediment samples (mud content higher than 80%). Figure 4 illustrates the critical shear stress for erosion $\tau_{e,mud}$, estimated to be 0.1 N·m^{-2}, suggesting a barely-consolidated easily erodible sediment.

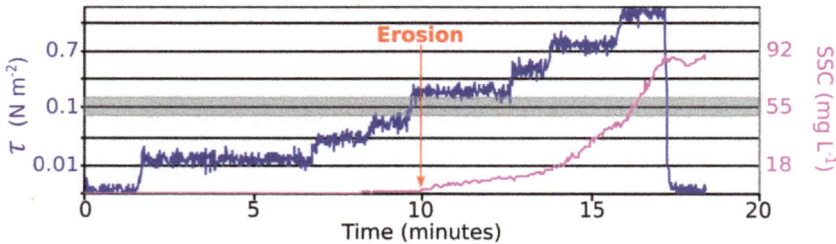

Figure 4. Erodimetry experiment conducted on a muddy sample of the BoBCS using the "erodimeter" device [7]. On the graph, the blue and pink curves represent time evolutions of bottom shear stress (N·m^{-2}) and suspended sediment concentration (SSC in mg·L^{-1}) during the few minutes of the experiment. The applied shear stress at which erosion begins (around 0.1 N·m^{-2}) is illustrated by a grey band.

Concerning the erosion coefficient $E_{0,mud}$, the range of values cited in the literature extends from 10^{-3} to 10^{-5} kg·m^{-2}·s^{-1} for natural mud beds in open water (e.g., [54]). Simulating fine sediment transport along the BoBCS, Tessier et al. [43] applied the Partheniades erosion law but used an even lower erosion constant ($E_0 = 1.3 \times 10^{-6}$ kg·m^{-2}·s^{-1}). As a first attempt in the present study, a low value $E_{0,mud} = 10^{-5}$ kg·m^{-2}·s^{-1} was used, and its appropriateness was demonstrated by comparing the computed erosion fluxes ($E_{modelled}$) from Equation (1) (with $\tau_e = \tau_{e,mud}$ and $n = n_{mud}$ otherwise) with measurements from erodimetry experiments ($E_{measured}$) conducted on three muddy sediment samples from the BoBCS (mud fraction higher than 70%) (Figure 5).

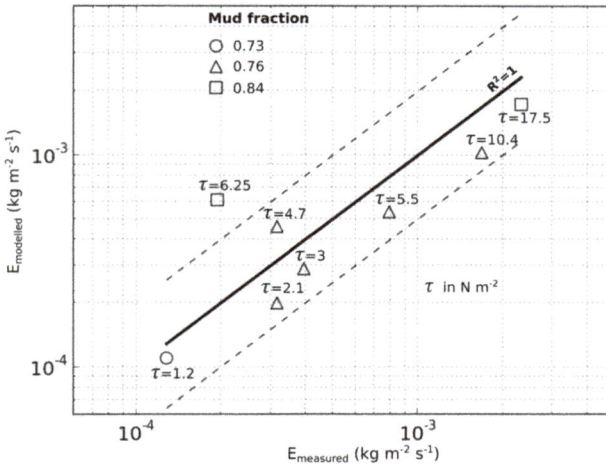

Figure 5. Comparisons of erosion fluxes deduced from erodimetry experiments conducted on muddy samples from the BoBCS and those computed from pure mud erosion law (Equation (1)) using similar shear stresses. Different symbols depict different sediment samples and labels refer to the applied shear stresses (τ in N·m^{-2}). The solid black line represents perfect agreement between modelled and measured fluxes, and the dotted lines delimit the range linked to model overestimation or underestimation by a factor 2.

The above-mentioned $E_{0,mud}$, $\tau_{e,mud}$, and n_{mud} values are summarized in Table 2 and constitute reference parameters characterizing pure mud erosion.

3.4. Erosion of Transitional Sand/Mud Mixtures: Selection of Transition Formulations to be Tested

Between critical mud fractions (f_{mcr1} and f_{mcr2}), E_0, τ_e, and n ranged between pure sand ($E_{0,sand}$, $\tau_{e,sand}$, and n_{sand}) and pure mud ($E_{0,mud}$, $\tau_{e,mud}$, and n_{mud}) parameters, following a transition trend which had to be specified. We defined several expressions of the erosion law for the transition between non-cohesive and cohesive behaviours as a function of the surficial sediment mud fraction (f_m).

First, we considered a linear transition type in which erosion-related parameters are linearly interpolated from the respective sand and mud parameters according to their respective concentrations in the mixture (default solution in the original paper by Le Hir et al. [1]).

Several experimental studies in the literature revealed that the transition from non-cohesive to cohesive could be more abrupt. For instance, Smith et al. [17] performed erodibility experiments on natural and artificial sand/mud mixtures, and showed a rapid decrease in erosion rates with increasing mud/clay contents. Nevertheless, the transition between the two regimes remains poorly documented. Here, we propose an exponential formulation that specifically enables adjustment of the sharpness of the transition, which is not possible with the few existing expressions (e.g., [1,19,20,23,24]). The exponential transition trend (Equation (10)) was applied to all erosion-related parameters (X_{exp}) as a function of mud content, with a coefficient C_{exp} allowing the adjustment of the sharpness of the transition, which becomes more abrupt with an increase in C_{exp}:

$$X_{exp} = (X_{sand} - X_{mud})e^{C_{exp} \cdot P_{exp}} + X_{mud} \tag{10}$$

where $P_{exp} = (f_{mcr1} - f_m)/(f_{mcr2} - f_{mcr1})$; $X_{sand} = \{E_{0,sand}; \tau_{e,sand}; n_{sand}\}$; and $X_{mud} = \{E_{0,mud}; \tau_{e,mud}; n_{mud}\}$.

Different settings of the erosion law were tested in 3D simulations to assess the correctness of the model response in terms of *SSC* and changes in the seabed. All settings are illustrated in Figure 6. Three kinds of transition, one linear, and two exponentials (with $C_{exp} = 10$ and 40 in Equation (10)), were defined to evaluate the effect of the transition trend only, using the reference critical mud fractions (f_{mcr1} = 20% and f_{mcr2} = 70%). The corresponding settings are named $S1_{LIN}$, $S1_{EXP1}$, and $S1_{EXP2}$, respectively.

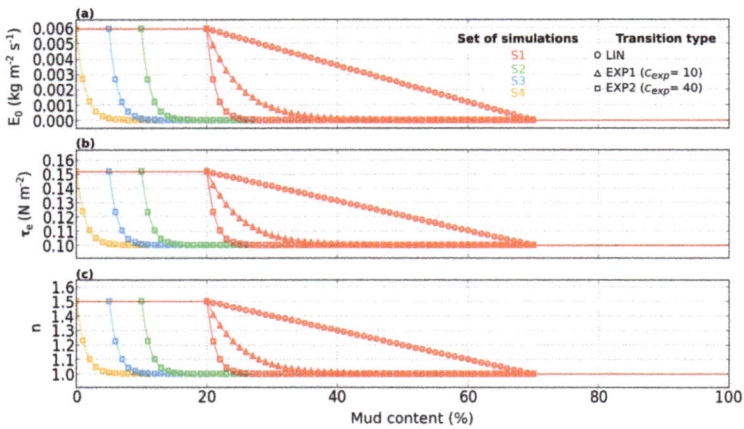

Figure 6. Variations in erosion-related parameters (erodibility parameter E_0 (**a**); critical shear stress τ_e (**b**); and exponent n (**c**); used in the erosion law (Equation (1)) as a function of the surficial sediment mud content in the different erosion settings tested.

A second series of simulations was run to evaluate the sensitivity of the results to critical mud fraction values. Thus, the f_{mcr1} value was successively reduced to 10%, 5%, and ~0% (with the corresponding f_{mcr2} = 60%, 55%, and 50%), but only the exponential transition regime (with C_{exp} = 40) was considered (simulations $S2_{EXP2}$, $S3_{EXP2}$, and $S4_{EXP2}$ respectively), as it produced better results than the other transitions (see results Section 4.1).

4. Results

4.1. Influence of the Transition Trend between Non-Cohesive and Cohesive Erosion Modes in the Erosion Law

The first step consisted in assessing the 3D model response in the model cell located closest to the LC station in terms of SSC and changes in the seabed (i.e., composition, thickness) by considering two transitions of the erosion law, one linear and one exponential (with C_{exp} = 40 in Equation (10)), between the non-cohesive and cohesive regimes (Figure 7). This first comparison was performed using the "reference" f_{mcr1} and f_{mcr2} values of 20% and 70%, respectively. The two erosion settings, $S1_{LIN}$ and $S1_{EXP2}$, are illustrated in Figure 6 (Section 3.4).

Total bottom shear stresses (i.e., those caused by waves and currents) and barotropic currents in Figure 7a illustrate changes in forcing throughout the period. In the water column, we successively represented SSC over the entire water column (Figure 7b–d) and at 1.67 m above the seabed (i.e., at the level of the first AWAC cell; Figure 7e) for the two simulations and for the AWAC measurements. Changes in the seabed in the two simulations are presented in Figure 7f,g. Lastly, a global sediment budget (in kg) was applied to the model cell used for the comparison, and the contribution of advection (hereafter referred to as F_{OBC} with $F_{OBC,\,mud}$ for mud and $F_{OBC,\,sand}$ for sand, representing the total amount (integrated) of sediment that crosses the borders of the cells along the water column, as net inflow if F_{OBC} increases or as net outflow if F_{OBC} decreases) is illustrated in Figure 7h.

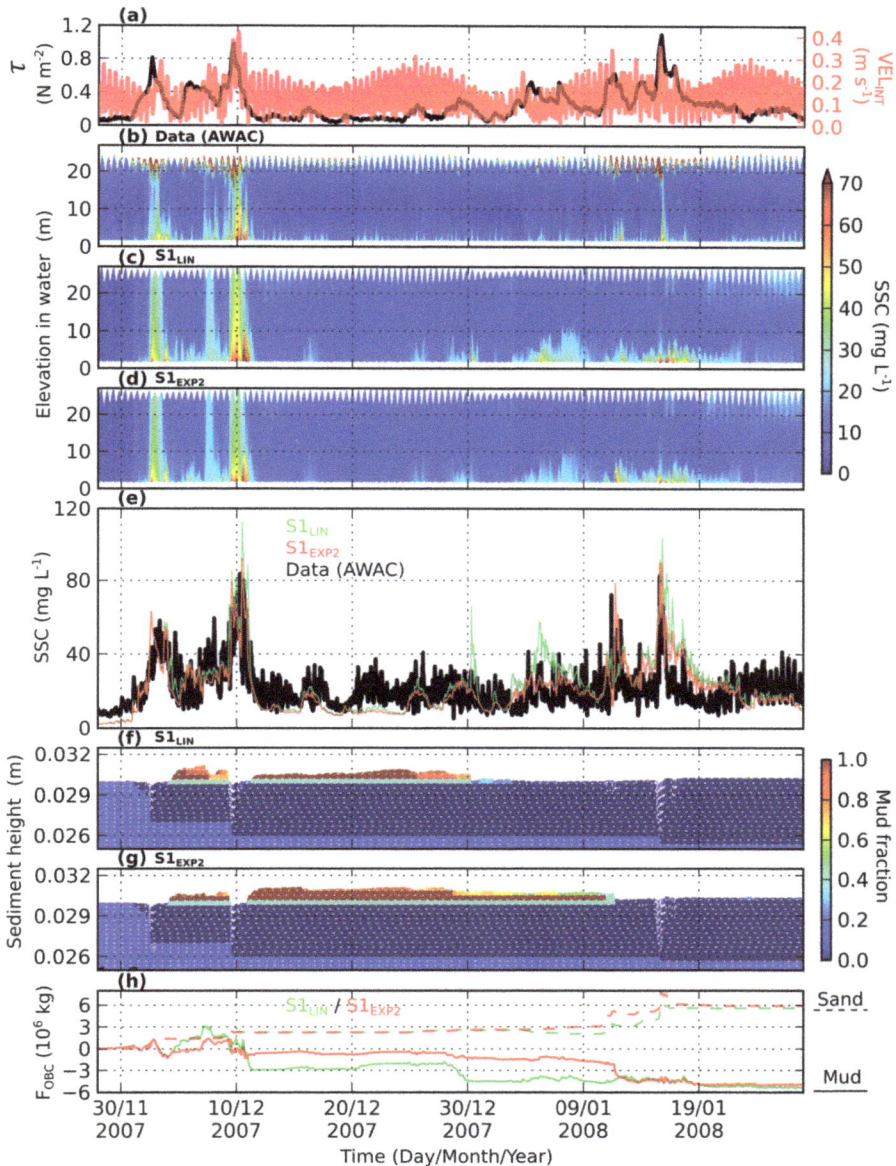

Figure 7. Comparisons of the results of the 3D model obtained from erosion settings $S1_{LIN}$ and $S1_{EXP2}$, and measurements made by the *AWAC* acoustic profiler. (**a**) Shear stresses τ and depth-integrated currents VEL_{INT}; (**b**) measured *SSC* over the entire water column; (**c**,**d**) computed *SSC* for $S1_{LIN}$ and $S1_{EXP2}$ simulations; (**e**) time series of measured and modelled *SSC* variations at the level of the *AWAC* first cell (1.67 m above the bottom); (**f**,**g**) changes in the seabed (mud fraction, thickness of the sediment layers) in the two simulations (white dotted lines represent the boundaries of the sediment layers); and (**h**) integrated amount of *SSC* advected through the water column (solid lines represent mud and dotted lines represent sand).

Regarding the dynamics of sediments in suspension, SSC from the $AWAC$ profiler ranged between 10 and 80 mg·L^{-1} over the study period. Four main resuspension events can be identified: the first event lasted from 1 to 5 December 2007 (Event 1), the second from 8 to 11 December 2007 (Event 2), the third from 11 to 12 January 2008 (Event 3), and the last from 15 to 17 January 2008 (Event 4). During these events, SSC values ranged between 20 and 80 mg·L^{-1} near the seabed, and did not exceed 40 mg·L^{-1} close to the surface (Figure 7). The rest of the time, a higher frequency turbidity signal linked to the semi-diurnal tide resuspension was recorded near the seabed with SSC in the range of 10–15 mg·L^{-1}. Turbidity peaks were regularly detected in the surface signal whereas there was no significant increase near the seabed: these peaks are probably due to the signal diffraction caused by wave-induced air bubbles (wave mixing; see Tessier [55]). Such a phenomenon also occurs during energetic events ($H_s > 2$ m) with a SSC signal in the surface higher than in the rest of the water column.

Despite the fact that the model response generally agreed with observations, computing SSC with the $S1_{LIN}$ erosion setting highlighted some periods during which turbidity was overestimated, for instance in the upper half of the water column during Event 2, and several times between 30 December and 9 January (Figure 7c,e). Overestimations were particularly noticeable in the SSC series at 1.67 m above the seabed (Figure 7e): modelled SSC regularly exceeded observed SSC by 20 to 40 mg·L^{-1} during Events 2 and 4, and even during calmer periods (e.g., between 30 December and 9 January). In contrast, modelled SSC were underestimated by a factor of 2 during Event 3. The $S1_{LIN}$ erosion setting led to a representation of observed SSC with a $RMSE$ of 14 mg·L^{-1} over the study period.

Using the $S1_{EXP2}$ erosion setting enabled a general improvement in modelled SSC with a $RMSE$ of 10.5 mg·L^{-1} over the period, and a correct response during the four energetic periods (Figure 7d,e). Differences in SSC between simulations during Events 1, 2, and 4 were mainly due to different erosion rates (especially due to E_0 in Equation (1)) prescribed for the same intermediate mud fraction in the surficial layer (in Figure 6a, this rate is clearly higher in the $S1_{LIN}$ setting). However, other differences, especially between 30 of December and 9 of January and during Event 3, are linked to the contrasted changes in the seabed between the two simulations.

One major difference in these changes occurs after Event 2 with more significant mud deposition in simulation $S1_{EXP2}$ (Figure 7f,g). This difference can also be seen in Figure 7h, which highlights a significant decrease in $F_{OBC, mud}$ (i.e., flow of mud out of the cell) in simulation $S1_{LIN}$, but not in $S1_{EXP2}$. The decrease in $F_{OBC, mud}$ in simulation $S1_{LIN}$ is probably due to less mud inputs from adjacent cells, resulting in relatively more mud exported by advection and thus less mud deposition during the decrease in shear stress following Event 2. Note that this difference in seabed changes influences the sediment dynamics in both simulations throughout the period. Following Event 2, a transition in surficial sediment from muddy to sandy occurs in both simulations but at different times. In the $S1_{LIN}$ simulation, the transition occurs half way through the period, around 30 December, and manifests itself as a SSC peak linked to mud resuspension near the bottom (Figure 7e), and by a decrease in $F_{OBC, mud}$ (relative loss by advection), which does not occur in the $S1_{EXP2}$ simulation. Following the transition in the nature of the seabed, the $S1_{LIN}$ simulation regularly gives incorrect SSC responses (e.g., overestimation around 6 January, underestimation during Event 3). These results underline the potential role of advection processes in the contrasted results of the two simulations, and the need for full 3D modelling to obtain a final fit of the erosion law. In the $S1_{EXP2}$ simulation, the transition occurs later in the period, around 11 of January, and enables a correct SSC response regarding Event 3, associated with a decrease in $F_{OBC, mud}$. It may mean that, on the one hand, setting $S1_{EXP2}$ allows a more accurate representation of resuspension dynamics in response to a given forcing, and on the other hand, it induces a more correct change in the nature of the seabed with respect to the variations in forcing over time. Note that despite contrasted sediment dynamics in the different simulations, the mud budget at the scale of the cell summed over the whole period led to similar trends corresponding to a relative loss by advection of around -5×10^6 kg (that is $-5/2.5^2$ kg·m^{-2}).

Sand contributes to turbidity over shorter periods than mud, mainly during Events 1, 2, and 4. Results provide evidence that sand is not subject to the same dynamics as mud. Until the transition in the nature of the surficial sediments in the middle of the period (30 December), sand dynamics appear to be quite similar in the two simulations (e.g., $F_{OBC, sand}$ in Figure 7h). Beyond this date, the contrasted nature of the seabed results in more regular sand resuspension in $S1_{LIN}$, with an advection component leading to a relative local sand loss (in the cell). Starting from Event 3, $F_{OBC, sand}$ increases (i.e., relative sand inflow into the cell) in both simulations while the advection flux of mud decreases (in $S1_{EXP2}$) or does not change (in $S1_{LIN}$). This highlights the fact that sand and mud dynamics are likely to differ depending on the nature of the seabed in adjacent cells.

The erosion setting $S1_{EXP1}$, which is characterized by a less abrupt exponential transition in erosion law ($C_{exp} = 10$ in Equation (10)), appears to be less accurate in terms of SSC (not illustrated here) with a $RMSE$ of 12 mg·L^{-1} over the study period (versus 10.5 mg·L^{-1} in $S1_{EXP2}$). The turbidity response provided by the model would be expected to be degraded while progressively reducing the decreasing trend of the transition (until a linear decrease is reached).

Considering the accurate representation of resuspension events and the lower $RMSE$ obtained with the erosion setting $S1_{EXP2}$, the latter was considered as an "optimum" setting, suggesting that the definition of an exponential transition to describe sand/mud mixture erosion between non-cohesive and cohesive erosion modes may be appropriate in hydro-sedimentary numerical models.

4.2. Influence of Critical Mud Fractions

The sensitivity of the SSC model results to critical mud fractions, f_{mcr1} and f_{mcr2}, was assessed, starting from the "optimum" erosion setting deduced in Section 4.1 and characterized by an exponential transition between $f_{mcr1} = 20\%$ and $f_{mcr2} = 70\%$ with $C_{exp} = 40$ in Equation (10) (i.e., $S1_{EXP2}$ setting). Both critical mud fractions were progressively reduced by 10% ($f_{mcr1} = 10\%$, $f_{mcr2} = 60\%$), 15% ($f_{mcr1} = 5\%$, $f_{mcr2} = 55\%$), and 20% ($f_{mcr1} \approx 0\%$, $f_{mcr2} = 50\%$). The corresponding settings, $S2_{EXP2}$, $S3_{EXP2}$, and $S4_{EXP2}$ are illustrated in Figure 6. Results linked to the application of these different settings are illustrated in Figure 8. Note that the second critical mud fraction f_{mcr2} appears in the extension of the exponential decay (Equation (10)), but, due to the shape of the exponential trend, it does not constitute a real critical mud fraction but rather an adjustment parameter for the transition. We can thus consider that this sensitivity analysis mainly deals with the setting of the first critical mud fraction f_{mcr1}.

First, it can be seen that the SSC linked to the four resuspension events progressively decrease with the decrease in f_{mcr1} (Figure 8b), which results in underestimation of turbidity with respect to observed values. While no clear differences in SSC appear between $S1_{EXP2}$ and $S2_{EXP2}$ (the latter is not illustrated in Figure 8), i.e., with a reduction of f_{mcr1} from 20% to 10%, significant SSC underestimations occur for $f_{mcr1} < 10\%$. The average turbidity during resuspension events is underestimated by 15–20% (respectively, 30%) in simulation $S3_{EXP2}$ (respectively, $S4_{EXP2}$). Regarding maximum SSC, underestimations of SSC peaks are around 15–30% (respectively, 40–50%) during Events 1 and 2 in simulation $S3_{EXP2}$ (respectively, $S4_{EXP2}$). In addition, SSC peaks during Events 3 and 4 are completely absent in these two simulations with an underestimation of about 60%. Other simulations with linear trend but low f_{mcr1} were tested and showed no improvement compared with the settings illustrated in Figures 7 and 8.

Changes in the seabed linked to the optimum erosion setting $S1_{EXP2}$ and the simulation $S3_{EXP2}$ ($f_{mcr1} = 5\%$) are illustrated in Figure 8c,d. Following Event 2, contrary to results in $S1_{EXP2}$, no drastic change in the nature of the seabed occurs in $S3_{EXP2}$ in the rest of the period, with a surficial sediment containing at least 30–40% of mud. This less dynamic change in the seabed is consistent with the lower SSC obtained in the water column. A reduction of f_{mcr1} led to the application of a pure mud erosion law starting from a lower mud content in the surficial sediment. This mostly resulted in less erosion ($E_{0,mud} << E_{0,sand}$) with weaker SSC and slower changes in the seabed. This is also visible in the variations in $F_{OBC, mud}$ (Figure 8e) which highlight the fact that the reduced sediment dynamics obtained by reducing f_{mcr1} results in weaker gradients (SSC, seabed nature) with adjacent cells, and a less dynamic advection term over the study period.

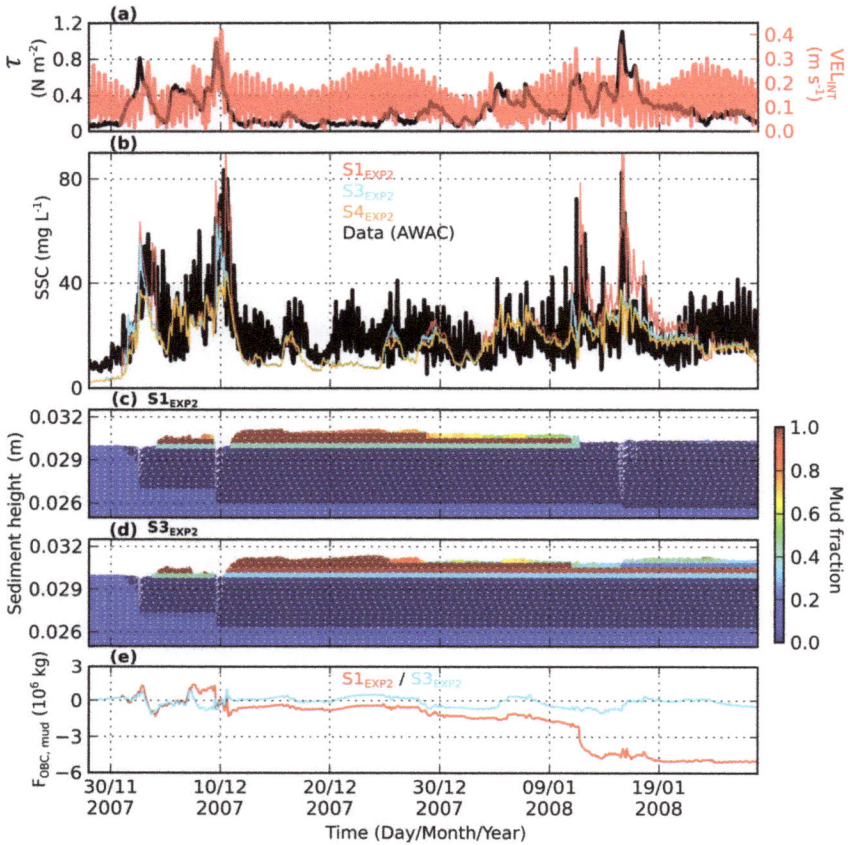

Figure 8. Comparisons of the results of the 3D model obtained from erosion settings $S1_{EXP2}$, $S3_{EXP2}$ and $S4_{EXP2}$, and measurements made with the *AWAC* acoustic profiler. (**a**) Shear stresses τ and depth-integrated currents VEL_{INT}; (**b**) time series of measured and modelled *SSC* at the level of the first *AWAC* cell (1.67 m above the bottom); (**c,d**) changes in the seabed (mud fraction, thickness of the sediment layers) in $S1_{EXP2}$ and $S3_{EXP2}$ simulations (the white dotted lines represent the boundaries of the sediment layers); and (**e**) integrated amount of mud advected through the water column in the $S1_{EXP2}$ and $S3_{EXP2}$ simulations.

5. Discussion

5.1. Setting Describing Erosion of a Sand/Mud Mixture

All experimental studies in the literature on the erosion of sand/mud mixtures mentioned a transition in the erosion mode when fine particles progressively fill the spaces between non-cohesive particles. Panagiotopoulos et al. [15] proposed a conceptual model showing the mechanism for the initiation of sediment motion for sand-mud mixtures, based on the forces acting on an individual grain and the associated angle of internal friction. When the mud content increases, clay minerals progressively fill the spaces between the sand particles, which slightly alters the pivoting characteristics and consequently the internal friction angle, and thus a slight change in erosion resistance. As soon as the sand particles are no longer in contact with one another, pivoting stops being the main mechanism behind the initiation of motion, and the resistance of the clay fraction mainly controls erosion. Depending on the authors, this transition is expressed by reasoning in terms of mud or clay content in the mixture. For instance, Mitchener and Torfs [14] proposed a transition between 3% and 15% mud content, and suggested using cohesive-type sediment transport equations above this transition value and sand transport theories below it. Other authors suggested that the transition occurs at much higher mud content, i.e., between 20% and 30% (e.g., [7,15]). More generally, previous investigations emphasized that only 2% to 10% of clay minerals (dry weight) added in a non-cohesive sediment matrix were sufficient to control the soil properties and increase the resistance of the bed to erosion (e.g., [17,24,45,56]).

Modelling results in the present study highlighted the fact that the critical mud fraction (f_{mcr1}), above which a transition toward a cohesive erosion mode would start, is at least 10% mud content. Grain size analyses of numerous sediment samples (from several locations, and at different depths in the sediment) from the BoBCS revealed that the clay content (per cent < 4 μm) corresponds to 30% ($\pm$3%, $R^2 = 0.96$) of the mud content (per cent < 63 μm). Such a constant ratio between the clay and mud fractions (or between the clay and silt fractions) in a given area has been observed in many sites worldwide (e.g., [13]). Thus, the critical clay content deduced from our modelling fitting would be around 3%. Therefore, our results are in agreement with experimental results of previous studies regarding the existence and the value of a critical mud/clay fraction indicating a transition in the mode of erosion.

Multiple transition trends of the erosion law (linear, exponentials) were tested to describe the erosion behaviour of transitional sand/mud mixtures, i.e., when the mud fraction exceeds the critical value of f_{mcr1} in the mixture. The quality of the model response was evaluated by comparing *SSC* results with turbidity measurements provided by the *AWAC* profiler over the entire water column. Based on *RMSE* and average or maximum *SSC* values reached during resuspension events, the results provided a more accurate representation of observations while considering an abrupt exponential transition of erosion parameters (i.e., E_0, τ_e, and n in the Partheniades form of the erosion law, see Equation (1)). Actually, changes in *SSC* produced by this transition formulation mainly hold in the contrasted E_0 values prescribed in erosion law depending on the seabed mud fraction ($E_{0,mud} \ll E_{0,sand}$; see Table 2). This result agrees with results recently obtained by Smith et al. [17] who performed erosion experiments on mixed sediment beds prepared in the laboratory (250–500 μm sands mixed with different clayey sediments corresponding to kaolinite or kaolinite/bentonite). In particular, they observed a rapid decrease in erosion rates, from 1.5 to 2.5 orders of magnitude, over a range of 2% to 10% clay content. In the present study, exponential transitions prescribed in settings $S1_{EXP1}$ and $S1_{EXP2}$ ($C_{exp} = 10$ and $C_{exp} = 40$ in Equation (10), Figure 6) led to a variation in the erodibility parameter E_0 of about 2.5 orders of magnitude over a mud content range of 10% and 40%, i.e., over a clay content ranging from 3% and 12% respectively. The best model results obtained from erosion setting $S1_{EXP2}$ are thus consistent with the findings of Smith et al. [17], and suggest that a rapid exponential transition may be appropriate to describe the erosion of a sand/mud mixture between non-cohesive and cohesive erosion modes in numerical hydro-sedimentary models.

5.2. Limitations of the Approach and Remaining Uncertainties

Despite successive assessments of model quality, some limitations and uncertainties concerning our modelling approach remain and have to be addressed.

5.2.1. Mud Erosion Law

A pure mud erosion law was set up based on erodimetry experiments performed on muddy sediment samples from the BoBCS. A critical shear stress for mud erosion ($\tau_{e,mud}$) of 0.1 N·m^{-2} was deduced. By combining this $\tau_{e,mud}$ value with the minimum erodibility parameter $E_{0,mud}$ recommended by e.g., Winterwerp [54], i.e., 10^{-5} kg·m^{-2}·s^{-1}, the application of the mud erosion law from the model (Section 3.3) led to good agreement between modelled erosion fluxes and those obtained in erodimetry experiments for comparable applied shear stresses (Figure 5). Such a lower critical stress for erosion when the mixture is muddier is opposite to trends most often published, characterized by an increase of the resistance to erosion when mud is added to sand (e.g., [6,7,14–17]). Other simulations were performed with higher $\tau_{e,mud}$ values, 0.15, 0.2, and 0.4 N·m^{-2}. As expected, modelled SSC was underestimated compared with observed SSC while $\tau_{e,mud}$ increased (even for 0.15 N·m^{-2}). Another assessment of $E_{0,mud}$ could have produced similar results, but we preferred to keep the shear strength provided by our experiments, the low value being justified by the fact that in our environment (erosion on a continental shelf with low bottom friction) the sediment is never remobilized at depth, and the surficial sediment remains unconsolidated.

5.2.2. Initial Condition of the Sediment and Time Variation of the Seabed

Seabed initialization was prescribed from the synthesis of sediment facies applied at the beginning of each simulation. To evaluate the influence of the sediment initialization on model results (SSC, seabed variations), the "optimum" model setting from the present study ($S1_{EXP2}$) was used again in a new simulation using the surficial sediment cover computed at the end of a one-year simulation used as spin up. We obtained similar SSC results with a $RMSE$ of 11.3 mg·L^{-1} over the study period (versus 10.5 mg·L^{-1} in the original $S1_{EXP2}$ simulation). Similarities in seabed variations (thickness and composition) in the two simulations were likewise remarkable (not illustrated here). Thus, the seabed initialization prescribed at the beginning of each simulation appears to be appropriate and does not correspond to a transitional state regarding the sediment dynamics.

Model results concerning changes in the seabed highlighted pronounced gradients in the nature of the surficial sediment in most simulations, with an alternation of muddy and sandy facies depending on the intensity of forcing (e.g., shear stress, advection). Such variations in the nature of the sediment are not unrealistic, since grain size analyses of a sediment core sampled at LC station revealed a layered bed, with alternating muddy and sandy layers at different depths in the sediment. These variations are also consistent with the geographical location of LC station (Figure 1), in a zone with horizontal gradients in sediment facies.

A further validation of the model in terms of thickness or elevation would require other measurements as in situ altimetry data (e.g., [57]). Simulations make it necessary to use altimetry data, which were not available in the present study.

5.2.3. Applicability of the Sand/Mud Mixture Erosion Law

Assessment of the effect of erosion settings on the quality of model results with respect to observations would require further comparisons in other study sites where the seabed consists of both sand and mud. This would make it possible to know if an abrupt transition between non-cohesive and cohesive erosion modes systematically improves model accuracy in terms of SSC.

The use of the sand/mud mixture erosion law derived from this study requires site-specific information beforehand, in particular grain size analyses for the assessment of the mud or the clay content. Since the critical mud fraction f_{mcr1}, above which erosion behaviour starts to change, mainly depends on clay content, the ratio between clay and mud fractions can be used. In future works, it would be interesting to explore other mud properties than grain size and sediment fractions, such as mineralogy, to represent more accurately the key role played by cohesive sediments in erosion process, especially for transitional sand/mud mixtures between the contrasted non-cohesive/cohesive regimes.

Lastly, the formulation of erosion was based on the Patheniades-Ariathurai law, with an erosion flux proportional to the normalized excess shear stress (τ/τ_e-1). Such a formulation is very sensitive to the value of the critical shear stress for erosion, which can be difficult to estimate and highly variable in the case of sand/mud mixtures. Alternatively, a formulation of the erosion flux proportional to the excess shear stress (τ-τ_e) would reduce the sensitivity of erosion to τ_e. It would also be in agreement with the Van Kesteren-Winterwerp-Jacobs erosion law [6,58], and deserves further investigations following the pioneering work of Jacobs et al. [6].

6. Conclusions

The aim of this study was to assess the influence of the erosion law prescribed in a 3D realistic hydro-sedimentary model on sediment dynamics in the case of a seabed composed of both fine sand and mud in a slightly energetic environment, representative of continental shelves. According to the sediment model described by Le Hir et al. [1], the sediment was eroded as a mixture and was assumed to behave as pure sand below a first critical mud fraction in the surficial sediment, and as pure mud above a second one. Following hydrodynamic validation of the model and rigorous assessment of pure sand and pure mud erosion dynamics, several transition trends of erosion-related parameters (erodibility parameter, critical shear stress, and exponent in the Partheniades erosion law) were tested to describe the erosion of transitional sand/mud mixtures between the two critical fractions. Different simulations were run using linear or exponential transitions, and different critical mud fractions. The accuracy of model results regarding suspended matter dynamics was evaluated at a single point, located on the Bay of Biscay shelf, by performing comparisons with turbidity observations provided by an acoustic profiler during two typical winter months. The main conclusions of this work are:

- Using an abrupt exponential transition, e.g., an erodibility parameter decrease of 2.5 orders of magnitude over a 10% (respectively, 3%) mud (respectively, clay) content range, improves *SSC* model results regarding measurements, compared to results obtained with linear or less abrupt exponential transitions. This conclusion agrees with recent experimental studies in the literature on the erosion of sand/mud mixtures, which mention a drastic change in erosion mode for only a small percentage of clay added in the mixture.
- A first critical mud fraction (above which the erosion mode begins to change) of 10–20% is required to ensure a relevant model response in turbidity. By reasoning in terms of the clay fraction, the corresponding critical clay fraction ranges between 3% and 6%. Once again, this conclusion agrees with experimental studies in the literature reporting that 2% to 10% of clay minerals in a sediment mixture are sufficient to control the soil properties.
- The erosion flux of mixed sediments appears to be very sensitive to the clay fraction of the surficial sediment, and then is likely to change considerably at a given location, according to erosion and deposition events.
- The need to perform 3D simulations to account for advection, which considerably influences sediment dynamics in terms of export of resuspended sediments, sediment inflows from adjacent cells, and consequent changes in the surficial seabed (nature and thickness of deposits).

Therefore, the optimal erosion law derived from this study to describe sand/mud mixture erosion led to model results consistent with measurements and with most of the conclusions deduced from experimental studies already published. This should encourage further similar comparisons and suggests that the application of this kind of erosion setting is appropriate for hydro-sedimentary models.

Acknowledgments: This study was supported by the SHOM (Service Hydrographique et Océanographique de la Marine) and IFREMER (Institut Français de Recherche pour l'Exploitation de la Mer). The authors would like to thank the SHOM for surficial mud content data. Lastly, the two anonymous reviewers are deeply thanked for their comments and suggestions that greatly improved the manuscript.

Author Contributions: All authors conceived the study; B.M., P.L.H., and F.C. designed the numerical experiments; T.G. performed the seabed initialization of the model describing the horizontal distribution of sediment facies; B.M. performed the simulations; B.M., P.L.H. and F.C. analysed the simulations; and B.M. and P.L.H. wrote the paper.

Conflicts of Interest: The authors declare no conflict of interest.

References

1. Le Hir, P.; Cayocca, F.; Waeles, B. Dynamics of sand and mud mixtures: A multiprocess-based modelling strategy. *Cont. Shelf Res.* **2011**, *31*, S135–S149. [CrossRef]

2. Righetti, M.; Lucarelli, C. May the Shields theory be extended to cohesive and adhesive benthic sediments? *J. Geophys. Res.* **2007**, *112*, C05039. [CrossRef]
3. Kimiaghalam, N.; Clark, S.P.; Ahmari, H. An experimental study on the effects of physical, mechanical, and electrochemical properties of natural cohesive soils on critical shear stress and erosion rate. *Int. J. Sediment Res.* **2016**, *31*, 1–15. [CrossRef]
4. Williamson, H.J.; Ockenden, M.C. Laboratory and field investigations of mud and sand mixtures. In Proceedings of the First International Conference on Hydro-Science and Engineering, Advances in Hydro-Science and Engineering, Washington, DC, USA, 7–11 June 1993; Wang, S.S.Y., Ed.; Volume 1, pp. 622–629.
5. Le Hir, P.; Monbet, Y.; Orvain, F. Sediment erodability in sediment transport modelling: Can we account for biota effects? *Cont. Shelf Res.* **2007**, *27*, 1116–1142. [CrossRef]
6. Jacobs, W.; Le Hir, P.; Van Kesteren, W.; Cann, P. Erosion threshold of sand–mud mixtures. *Cont. Shelf Res.* **2011**, *31*, S14–S25. [CrossRef]
7. Le Hir, P.; Cann, P.; Waeles, B.; Jestin, H.; Bassoullet, P. Erodibility of natural sediments: Experiments on sand/mud mixtures from laboratory and field erosion tests. In *Sediment and Ecohydraulics: INTERCOH 2005 (Proceedings in Marine Science)*; Kusuda, T., Yamanishi, H., Spearman, J., Gailani, J.Z., Eds.; Elsevier: Amsterdam, The Netherlands, 2008; Volume 9, pp. 137–153.
8. Van Rijn, L.C. Sediment pick-up functions. *J. Hydraul. Eng.* **1984**, *110*, 1494–1502. [CrossRef]
9. Emadzadeh, A.; Cheng, N.S. Sediment pickup rate in uniform open channel flows. In Proceedings of the River Flow 2016, Iowa City, IA, USA, 11–14 July 2016; Constantinescu, G., Garcia, M., Hanes, D., Eds.; Taylor & Francis Group: London, UK, 2016; Volume 1, pp. 450–457.
10. Partheniades, E. A Study of Erosion and Deposition of Cohesive Soils in Salt Water. Ph.D. Thesis, University of California, Berkeley, CA, USA, 1962.
11. Ariathurai, C.R. A Finite Element Model of Cohesive Sediment Transportation. Ph.D. Thesis, University of California, Davis, CA, USA, 1974.
12. Van Ledden, M.; Van Kesteren, W.G.M.; Winterwerp, J.C. A conceptual framework for the erosion behaviour of sand–mud mixtures. *Cont. Shelf Res.* **2004**, *24*, 1–11. [CrossRef]
13. Flemming, B.W. A revised textural classification of gravel-free muddy sediments on the basis of ternary diagrams. *Cont. Shelf Res.* **2000**, *20*, 1125–1137.
14. Mitchener, H.; Torfs, H. Erosion of mud/sand mixtures. *Coast. Eng.* **1996**, *29*, 1–25. [CrossRef]
15. Panagiotopoulos, I.; Voulgaris, G.; Collins, M.B. The influence of clay on the threshold of movement of fine sandy beds. *Coast. Eng.* **1997**, *32*, 19–43. [CrossRef]
16. Ye, Z.; Cheng, L.; Zang, Z. Experimental study of erosion threshold of reconstituted sediments. In Proceedings of the ASME 2011 30th International Conference on Ocean, Offshore and Arctic Engineering, Rotterdam, The Netherlands, 19–24 June 2011; American Society of Mechanical Engineers: New York, NY, USA; Volume 7, pp. 973–983. [CrossRef]
17. Smith, S.J.; Perkey, D.; Priestas, A. Erosion thresholds and rates for sand-mud mixtures. In Proceedings of the 13th International Conference on Cohesive Sediment Transport Processes (INTERCOH), Leuven, Belgium, 7–11 September 2015; Toorman, E., Mertens, T., Fettweis, M., Vanlede, J., Eds.;
18. Gailani, J.Z.; Jin, L.; McNeil, J.; Lick, W. Effects of Bentonite Clay on Sediment Erosion Rates. DOER Technical Notes Collection. Available online: http://www.dtic.mil/docs/citations/ADA390214 (accessed on 21 July 2017).
19. Waeles, B.; Le Hir, P.; Lesueur, P. A 3D morphodynamic process-based modelling of a mixed sand/mud coastal environment: The Seine estuary, France. In *Sediment and Ecohydraulics: INTERCOH 2005, Proceedings in Marine Science*; Kusuda, T., Yamanishi, H., Spearman, J., Gailani, J.Z., Eds.; Elsevier: Amsterdam, The Netherlands, 2008; Volume 9, pp. 477–498.
20. Bi, Q.; Toorman, E.A. Mixed-sediment transport modelling in Scheldt estuary with a physics-based bottom friction law. *Ocean Dyn.* **2015**, *65*, 555–587. [CrossRef]
21. Migniot, C. Tassement et rhéologie des vases—Première partie. *La Houille Blanche* **1989**, *1*, 11–29. (In French) [CrossRef]
22. Dickhudt, P.J.; Friedrichs, C.T.; Sanford, L.P. Mud matrix solids fraction and bed erodibility in the York River estuary, USA, and other muddy environments. *Cont. Shelf Res.* **2011**, *31*, S3–S13. [CrossRef]

23. Carniello, L.; Defina, A.; D'Alpaos, L. Modeling sand-mud transport induced by tidal currents and wind waves in shallow microtidal basins: Application to the Venice Lagoon (Italy). *Estuar. Coast. Shelf Sci.* **2012**, *102*, 105–115. [CrossRef]

24. Van Ledden, M. Sand-Mud Segregation in Estuaries and Tidal Basins. Ph.D. Thesis, Delft University of Civil Engineering, Delft, The Netherlands, 2003.

25. Ahmad, M.F.; Dong, P.; Mamat, M.; Wan Nik, W.B.; Mohd, M.H. The critical shear stresses for sand and mud mixture. *Appl. Math. Sci.* **2011**, *5*, 53–71.

26. Wiberg, P.L.; Drake, D.E.; Harris, C.K.; Noble, M. Sediment transport on the Palos Verdes shelf over seasonal to decadal time scales. *Cont. Shelf Res.* **2002**, *22*, 987–1004. [CrossRef]

27. Ulses, C.; Estournel, C.; Durrieu de Madron, X.; Palanques, A. Suspended sediment transport in the Gulf of Lions (NW Mediterranean): Impact of extreme storms and floods. *Cont. Shelf Res.* **2008**, *28*, 2048–2070. [CrossRef]

28. Fard, I.K.P. Modélisation des Échanges Dissous Entre L'estuaire de la Loire et les Baies Côtières Adjacentes. Ph.D. Thesis, University of Bordeaux, Bordeaux, France, 2015.

29. Lurton, X. *An Introduction to Underwater Acoustics: Principles and Applications*; Springer: Berlin, Germany, 2002.

30. Tessier, C.; Le Hir, P.; Lurton, X.; Castaing, P. Estimation de la matière en suspension à partir de l'intensité rétrodiffusée des courantomètres acoustiques à effet Doppler (ADCP). *C. R. Geosci.* **2008**, *340*, 57–67. (In French) [CrossRef]

31. Lazure, P.; Dumas, F. An external–internal mode coupling for a 3D hydrodynamical model for applications at regional scale (MARS). *Adv. Water Resour.* **2008**, *31*, 233–250. [CrossRef]

32. Song, Y.; Haidvogel, D. A semi-implicit ocean circulation model using a generalized topography-following coordinate system. *J. Comput. Phys.* **1994**, *115*, 228–244. [CrossRef]

33. Rodi, W. *Turbulence Models and Their Application in Hydraulics*, 3rd ed.; IAHR Monograph: Delft, The Netherlands, 1993.

34. Ferry, N.; Parent, L.; Garric, G.; Barnier, B.; Jourdain, N.C. Mercator global Eddy permitting ocean reanalysis GLORYS1V1: Description and results. *Mercator-Ocean Q. Newsl.* **2010**, *36*, 15–27.

35. Boudière, E.; Maisondieu, C.; Ardhuin, F.; Accensi, M.; Pineau-Guillou, L.; Lepesqueur, J. A suitable metocean hindcast database for the design of Marine energy converters. *Int. J. Mar. Energy* **2013**, *3–4*, e40–e52. [CrossRef]

36. Déqué, M.; Dreveton, C.; Braun, A.; Cariolle, D. The ARPEGE/IFS atmosphere model: A contribution to the French community climate modelling. *Clim. Dyn.* **1994**, *10*, 249–266. [CrossRef]

37. Lyard, F.; Lefèvre, F.; Letellier, T.; Francis, O. Modelling the global ocean tides: Modern insights from FES2004. *Ocean Dyn.* **2006**, *56*, 394–415. [CrossRef]

38. Jonsson, I.G. Wave boundary layers and friction factors. In Proceedings of the 10th International Conference on Coastal Engineering, Tokyo, Japan, September 1966; American Society of Civil Engineers: New York, NY, USA, 1966; pp. 127–148.

39. Soulsby, R.L.; Hamm, L.; Klopman, G.; Myrhaug, D.; Simons, R.R.; Thomas, G.P. Wave-current interaction within and outside the bottom boundary layer. *Coast. Eng.* **1993**, *21*, 41–69. [CrossRef]

40. Soulsby, R. *Dynamics of Marine Sands: A Manual for Practical Applications*; Thomas Telford: London, UK, 1997.

41. Grasso, F.; Le Hir, P.; Bassoullet, P. Numerical modelling of mixed-sediment consolidation. *Ocean Dyn.* **2015**, *65*, 607–616. [CrossRef]

42. Van Leussen, W. Estuarine Macroflocs and Their Role in Fine-Grained Sediment Transport. Ph.D. Thesis, University of Utrecht, Utrecht, The Netherlands, 1994.

43. Tessier, C.; Le Hir, P.; Dumas, F.; Jourdin, F. Modélisation des turbidités en Bretagne sud et validation par des mesures in situ. *Eur. J. Environ. Civ. Eng.* **2008**, *12*, 179–190. [CrossRef]

44. Verney, R.; Gangloff, A.; Chapalain, M.; Le Berre, D.; Jacquet, M. Floc features in estuaries and coastal seas. In Proceedings of the 5th Particles in Europe Conference, Budapest, Hungary, 3–5 October 2016.

45. Dyer, K.R. *Coastal and Estuarine Sediment Dynamics*; John Wiley & Sons: New York, NY, USA, 1986.

46. Mengual, B. Variabilité Spatio-Temporelle des Flux Sédimentaires Dans le Golfe de Gascogne: Contributions Relatives des Forçages Climatiques et des Activités De Chalutage. Ph.D. Thesis, University of Western Brittany, Brest, France, 2016.

47. Bouysse, P.; Lesueur, P.; Klingebiel, A. Carte Des Sédiments Superficiels du Plateau Continental du Golfe de Gascogne: Partie Septentrionale au 1/500 000. Co-Éditée par le BRGM Et l'IFREMER, 1986. Available online: http://sextant.ifremer.fr/record/ea0b61b0-71c6-11dc-b1e4-000086f6a62e/ (accessed on 21 July 2017).

48. Mengual, B.; Cayocca, F.; Le Hir, P.; Draye, R.; Laffargue, P.; Vincent, B.; Garlan, T. Influence of bottom trawling on sediment resuspension in the "Grande-Vasière" area (Bay of Biscay, France). *Ocean Dyn.* **2016**, *66*, 1181–1207. [CrossRef]

49. Van Rijn, L.C. Unified view of sediment transport by currents and waves. II: Suspended transport. *J. Hydraul. Eng.* **2007**, *133*, 668–689. [CrossRef]

50. Van Rijn, L.C. Sediment transport, part II: Suspended load transport. *J. Hydraul. Eng.* **1984**, *110*, 1613–1641. [CrossRef]

51. Engelund, F.; Hansen, E. *A Monograph on Sediment Transport in Alluvial Streams*; Teknish Forlag, Technical Press: Copenhagen, Denmark, 1967.

52. Yang, C.T. Incipient motion and sediment transport. *J. Hydraul. Div.* **1973**, *99*, 1679–1704.

53. Dufois, F.; Le Hir, P. Formulating Fine to Medium Sand Erosion for Suspended Sediment Transport Models. *J. Mar. Sci. Eng.* **2015**, *3*, 906–934. [CrossRef]

54. Winterwerp, J.C. Flow-Induced Erosion of Cohesive Beds; A Literature Survey. Rijkswaterstaat—Delft Hydraulics, Cohesive Sediments Report 25, February 1989. Available online: http://publicaties.minienm.nl/download-bijlage/45703/164198.pdf (accessed on 21 July 2017).

55. Tessier, C. Caractérisation et Dynamique des Turbidités en Zone Côtière: L'exemple de la Région Marine Bretagne Sud. Ph.D. Thesis, University of Bordeaux 1, Bordeaux, France, 2006.

56. Raudkivi, A.J. *Loose Boundary Hydraulics*, 3rd ed.; Pergamon Press: Oxford, UK, 1990.

57. Bassoullet, P.; Le Hir, P.; Gouleau, D.; Robert, S. Sediment transport over an intertidal mudflat: Field investigations and estimation of fluxes within the "Baie de Marennes-Oléron" (France). *Cont. Shelf Res.* **2000**, *20*, 1635–1653. [CrossRef]

58. Winterwerp, J.C.; Van Kesteren, W.G.M. *Introduction to the Physics of Cohesive Sediment Dynamics in the Marine Environment*; Developments in Sedimentology; Elsevier: Amsterdam, The Netherlands, 2004.

water

MDPI

Article

Spatial and Seasonal Variation of Biomineral Suspended Particulate Matter Properties in High-Turbid Nearshore and Low-Turbid Offshore Zones

Michael Fettweis [1] and Byung Joon Lee [2,*]

[1]   Operational Directorate Natural Environment, Royal Belgian Institute of Natural Sciences, Gulledelle 100,
      B-1200 Brussels, Belgium; mfettweis@naturalsciences.be
[2]   Department of Disaster Prevention and Environmental Engineering, Kyungpook National University,
      2559 Gyeongsang-daero, Sangju, Gyeongbuk 742-711, Korea
*   Correspondence: bjlee@knu.ac.kr; Tel.: +82-54-530-1444

Received: 3 August 2017; Accepted: 11 September 2017; Published: 12 September 2017

Abstract: Suspended particulate matter (SPM) is abundant and essential in marine and coastal waters, and comprises a wide variety of biomineral particles, which are practically grouped into organic biomass and inorganic sediments. Such biomass and sediments interact with each other and build large biomineral aggregates via flocculation, therefore controlling the fate and transport of SPM in marine and coastal waters. Despite its importance, flocculation mediated by biomass-sediment interactions is not fully understood. Thus, the aim of this research was to explain biologically mediated flocculation and SPM dynamics in different locations and seasons in marine and coastal waters. Field measurement campaigns followed by physical and biochemical analyses had been carried out from 2004 to 2011 in the Belgian coastal area to investigate bio-mediated flocculation and SPM dynamics. Although SPM had the same mineralogical composition, it encountered different fates in the turbidity maximum zone (TMZ) and in the offshore zone (OSZ), regarding bio-mediated flocculation. SPM in the TMZ built sediment-enriched, dense, and settleable biomineral aggregates, whereas SPM in the OSZ composed biomass-enriched, less dense, and less settleable marine snow. Biological proliferation, such as an algal bloom, was also found to facilitate SPM in building biomass-enriched marine snow, even in the TMZ. In short, bio-mediated flocculation and SPM dynamics varied spatially and seasonally, owing to biomass-sediment interactions and bio-mediated flocculation.

Keywords: suspended particulate matter; aggregates; flocculation; biomass; sediment

---

1. Introduction

Suspended particulate matter (SPM), produced by biological and geophysical actions on the Earth's crust, enters into marine and coastal waters and is dispersed by flow-driven transportation, such as advection and dispersion [1–3]. The SPM concentration is an important parameter to understand the marine ecosystem as it controls the water turbidity and mediates many physical and biochemical processes [4–6].

SPM comprises a wide variety of biomineral clay to sand sized particles, comprising living (microbes, phyto- and zooplankton) and non-living organic matter (fecal and pseudo-fecal pellets, detritus and its decomposed products from microbial activity such as mucus, exopolymers), and minerals from a physico-chemical (e.g., clay minerals, quartz, feldspar) and biogenic origin (e.g., calcite, aragonite, opal), which are practically grouped into organic biomass and inorganic sediments [7]. It is important to note that when clays or other charged particles and polymers are in suspension, they become attached to each other and form fragile structures or flocs with compositions, sizes, densities, and structural complexities that vary as a function of turbulence and biochemical composition [3,8–11]. Flocculation combines biomass and sediments into larger aggregates (i.e., flocs) that can be classified as either mineral, biomineral, or biological aggregates. Flocculation usually integrates aggregation and disaggregation (i.e., breakup) kinetics, depending on the hydrodynamics of a suspension. Electrostatic and colloidal chemistry is the fundamental driver for flocculation in a cohesive suspension. For example, high ionic strength reduces the electrostatic repulsion between colloidal particles, thereby increasing the aggregation of colloidal suspension. Also, regarding the heterogeneity of a natural suspension with various biomass and sediments, physical and biochemical conditions are favorable for flocculation, like low turbulence intensity, high ionic strength, and sticky polymeric substances, which help individual biomineral particles to build large aggregates. Clay mineralogy is also important for determining electrochemistry and flocculation capability. Depending on the biomass composition, such aggregates are classified into mineral, biomineral, and biological aggregates [12,13]. Mineral and biomineral aggregates form in the sediment-enriched environment, such as a turbidity maximum zone (TMZ) or a nearshore area [6,14,15], while biological aggregates (i.e., marine snow) form in the mineral-depleted environment typically found in an offshore zone (OSZ) [16].

Flocculation mediated by biological composition determines the size, density, and settling velocity of aggregates [12,14,17]. For example, in a tidal cycle, low flow intensity during slack water enhances flocculation capability, building large, settleable aggregates, whereas high flow intensity at peak flow reduces flocculation capability, breaking down aggregates to small, less (or hardly) settleable aggregates or primary particles [5,18]. Moreover, sticky biomass (e.g., extracellular polymeric substances (EPSs) or transparent extracellular polymers (TEPs)) helps build large biomineral aggregates [7,19–22]. Flocculation which can be mediated by biological factors consequently controls sedimentation, resuspension, deposition, and erosion, and determines the overall SPM dynamics in marine and coastal waters [12,23].

Bio-medicated flocculation and SPM dynamics are important in science and engineering because they eventually control the sediment, carbonaceous, and nitrogenous mass balances at the regional or global scale [24,25]. Despite their importance, bio-mediated flocculation and SPM dynamics are not fully understood in coastal and marine waters. Geologists and hydraulic engineers have focused more on sediments and less on biomass [3,14], and marine biologists vice versa [16]. In our opinion, the biomass-sediment interactions in coastal and marine waters have only recently been studied in a systematic and quantitative way [7,9,13,26–29], and mathematical models which can take into account the heterogeneous composition/morphology of biomineral aggregates were developed only a few years ago [12,30]. These efforts should be paid more attention.

Therefore, the aim of the study was to add to our current understanding of bio-mediated flocculation and its impact on the SPM in marine and coastal waters. First, we investigated the spatial variation of SPM dynamics in a sediment-enriched TMZ and a mineral-depleted OSZ, especially concerning bio-mediated flocculation. Second, we investigated the seasonal variation of SPM dynamics in a TMZ to understand how seasonal changes in biological activity, especially algae blooms, affect bio-mediated flocculation and SPM dynamics. This paper describes and discusses bio-mediated flocculation and SPM dynamics for different locations and seasons.

2. Materials and Methods

2.1. Site Description

The study area is situated in the Southern Bight of the North Sea, specifically in the Belgian coastal zone. Measurements have indicated SPM concentrations of 20–70 mg/L in the nearshore area; reaching 100 to more than a few g/L near the bed; lower values (<10 mg/L) occur in the offshore [31]. As shown in Figure 1, the MOW1 measurement site is located in the TMZ. The Gootebank (G-Bank), Hinderbank (H-Bank), and Kwintebank (K-Bank) sites are in the OSZ, out of or at the edge of the turbidity maximum. Satellite images of surface SPM and chlorophyll-a (Chl) concentrations in the study area show clear spatial and seasonal changes. Regarding the seasonality, the annual cycle of SPM concentration in the high turbidity area off the Belgian coast is mainly caused by the seasonal biological cycle, rather than wind and waves. Wind strengths and wave heights have a seasonal signal, but these do not explain the large differences observed in SPM concentration [31,32]. This seasonality is linked with the seasonal changes in aggregate size and thus settling velocity due to biological effects. The aggregate sizes and settling velocities are smaller in winter and larger in summer. As a result, the SPM is more concentrated in the near-bed layer, whereas in winter, the SPM is better mixed throughout the water column. This explains the inverse correlation found between the surface SPM and the Chl concentrations in Figure 1. Water depths of the measuring area vary between 5 and 35 m. The mean tidal ranges at Zeebrugge are 4.3 and 2.8 m at spring and neap tides, respectively. The tidal current ellipses are elongated in the nearshore area and become gradually more semicircular towards the offshore area. The current velocities near Zeebrugge (nearshore) vary from 0.2 to 1.5 m/s during spring tide and from 0.2 to 1.0 m/s during neap tide. Salinity varies between 28 and 34 practical salinity units (PSU) in the coastal zone, because of the wind-induced advection of water masses and river discharge [33,34]. The most important sources of SPM are from the erosion and resuspension of the Holocene mud deposits outcropping in the Belgian nearshore area; the French rivers discharging into the English Channel, and the coastal erosion of the Cretaceous cliffs at Cap Gris-Nez and Cap Blanc-Nez (France) are only minor sources [35,36].

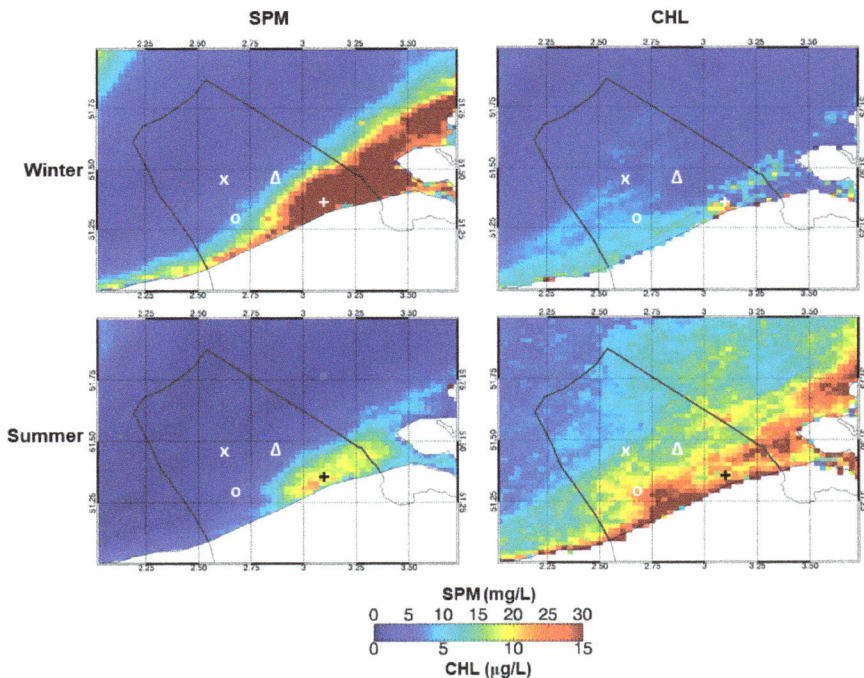

Figure 1. Mean surface suspended particulate matter (SPM) and chlorophyll-a (CHL) concentrations in the southern North Sea during the winter (October–March, top) and summer season (April–September, bottom) derived from the MERIS satellite. The +, Δ, X, and O symbols indicate the measurement sites of MOW1, Gootebank (G-Bank), Hilnderbank (H-Bank), and Kwintebank (K-Bank), respectively. Turbidity maximum zone (TMZ) has SPM concentrations above 15–20 mg/L and the offshore zone (OSZ) below 10 mg/L.

2.2. Tidal Measurements

Field measurements in the TMZ (MOW1) and in the OSZ (G-Bank, H-Bank, and K-Bank) were carried out about four times a year from February 2004 until 2011. During each campaign, sensor measurements (flow, SPM dynamics) and water sampling (SPM properties) were executed, while the research vessel was moored to maintain a specific measuring position for a 13-h tidal cycle. A Sea-Bird SBE09 SCTD carousel sampling system (containing twelve 10 L Niskin bottles) (Sea-Bird Electronics Inc., Bellevue, WA, USA) was kept at least 4.5 m below the surface and about 3 m above the bottom. A LISST 100X (Laser In-Situ Scattering and Transmissometry, range 2.5–500 μm) (Sequoia Scientific Inc., Bellevue, WA, USA) was attached directly to the carousel sampling system to measure particle size distribution (PSD) at the same location as the water sampling system [37]. The volume concentration of each size group was estimated with an empirical volume calibration constant, which was obtained under a presumed sphericity of particles [37–39]. The LISST has a sampling volume which permits it to statistically sample the less numerous large aggregates, but it cannot detect aggregates larger than 500 μm or smaller than 2.5 μm. Particles smaller than the size range affect the entire PSD, with an increase in the volume concentration of the smallest two size classes, a decrease in the next size classes and, an increase in the largest size classes [40]. Similar remarks have been formulated by Graham and coworkers [41], who observed an overestimation of one or two orders of magnitude in the number of fine particles measured by the LISST. A rising tail in the lowest size classes of the LISST occurs regularly in the data during highly turbulent conditions and is interpreted as an indication of the presence of very fine particles and thus a break-up of the aggregates. Particles exceeding the LISST size range of 500 μm also contaminate the PSD. The large out of range particles increase the volume concentration of particles in multiple size classes in the range between 250 and 500 μm and in the smaller size classes [42–44]. The occurrence of rising tail in the largest size classes indicates the occurrence of large particles rather than an absolute value. Other uncertainties of the LISST-100C are related to the often non-spherical shape of the particles occurring in nature [40,41,44]. A hull mounted, acoustic Doppler current profiler (ADCP) type, Workhorse Mariner 300 kHz (RD Instruments, Poway, CA, USA), was used to determine the velocity profiles.

2.3. Water Samples and Analysis

A Niskin bottle of the carousel sampling system was closed every 20 min, thus collecting about 40 samples during a 13 h flood-ebb tidal cycle. Note that the carousel sampling system was deployed to take water samples in the middle of the water column, at least 3 m above the bed layer. The carousel was brought aboard every hour. Three sub-samples from each water sample were then filtered on board using pre-weighed filter papers (Whatman GF/C, Sigma-Aldrich, St. Louis, MO, USA). In total, 120 filtrations were thus carried out per tidal cycle. After filtration, the filter papers were rinsed with demineralized water (±50 mL) to remove the salt, dried at 105 °C, and weighed again to determine the SPM concentrations. Every hour, a fourth sub-sample was filtered on board to determine particulate organic carbon (POC) and particulate organic nitrogen (PON) concentrations. The residues on the filter paper were carefully collected and acidified with 1 N HCl. Then, the POC and PON of the residues were quantified with a Carbon Nitrogen elemental analysis.

2.4. Grain Size and Mineralogical Analysis

Primary grain size and mineralogical analyses were performed to determine the mineralogical composition of the SPM samples. Suspension samples were obtained by the centrifugation of seawater collected by an ALFA Laval MMB 304S flow-through centrifuge (Alfa Laval Corp., Lund, Sweden), while the bed samples have been taken with a Van Veen grab sampler. Collected and stored samples were dried at 105 °C and chemically treated by adding HCl and H_2O_2 in order to remove the organic and carbonate fractions. The pretreated samples were rinsed with demineralized water, dried at 105 °C, and added to 100 mL demineralized water with 5 mL of peptizing agent (a mixture of

NaCO$_3$ and Na-oxalate). The suspension was dispersed and disaggregated using a magnetic stirrer and an ultrasonic bath. The grain size distribution and clay-silt-sand fractions of the SPM sample were analyzed with a Sedigraph 5100 (Micrometrics Instrument Corp., Norcross, GA, USA) for the fraction < 75 μm and sieved for the coarser fraction. The mineralogical composition of the clay fraction of the samples was determined with a Seifert 3003 theta-theta X-ray diffractometer (GE Measurement & Control, Billerica, MA, USA). Details of the analytical methods are documented in the earlier dissertation [36].

3. Results and Discussion

3.1. Mineralogical Characteristics of TMZ and OSZ

The mineralogical composition of the bed materials in the TMZ and OSZ are shown in Table 1. The respective clay and quartz contents of the bed materials in the TMZ were 25.0% and 39.6%, respectively, whereas those in the OSZ were 12.4% and 66.7%. Thus, the bed materials in the TMZ were found to be a mud-sand mixture, while the bed materials in the OSZ were sandy. Carbonates, such as calcite, Mg-calcite, aragonite, and dolomite, comprise about 20% and 10% of the TMZ and OSZ, respectively. Feldspar (i.e., K-feldspar and plagioclase) was also found to be an important content of the bed materials of the TMZ and OSZ, with a value of about 8%. Amorphous species in the TMZ and OSZ comprised 4.2% and 1.1%, respectively. Amorphous species are considered biogenic minerals influenced by biogeochemical actions [36,45]. The clay minerals at both sites comprised about 5% Kaolinite, 10% Chlorite, and 85% 2:1 layered silicates.

Table 1. Average mineralogical fractions (%) of the bulk deposits and suspended particulate matters (SPM) in the turbidity maximum zone (TMZ) and offshore zone (OSZ) measuring sites.

Material	Location	Clays	Quartz	Carbonates	Amorphous	Feldspar	Others
Bed Materials	TMZ	25.0	39.6	21.1	4.2	8.0	2.1
	OSZ	12.4	66.7	10.7	1.1	8.1	0.9
SPM	TMZ	36.2	14.6	29.9	12.7	4.2	2.3
	OSZ	31.3	20.6	29.7	10.1	6.4	1.8

In contrast to the bed materials, the SPM in the TMZ and OSZ had a similar mineralogical composition. For instance, the clays and quartz contents of the SPM only differed by 5% between the TMZ and OSZ, and the contents of carbonates, amorphous, feldspar, and others differed by less than 2.5%. This happened because the SPM samples do not contain coarser bed material, as the sand grains in suspension are seldom found above the near-bed layer. This also shows that the SPM in the TMZ and OSZ has the same origin, as suggested by the earlier geological survey in this area [35,36]. It is also important to note that the respective fractions of carbonate and amorphous species are large, at about 30% and 11% for both the TMZ and OSZ, thereby indicating high biological activity in the measuring area.

3.2. Spatial Variation of SPM Dynamics in the TMZ and OSZ

During the entire measurement period (2004–2011), the POC/SPM ratios in the OSZ were substantially higher than those in the TMZ (Figure 2a). This observation indicates that the SPM in the OSZ comprises more biomass and less sediments, and vice versa for the SPM in the TMZ. A scatter plot with POC content and SPM concentration shows the transition from a high mineral to the low-mineral SPM, when shifting from the TMZ to the OSZ (Figure 3). Generally, POC content increased with a decreasing SPM concentration (i.e., mineral-depleted condition). The mineral-depleted SPM in the OSZ seemed analogous to the muddy marine snow from an Australian coastal area where minerals were bound together with planktonic and transparent exopolymer particulate matter [46]. However, the PON/POC ratios of the TMZ and OSZ did not show such a clear difference during the entire study period, and their 95% confidence levels overlapped (Figure 2b).

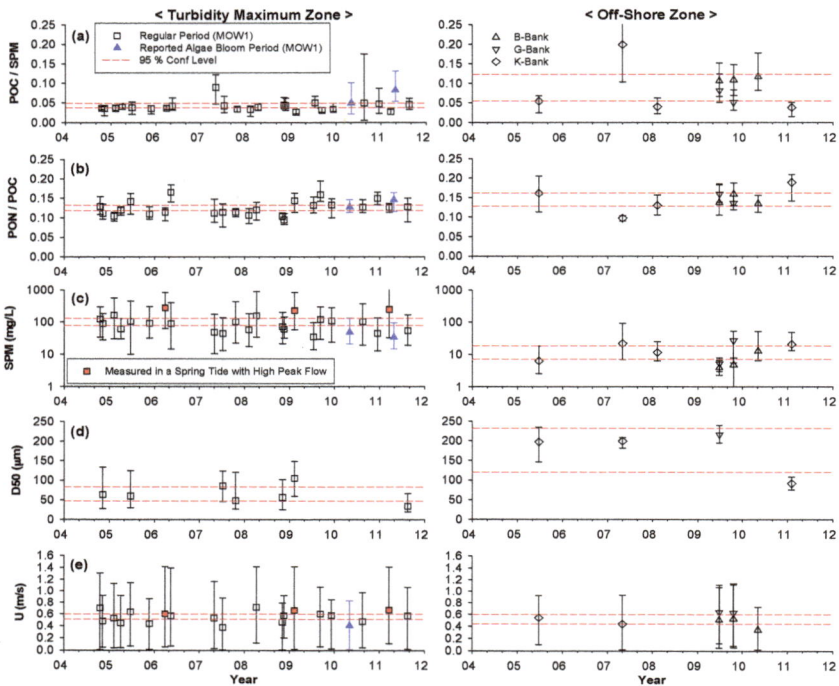

Figure 2. Spatial and seasonal variation of experimental indices in the measurement sites, during the entire measurement period, from 2004 to 2011. The left and right panels illustrate the data obtained from the turbidity maximum zone (TMZ) and the offshore zone (OSZ), respectively. (**a**) POC content in the SPM; (**b**) POC/PON ratio; (**c**) SPM concentration; (**d**) D50: median of the volumetric particle/aggregate size distribution; (**e**) U: flow velocity; MOW1: measurement site in the TMZ; B-Bank, G-Bank, and K-Bank: measurement sites in the OSZ.

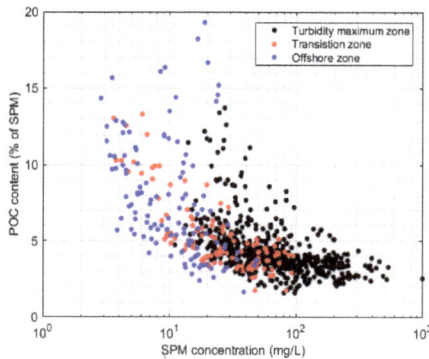

Figure 3. Scatter plot of POC content (% of SPM) versus SPM concentration. The coupled data sets of POC content and SPM concentration were obtained from all the 13 h measurement campaigns.

SPM concentrations in the TMZ are about an order magnitude higher than those in the OSZ (Figure 2c), similar to the satellite images of SPM concentration in Figure 1. This observation suggests that a substantial amount of sediment resides in the TMZ, which is transported back and forth in the flood and ebb tides. In addition, SPM concentrations in the TMZ were more vulnerable to flow intensity. High flow velocity (U) was found to increase SPM concentrations (e.g., 29 March 2006, 7 February 2008, 10 February 2009 and 21 March 2011, in Figure 2), because it increases sediment erosion and resuspension from the sea floor. It is also important to note that the TMZ had a two to three times smaller aggregate size (D50; median of the volumetric particle/aggregate size distribution) than the OSZ (Figure 2d) [15]. Thus, the TMZ enriched with sediments (i.e., higher SPM concentration and lower POC/SPM) had a lower flocculation capability (i.e., lower D50) than the OSZ. Tang and Maggi reported that small, dense aggregates are formed in sediment(mineral)-enriched environments, such as the TMZ in this study, whereas large, fluffy aggregates are formed in biomass-enriched environments [13,29]. The former was defined as mineral or biomineral aggregates, and the latter as biological aggregates.

SPM and POC concentrations and D50 in the TMZ were subject to ups and downs during the 13-h tidal cycle (Figure 4a). Generally, SPM and POC concentrations increased to their maximum around the peak flows. Biomass and minerals were likely combined in large, settleable biomineral aggregates, because SPM and POC concentrations had the same up-and-down movement during a tidal cycle. Such biomineral aggregates in the TMZ are vulnerable to aggregation and disaggregation (i.e., breakup), depending on the flow intensity (turbulence), available aggregation time to reach the equilibrium aggregate size, and organic matter content, therefore changing D50 in a flow-varying tidal cycle [5,18]. D50 increased to the maximum when approaching slack water, but decreased to a minimum around peak flow. Regarding flocculation kinetics, aggregation kinetics dominated over disaggregation kinetics for the slack water, and vice versa for the peak flow [5,18]. In contrast, SPM and POC concentrations and D50 in the OSZ were rather constant, randomly scattered without apparent ups and downs (Figure 4b), showing that aggregation kinetics dominate over disaggregation kinetics for the entire period. SPM in the OSZ might be mainly composed of biomass and some mineral particles, building more shear-resistant and less settleable marine snow [46]. Although biological aggregates (i.e., marine snow) are usually much larger, up to several millimeters, than mineral or biomineral aggregates, they settle more slowly because of their low density and fluffy structure [30]. The latter is confirmed by an earlier study [38], where the excess density of aggregates has been calculated for some of the tidal cycles investigated here; the mean excess density was 550 kg/m^3 and the mean D50 of the aggregates 65 μm (five tidal cycles) in the TMZ versus 180 kg/m^3 and 115 μm (three tidal cycles) in the OSZ. Although both the TMZ and the OSZ are governed by tidal dynamics, small differences in the current regime occur between both areas [15], as is also shown in Figure 4. The TMZ is situated in the nearshore, where the current ellipses are more elongated, whereas more offshore, the ellipses tend to be more spherical. This will cause higher velocity gradients, stronger turbulence, more stress exerted on the aggregates, and a reduction of the time needed for the aggregates to reach equilibrium size in the TMZ. Considering these differences in hydrodynamics, the mineral and biomineral aggregates in the TMZ are more susceptible to the hydrodynamics than the biological aggregates in the OSZ.

Time series of the PSDs during the 13 h tidal cycles are shown in Figure 5, for the TMZ and OSZ, respectively. PSDs in the TMZ skewed toward a smaller size around peak flow (e.g., t = 3, 4 h at location MOW1 on 10 July 2007) and then to a larger size around slack water (e.g., t = 6, 7 h). Except for the PSDs in 23 October 2007, the other PSDs in the TMZ showed bimodality, comprising microflocs (20–200 μm) and macroflocs (>200 μm), as reported in the earlier studies [5]. The primary peak of microflocs in a PSD was prominent around the peak flow. However, while approaching the slack water, the secondary peak of macroflocs became dominant over the primary peak. Low flow/turbulence intensity might promote the aggregation of microflocs (i.e., mineral, biomineral aggregates) to macroflocs (i.e., biological aggregates) [4,5,18]. On the other hand, large hardly-settleable biological aggregates which were suspended in the water column might dominate in the slack water. Maggi and Tang recently reported that larger biological aggregates can be lighter and even settle slower than smaller mineral, bio-mineral aggregates [13]. Here, larger biological aggregates can be suspended in the slack water, while smaller mineral, bio-mineral aggregates settle and deposit, thereby developing the secondary peak of biological aggregates.

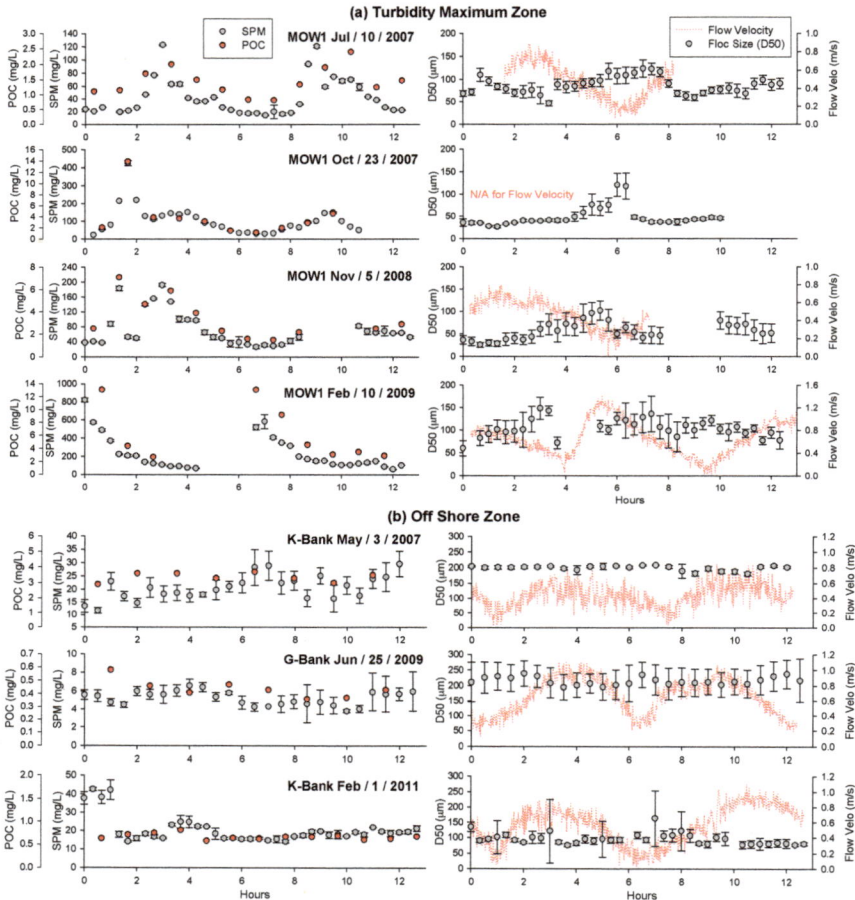

Figure 4. Dynamic behaviors of suspended particulate matter (SPM) and particulate organic carbon (POC) concentrations and aggregate size (D50) in 13-h tidal cycles, in (**a**) the turbidity maximum zone (TMZ) and (**b**) the offshore zone (OSZ). Each set of the SPM/POC and D50 data was measured on a specific date of a field campaign. MOW1: measurement site in the TMZ; K-Bank and G-Bank: measurement sites in the OSZ.

However, PSDs in the OSZ remained rather constant during the entire tidal cycle, consistently skewing toward a larger size (Figure 5b). A substantial fraction of the PSDs occupied the upper most measuring bin of the LISST-100X instrument (i.e., 500 μm). Aggregates in the OSZ, even with such a large size, apparently did not properly settle but floated in the water column (see also the previous paragraph and Figure 4). Thus, SPM in the OSZ is likely composed of large but light, fluffy, and hardly-settleable biological aggregates (i.e., marine snow), whereas SPM in the

TMZ comprises dense, compact, and readily-setteable mineral, biomineral aggregates, as well as biological aggregates [13,29,30,46]. However, note that this argument is supported by a rather indirect measurement of SPM dynamics in this research and observations from earlier studies. Direct ways of measuring aggregate morphology might be required in the future to explain realistic structures and behaviors of mineral, biomineral, and biological aggregates.

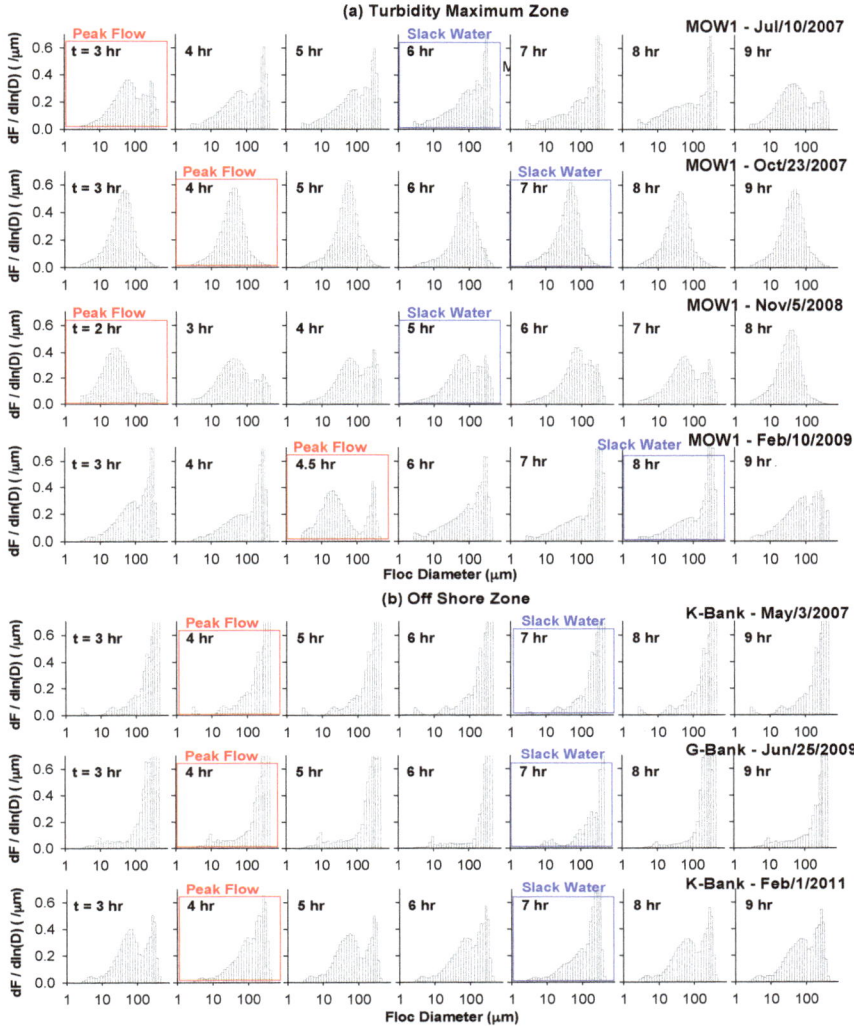

Figure 5. Particle size distributions (PSDs) of suspended particulate matter (SPM) in 13-h tidal cycles, for (**a**) the turbidity maximum zone (TMZ) (MOW1) and (**b**) the offshore zone (OSZ) (K-Bank and G-Bank). Each set of the PSDs was measured on a specific date of a field campaign. Each PSD was plotted on a logarithmic scale, and the fraction of a size bin was normalized by the width of the size bin in y-axis. Thus, dF/dln (D) is the normalized volumetric fraction by the width of the size interval in the log scale, in accordance with the lognormal distribution function [5,47].

3.3. SPM Dynamics during the Algae Bloom and Normal Periods in the TMZ

Flow intensity of the spring and neap tides was found to alter SPM properties (e.g., aggregate size and settling velocity) and SPM dynamics (e.g., flocculation, sedimentation, and deposition) in the TMZ (i.e., the MOW1 site). A spring tide, associated with a strong peak flow (up to 1.5 m/s), increased SPM concentrations substantially, compared to a neap tide with a weak peak flow (up to 1.0 m/s). For example, SPM concentrations increased up to 800 mg/L during a spring tide (e.g., MOW1—10 February 2009 in Figure 4a), whereas they remained under 120 mg/L during a neap tide (e.g., MOW1—10 July 2007). When the pairs of the maximum SPM concentration (SPM$_{max}$) and peak flow velocity (U$_{max}$) in each 13-h tidal cycle are plotted (Figure 6), they are proportional. A spring tide with high U$_{max}$ resulted in high SPM$_{max}$, because it enhanced the disaggregation, erosion, and resuspension of sediment particles/aggregates. However, a neap tide with low U$_{max}$ resulted in low SPM$_{max}$, because it enhanced aggregation, sedimentation, and deposition. Thus, the fate and transport of SPM in the TMZ, which was governed by aggregation-disaggregation, sedimentation-resuspension, and erosion-deposition, highly depended on flow intensity. However, an exception against the SPM-flow intensity relation was found during an algae bloom period.

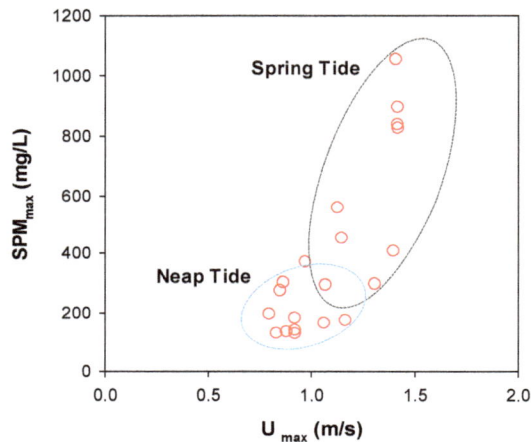

Figure 6. Plots of maximum suspended particulate matter concentration (SPMmax) versus maximum flow velocity (Umax). SPM concentration and flow velocity were measured in the middle of the water column. Each point represents a pair of SPMmax and Umax measured in a 13-h tidal cycle. All the data were measured in the turbidity maximum zone (TMZ) from 2004 to 2011.

During the reported spring algae bloom (MOW1—26 April 2011 in Figure 7a), SPM and POC concentrations did not show a clear up-and-down trend with tide, but behaved similar to those in the OSZ. The aggregate sizes during the algae bloom period (26 April 2011 in Figure 7a) were two to three times larger than aggregates during the normal period, measured at the same site four months later (18 August 2011 in Figure 7b). Although aggregates were enlarged (>100 μm) during the algae bloom period, they did not show a clear sign of downward settling. Considering that such large aggregates during the algae bloom period were subject to floatation without a clear sign of sedimentation and resuspension, they were found to be lighter and less settleable than during a regular period, and thus more similar to the marine snow (i.e., biological aggregates) found in the OSZ (see Section 3.2). In the TMZ, two different aggregates may thus occur: (1) sediment-enriched, dense, and settleable biomineral aggregates during normal periods; and (2) biomass-enriched, light, and less settleable marine snow during algal bloom periods (Figure 8). The latter type of aggregate corresponds better to the one observed in the OSZ. The aggregates occurring during algae bloom periods or in the OSZ have a lower settling velocity as a larger fraction is composed of organic matter and sticky bio-polymers organized in a fluffy structure [16,48].

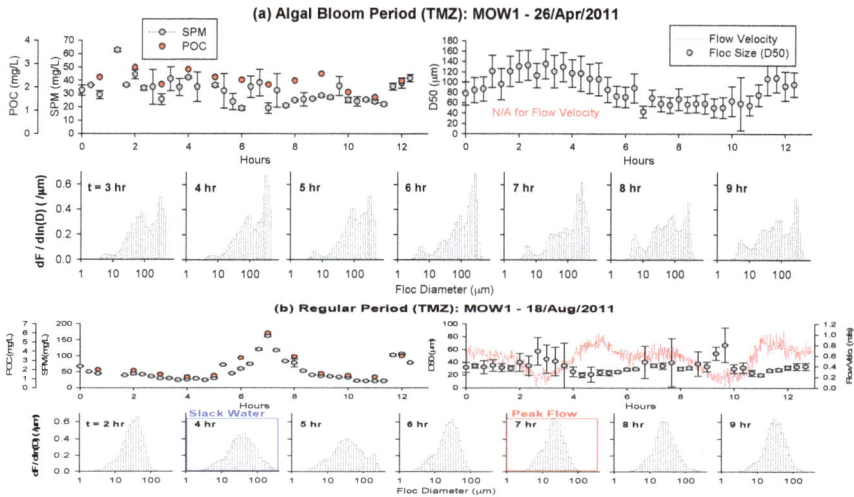

Figure 7. Dynamic behaviors of suspended particulate matter (SPM) and particulate organic carbon (POC) concentrations, aggregate size (D50), and particle size distribution (PSD) in a 13-h tidal cycle. The two data sets were collected in the TMZ (i.e., the MOW1 site) on different dates in 2011, representing (**a**) algal bloom period and (**b**) regular periods.

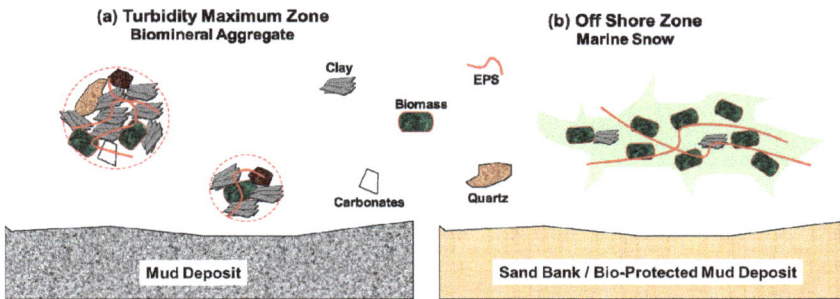

Figure 8. Schematic diagrams of (**a**) biomineral aggregates in the turbidity maximum zone (TMZ) and (**b**) biological marine snow in the offshore zone (OSZ). EPS: extracellular polymeric substances.

Previous studies, carried out in the same TMZ, reported that large and settleable biomineral aggregates were dominant SPM species during bio-enriched spring and summer periods [31,32]. These large, settleable biomineral aggregates are contrary to less settleable biological aggregates observed in this current study. However, it is important to note that the SPM samples in this study were taken in the middle of the water column, well above the near-bed layer. Dense, compacted, and settleable biomineral aggregates might be stored in the near-bed layer, causing mineral-depletion in the water column, and hence less dense, fluffy, and hardly settleable biological aggregates might be formed and float around in the water column. Enhanced primary production during an algal bloom period generates more sticky, particle-binding polymeric substances, such as EPSs and TEPs. These sticky polymeric substances can not only enhance flocculation, but also reduce the erosion and resuspension of muddy deposits from the seabed to the water column [49]. A large amount of cohesive sediments are thus stored in or on the seafloor as a fluid-mud layer or a muddy deposit, and the marine snow with more biomass and less sediments is suspended in the water column [3]. This SPM behavior during an algal bloom period with high primary production agrees with the satellite images of low SPM and high Chl concentrations in summer (Figure 1). Similar observations were made in the port of Zeebrugge. High primary production and low turbulence in summer provoked a large amount of mud deposition in the near-bed layer (or formation of a fluid mud layer) and reduction of the SPM concentration in the water column.

Reviewing other studies [28,50] revealed similar SPM dynamics around an algal bloom period. Proliferation of a specific algae group could enhance flocculation and store sediments in the near-bed layer, and hence could cause large but suspended biological aggregates and a low SPM concentration in the water column. Maerz and co-workers [51] have been looking at the whole gradient from the nearshore TMZ to the OSZ; they have found a maximum settling velocity in the transition zone between the TMZ and the OSZ where the aggregates are larger as compared to near-coast TMZ and denser as compared to the low turbid OSZ. This maximum in settling velocity is caused by similar gradients in aggregate size, POC content, density, and chlorophyll concentration than found in our data. The fact that algae are involved in these observed gradients points to seasonal influences. Another study [52], however, does not confirm the leading role of the algae bloom on SPM dynamics. The reason for these different findings may be due to differences in, amongst others, hydrodynamics, wave climate, nutrient availability, and algae species at the different study sites. The importance of each of these parameters will explain to a smaller or larger part the observed seasonal variations in SPM dynamics.

Biomineral and biological aggregates are often approximated by a single parameter (e.g., a characteristic diameter) in practical applications, although they are very different in composition and mechanical property. For example, a traditional aggregate structure model, based on fractal theory, includes only mineral particles and disregards organic matter, which is instead assumed to be part of the pore space for simplicity and ease [53,54]. This approximation might not be valid for biological aggregates (i.e., marine snow) with a high content of organic matter or in environments where aggregate properties change in time (regular versus algae bloom period) or space (inside and outside harbours). Thus, the heterogeneity of aggregates, at least the two fractions of biomass and sediments, should be considered when developing a rigorous aggregate structure model and accurately predicting the fate and transport of biomass and sediments in marine and coastal waters [30].

A higher biomass content (indicated by a higher POC/SPM ratio) was generally found to enhance flocculation, thereby increasing aggregate size. However, the quantity of biomass is not the only factor determining the flocculation capability. For example, in June 2009 at G-Bank (Figure 2a), aggregate size increased to over 200 μm, even with a low POC/SPM. Besides the quantity of biomass, the quality, such as stickiness, is important for controlling flocculation kinetics, as reported in previous research [50,55]. Specifically, extracellular polymeric substances (EPSs) or transparent extracellular polymers (TEPs) are sticky and increase flocculation [19,20,22,23]. Long polymeric chain structures of EPSs and TEPs, which are produced by aquatic microorganisms (e.g., algae), can bind biomass and sediment particles to large mineral, biomineral, and biological aggregates. Even in an unfavorable chemical condition for flocculation (e.g., terrestrial water with low ionic strength), a small amount of EPSs and TEPs can cause substantial flocculation, because they can overcome the electrostatic repulsive force of negatively-charged colloidal particles and bind such particles to large aggregates [56,57]. Therefore, qualitative measures of biomass, such as EPS and/or TEP concentration, likely need to be included to explain bio-mediated flocculation and SPM dynamics in marine and coastal waters.

4. Conclusions

The monitoring and analysis of SPM dynamics explained how organic biomass and inorganic sediment interact with each other to build large biomineral aggregates or marine snow in marine and coastal waters. SPM in the TMZ and OSZ had a similar mineralogical composition, but encountered different fates in association with biomass. SPM in the TMZ built sediment-enriched, dense, and settleable biomineral aggregates, whereas SPM in the OSZ was composed of biomass-enriched, light, and less settleable marine snow. Biological proliferation, such as an algae bloom, also facilitated the occurrence of marine snow in the water column, even in the TMZ. Enhanced flocculation in summer could also scavenge SPM in the water column down to the sea bed, resulting in a low SPM concentration in the water column. In short, bio-mediated flocculation and SPM dynamics were found to vary spatially and seasonally, affected by the biota. The proposed concept to combine organic and mineral particles in aggregates will help us to better understand and predict bio-mediated flocculation and SPM dynamics in marine and coastal waters.

Acknowledgments: This research was supported by the Basic Science Research Program through the National Research Foundation of Korea (NRF) funded by the Ministry of Education (No: NRF-2017R1D1A3B03035269), the Maritime Access Division of the Flemish Ministry of Mobility and Public Works (MOMO project), and the Belgian Science Policy (BELSPO) within the BRAIN-be program (INDI67 project). The ship time RV Belgica was provided by BELSPO and the RBINS–Operational Directorate Natural Environment.

Author Contributions: M.F. conceived, designed, and performed the experiments; B.J.L. analyzed the experimental data; and M.F. and B.J.L. wrote the paper.

Conflicts of Interest: The authors declare no conflict of interest.

References

1. Ouillon, S.; Douillet, P.; Andrefouet, S. Coupling satellite data with in situ measurements and numerical modeling to study fine suspended-sediment transport: A study for the lagoon of New Caledonia. *Coral Reefs* **2004**, *23*, 109–122.
2. Perianez, R. Modelling the transport of suspended particulate matter by the Rhone River plume (France). Implications for pollutant dispersion. *Environ. Pollut.* **2005**, *133*, 351–364. [CrossRef] [PubMed]
3. Winterwerp, J.; van Kesteren, W. *Introduction to the Physics of Cohesive Sediment in the Marine Environment*; Elsevier B.V.: Amsterdam, The Netherlands, 2004.
4. Lee, B.J.; Toorman, E.; Molz, F.J.; Wang, J. A two-class population balance equation yielding bimodal flocculation of marine or estuarine sediments. *Water Res.* **2011**, *45*, 2131–2145. [CrossRef] [PubMed]
5. Lee, B.J.; Fettweis, M.; Toorman, E.; Molz, F.J. Multimodality of a particle size distribution of cohesive suspended particulate matters in a coastal zone. *J. Geophys. Res. Oceans* **2012**, *117*, C03014. [CrossRef]
6. Chen, M.S.; Wartel, S.; Temmerman, S. Seasonal variation of floc characteristics on tidal flats, the Scheldt estuary. *Hydrobiologia* **2005**, *540*, 181–195. [CrossRef]
7. Droppo, I.G. Rethinking what constitutes suspended sediment. *Hydrol. Process.* **2001**, *15*, 1551–1564. [CrossRef]
8. Eisma, D. Flocculation and de-flocculation of suspended matter in estuaries. *Neth. J. Sea Res.* **1986**, *20*, 183–199. [CrossRef]
9. Droppo, I.; Leppard, G.; Liss, S.; Milligan, T. *Flocculation in Natural and Engineered Environmental Systems*; CRC Press Inc.: Boca Raton, FL, USA, 2005.
10. Jago, C.F.; Kennaway, G.M.; Novarino, G.; Jones, S.E. Size and settling velocity of suspended flocs during a phaeocystis bloom in the tidally stirred Irish Sea, NW European Shelf. *Mar. Ecol. Prog. Ser.* **2007**, *345*, 51–61. [CrossRef]
11. Tan, X.L.; Zhang, G.P.; Yi, H.; Reed, A.H.; Furukawa, Y. Characterization of particle size and settling velocity of cohesive sediments affected by a neutral exopolymer. *Int. J. Sediment Res.* **2012**, *27*, 473–485. [CrossRef]
12. Maggi, F. Biological flocculation of suspended particles in nutrient-rich aqueous ecosystems. *J. Hydrol.* **2009**, *376*, 116–125. [CrossRef]
13. Maggi, F.; Tang, F.H.M. Analysis of the effect of organic matter content on the architecture and sinking of sediment aggregates. *Mar. Geol.* **2015**, *363*, 102–111. [CrossRef]
14. Van Leussen, W. Estuarine Macroflocs: Their Role in Fine-Grained Sediment Transport. Ph.D. Thesis, Utrecht University, Utrecht, The Netherlands, February 1994.
15. Fettweis, M.; Francken, F.; Pison, V.; Van den Eynde, D. Suspended particulate matter dynamics and aggregate sizes in a high turbidity area. *Mar. Geol.* **2006**, *235*, 63–74. [CrossRef]

16. Alldredge, A.; Silver, M. Characteristics, dynamics and significance of marine snow. *Prog. Oceanogr.* **1988**, *20*, 41–82. [CrossRef]

17. Markussen, T.N.; Andersen, T.J. A simple method for calculating in situ floc settling velocities based on effective density functions. *Mar. Geol.* **2013**, *344*, 10–18. [CrossRef]

18. Lee, B.J.; Toorman, E.; Fettweis, M. Multimodal particle size distributions of fine-grained sediments: Mathematical modeling and field investigation. *Ocean Dyn.* **2014**, *64*, 429–441. [CrossRef]

19. Passow, U. Transparent exopolymer particles (TEP) in aquatic environments. *Prog. Oceanogr.* **2002**, *55*, 287–333. [CrossRef]

20. Engel, A.; Thoms, S.; Riebesell, U.; Rochelle-Newall, E.; Zondervan, I. Polysaccharide aggregation as a potential sink of marine dissolved organic carbon. *Nature* **2004**, *428*, 929–932. [CrossRef] [PubMed]

21. Sahoo, G.B.; Nover, D.; Schladow, S.G.; Reuter, J.E.; Jassby, D. Development of updated algorithms to define particle dynamics in Lake Tahoe (CA-NV) USA for total maximum daily load. *Water Resour. Res.* **2013**, *49*, 7627–7643. [CrossRef]

22. Mari, X.; Passow, U.; Migon, C.; Burd, A.; Legendre, L. Transparent Exopolymer Particles: Effects on carbon cycling in the ocean. *Prog. Oceanogr.* **2017**, *151*, 13–37. [CrossRef]

23. Jouon, A.; Ouillon, S.; Douillet, P.; Lefebvre, J.P.; Fernandez, J.M.; Mari, X.; Froidefond, J. Spatio-temporal variability in suspended particulate matter concentration and the role of aggregation on size distribution in a coral reef lagoon. *Mar. Geol.* **2008**, *256*, 36–48. [CrossRef]

24. Tranvik, L.J.; Downing, J.A.; Cotner, J.B.; Loiselle, S.A.; Striegl, R.G.; Ballarore, T.J.; Dillon, P.; Finlay, K.; Fortino, K.; Knoll, L.B.; et al. Lakes and reservoirs as regulators of carbon cycling and climate. *Limnol. Oceanogr.* **2009**, *54*, 2298–2314. [CrossRef]

25. Gudasz, C.; Bastviken, D.; Premke, K.; Steger, K.; Tranvik, L.J. Constrained microbial processing of allochthonous organic carbon in boreal lake sediments. *Limnol. Oceanogr.* **2012**, *57*, 163–175. [CrossRef]

26. Barkmann, W.; Schafer-Neth, C.; Balzer, W. Modelling aggregate formation and sedimentation of organic and mineral particles. *J. Mar. Syst.* **2010**, *82*, 81–95. [CrossRef]

27. Burd, A.; Jackson, G. Modeling steady-state particle size spectra. *Environ. Sci. Technol.* **2002**, *36*, 323–327. [CrossRef] [PubMed]

28. De Lucas Pardo, M.A.; Sarpe, D.; Winterwerp, J.C. Effect of algae on flocculation of suspended bed sediments in a large shallow lake. Consequences for ecology and sediment transport processes. *Ocean Dyn.* **2015**, *65*, 889–903. [CrossRef]

29. Tang, F.H.M.; Maggi, F. A mesocosm experiment of suspended particulate matter dynamics in nutrient- and biomass-affected waters. *Water Res.* **2016**, *89*, 76–86. [CrossRef] [PubMed]

30. Maggi, F. The settling velocity of mineral, biomineral, and biological particles and aggregates in water. *J. Geophys. Res. Oceans* **2013**, *118*, 2118–2132. [CrossRef]

31. Fettweis, M.; Baeye, M.; Van der Zande, D.; Van den Eynde, D.; Lee, B.J. Seasonality of floc strength in the southern North Sea. *J. Geophys. Res. Oceans* **2014**, *119*, 1911–1926. [CrossRef]

32. Fettweis, M.; Baeye, M. Seasonal variation in concentration, size and settling velocity of muddy marine flocs in the benthic boundary layer. *J. Geophys. Res. Oceans* **2015**, *120*, 5648–5667. [CrossRef]

33. Fettweis, M.; Francken, F.; Van den Eynde, D.; Verwaest, T.; Janssens, J.; Van Lancker, V. Storm influence on SPM concentrations in a coastal turbidity maximum area with high anthropogenic impact (southern North Sea). *Cont. Shelf Res.* **2010**, *30*, 1417–1427. [CrossRef]

34. Lacroix, G.; Ruddick, K.; Ozer, J.; Lancelot, C. Modelling the impact of the Scheldt and Rhine/Meuse plumes on the salinity distribution in Belgian waters (southern North Sea). *J. Sea Res.* **2004**, *52*, 149–163. [CrossRef]

35. Fettweis, M.; Nechad, B.; Van den Eynde, D. An estimate of the suspended particulate matter (SPM) transport in the southern North Sea using SeaWiFS images, in situ measurements and numerical model results. *Cont. Shelf Res.* **2007**, *27*, 1568–1583. [CrossRef]

36. Zeelmaekers, E. Computerized Qualitative and Quantitative Clay Minerology: Introduction and Application to Known Geological Cases. Ph.D. Thesis, Katholieke Universiteit Leuven, Leuven, Belgium, April 2011.

37. Agrawal, Y.; Pottsmith, H. Instruments for particle size and settling velocity observations in sediment transport. *Mar. Geol.* **2000**, *168*, 89–114. [CrossRef]

38. Fettweis, M. Uncertainty of excess density and settling velocity of mud flocs derived from in situ measurements. *Estuar. Coast. Shelf Sci.* **2008**, *78*, 426–436. [CrossRef]

39. Mikkelsen, O.; Curran, K.; Hill, P.; Milligan, T. Entropy analysis of in situ particle size spectra. *Estuar. Coast. Shelf Sci.* **2007**, *72*, 615–625. [CrossRef]

40. Andrews, S.; Nover, D.; Schladow, S. Using laser diffraction data to obtain accurate particle size distributions: The role of particle composition. *Limnol. Oceanogr. Methods* **2010**, *8*, 507–526. [CrossRef]

41. Graham, G.W.; Davies, E.; Nimmo-Smith, A.; Bowers, D.G.; Braithwaite, K.M. Interpreting LISST-100X measurements of particles with complex shape using digital in-line holography. *J. Geophys. Res. Oceans* **2012**, *117*, C05034. [CrossRef]

42. Mikkelsen, O.A.; Hill, P.S.; Milligan, T.; Chant, R.J. In situ particle size distributions and volume concentrations from a LISST-100 laser particle sizer and a digital floc camera. *Cont. Shelf Res.* **2005**, *25*, 1959–1978. [CrossRef]

43. Smith, S.J.; Friedrichs, C.T. Size and settling velocities of cohesive flocs and suspended sediment aggregates in a trailing suction hopper dredge plume. *Cont. Shelf Res.* **2011**, *31*, S50–S63. [CrossRef]

44. Davies, E.; Nimmo-Smith, A.; Agrawal, Y.; Souza, A. LISST-100 response to large particles. *Mar. Geol.* **2012**, *307–311*, 117–122. [CrossRef]

45. Kastner, M. Oceanic minerals: Their origin, nature of their environment, and significance. *Proc. Natl. Acad. Sci. USA* **1999**, *96*, 3380–3387. [CrossRef] [PubMed]

46. Bainbridge, Z.; Wolanski, E.; Alvarez-Romero, J.G.; Lewis, S.E.; Brodie, J.E. Fine sediment and nutrient dynamics related to particle size and floc formation in a Burdekin River flood plume, Australia. *Mar. Pollut. Bull.* **2012**, *65*, 236–248. [CrossRef] [PubMed]

47. Hinds, W. *Aerosol Technology: Properties, Behavior, and Measurement of Airborne Particles*, 2nd ed.; John Wiley: New York, NY, USA, 1999.

48. Fennessy, M.; Dyer, K.; Huntley, D. INSSEV: An instrument to measure the size and settling velocity of flocs in situ. *Mar. Geol.* **1994**, *117*, 107–117. [CrossRef]

49. Vos, P.; De Boer, P.; Misdorp, R. Sediment stabilization by benthic diatoms in intertidal sandy shoals: Qualitative and quantitative observations. In *Tide-Influenced Sedimentary Environments and Facies*; D. Reidel Publishing: Dordrecht, The Netherlands, 1988; pp. 511–526.

50. Van der Lee, W.T.B. Temporal variation of floc size and settling velocity in the Dollard estuary. *Cont. Shelf Res.* **2000**, *20*, 1495–1511. [CrossRef]

51. Maerz, J.; Hofmeister, R.; van der Lee, E.M.; Grawe, U.; Riethmuller, R.; Wirtz, K.W. Maximum sinking velocities of suspended particulate matter in a coastal transition zone. *Biogeosciences* **2016**, *13*, 4863–4876. [CrossRef]

52. Van der Hout, C.M.; Wittbaard, R.; Bergman, M.J.M.; Duineveld, G.C.A.; Rozemeijer, M.J.C. The dynamics of suspended particulate matter (SPM) and chlorophyll-a from intratidal to annual time scales in a coastal turbidity maximum. *J. Sea Res.* **2017**. [CrossRef]

53. Khelifa, A.; Hills, P.S. Models for effective density and settling velocity of flocs. *J. Hydraul. Res.* **2006**, *44*, 390–401. [CrossRef]

54. Maggi, F. Variable fractal dimension: A major control for floc structure and flocculation kinematics of suspended cohesive sediment. *J. Geophys. Res. Oceans* **2007**, *112*, C07012. [CrossRef]

55. Van der Lee, W.T.B. Parameters affecting mud floc size on a seasonal time scale: The impact of a phytoplankton bloom in the Dollard estuary, The Netherlands. In *Coastal and Estuarine Fine Sediment Transport Processes*; McAnally, W.H., Mehta, A.J., Eds.; Elsevier: Amsterdam, The Netherlands, 2001; Volume 3, pp. 403–421.

56. Furukawa, Y.; Reed, A.H.; Zhang, G. Effect of organic matter on estuarine flocculation: A laboratory study using montmorillonite, humic acid, xanthan gum, guar gum and natural estuarine flocs. *Geochem. Trans.* **2014**, *15*, 1–9. [CrossRef] [PubMed]

57. Lee, B.J.; Hur, J.; Toorman, E. Seasonal Variation in Flocculation Potential of River Water: Roles of the Organic Matter Pool. *Water* **2017**, *9*, 335. [CrossRef]

water

MDPI

Article

Shoreline Changes on the Wave-Influenced Senegal River Delta, West Africa: The Roles of Natural Processes and Human Interventions

Mamadou Sadio [1,2], Edward J. Anthony [1,*], Amadou Tahirou Diaw [2], Philippe Dussouillez [1], Jules T. Fleury [1], Alioune Kane [3], Rafael Almar [4] and Elodie Kestenare [4]

[1] Aix-Marseille University, CEREGE UM 34, Europôle de l'Arbois, 13545 Aix en Provence Cedex 04, France; sadio@cerege.fr (M.S.); dussou@cerege.fr (P.D.); fleury@cerege.fr (J.T.F.)
[2] Laboratoire d'Enseignement et de Recherche en Géomatique, Ecole Supérieure Polytechnique, Université Cheikh Anta Diop, Dakar, Sénégal; guede1914@gmail.com
[3] Laboratoire de Morphologie et d'Hydrologie, Université Cheikh Anta Diop, Dakar, Sénégal; akane@ucad.sn
[4] LEGOS (CNRS-IRD-CNES-University of Toulouse), 31400 Toulouse, France; rafael.almar@ird.fr (R.A.); Elodie.Kestenare@legos.obs-mip.fr (E.K.)
* Correspondence: anthony@cerege.fr

Academic Editors: Sylvain Ouillon and John W. Day
Received: 24 December 2016; Accepted: 12 May 2017; Published: 19 May 2017

Abstract: The Senegal River delta in West Africa, one of the finest examples of "wave-influenced" deltas, is bounded by a spit periodically breached by waves, each breach then acting as a shifting mouth of the Senegal River. Using European Re-Analysis (ERA) hindcast wave data from 1984 to 2015 generated by the Wave Atmospheric Model (WAM) of the European Centre for Medium-Range Weather Forecasts (ECMWF), we calculated longshore sediment transport rates along the spit. We also analysed spit width, spit migration rates, and changes in the position and width of the river mouth from aerial photographs and satellite images between 1954 and 2015. In 2003, an artificial breach was cut through the spit to prevent river flooding of the historic city of St. Louis. Analysis of past spit growth rates and of the breaching length scale associated with maximum spit elongation, and a reported increase in the frequency of high flood water levels between 1994 and 2003, suggest, together, that an impending natural breach was likely to have occurred close to the time frame of the artificial 2003 breach. Following this breach, the new river mouth was widened rapidly by flood discharge evacuation, but stabilised to its usual hydraulic width of <2 km. In 2012, severe erosion of the residual spit downdrift of the mouth may have been due to a significant drop (~15%) in the longshore sand transport volume and to a lower sediment bypassing fraction across the river mouth. This wave erosion of the residual spit led to rapid exceptional widening of the mouth to ~5 km that has not been compensated by updrift spit elongation. This wider mouth may now be acting as a large depocentre for sand transported alongshore from updrift, and has contributed to an increase in the tidal influence affecting the lower delta. Wave erosion of the residual spit has led to the destruction of villages, tourist facilities and infrastructure. This erosion of the spit has also exposed part of the delta plain directly to waves, and reinforced the saline intrusion within the Senegal delta. Understanding the mechanisms and processes behind these changes is important in planning of future shoreline management and decision-making regarding the articulations between coastal protection offered by the wave-built spit and flooding of the lower delta plain of the Senegal River.

Keywords: Senegal River delta; Langue de Barbarie spit; delta vulnerability; river-mouth migration; spit breaching; ERA hindcast waves; longshore sediment transport

1. Introduction

The impacts of human activities on coasts are often accompanied by a lack of understanding of the consequences of these activities on the hydrodynamic and sediment redistribution processes that shape coasts [1,2]. Alongshore sediment transport, gradients in transport, and interception of drifting sediment by natural or artificial (man-made) boundaries, including river mouths and inlets, are, from a coastal management point of view, very important, as these processes are significant drivers of short-term to medium-term (days to years) shoreline change. Much of the West African coast (Figure 1) is wave-dominated, and is classified as a cyclone- and storm-free "West Coast Swell Environment" in the global wave classification scheme of Davies [3], with a subsidiary contribution from shorter-period trade-wind waves from the Atlantic. In Figure 1, continental shelf width (clearer hue along the coast) is a fine indicator of the distribution of long stretches of wave-dominated coast (narrow shelf) and the much more limited, predominantly tidal, estuarine sector between Sierra Leone and Guinea-Bissau (broad, low-gradient shelf), subject to significant wave energy dissipation [4,5]. The West African coast is also characterised by a plethora of river deltas, the largest of which are those of the Niger, Senegal and Volta (Figure 1). Abundant sand supplies and strong wave-induced longshore drift have favoured the construction of numerous sand barriers, including at the mouths of these three deltas. These barriers are major settlement sites on the coast as they provide higher-lying areas above lagoons and wetlands, while acting as valuable aquifers. On the coast of Senegal, the barriers are generally elongate to curvilinear spits formed at the mouths of tidal or fluvial ria-like embayments. These spits are commonly capped by dunes, but individual beach ridges are visible in some of the more southern ones. These spits have a protective role on the back-barrier wetlands and lagoons by buffering wave energy. By forming alongshore barriers, they are also important in the regulation of the freshwater-saltwater balance and ecology of these lagoons and back-barrier wetlands, both of which can be considerably altered by breaches in the spits or by spit erosion [6,7]. This is particularly the case of the largest of the Senegal wetland systems, that of the lower Senegal River and delta plain (Figure 2).

The Senegal River delta is an iconic example of a delta subject to strong wave action [8–12]. This delta is often represented in the ternary (river-wave-tide) classification of Galloway [13] at the wave apex. Using a fluvial dominance ratio—defined as river sediment input versus the potential maximum alongshore sediment transport away from the delta mouth—to quantify the balance between river inputs and the ability of waves to spread sediments along the coast, Nienhuis et al. [12] computed a value of 0.04 for the Senegal, which highlights the strong role of wave-induced longshore transport along this delta's shoreline. A manifestation of this strong longshore transport potential is a long narrow sand spit presently fronting the delta plain, the Langue de Barbarie [10]. This spit has historically played an important role not only in the protection of the lower Senegal delta plain but also in regulating saltwater intrusion by diverting the mouth of the river several kilometres southwards. Of particular significance, in terms of long-term flood-risk and coastal management, is the historic and picturesque city of St. Louis (population in 2013: 300,000), a UNESCO world heritage site located in the proximal part of the delta (Figure 2). The cultural attractiveness of St. Louis, a French colonial city, and the biodiversity of the deltaic wetlands and lagoon bound by the Langue de Barbarie spit have also generated a substantial rise in tourism. Much of St. Louis, which has undergone a rapid growth in population over the last 50 years, lies at an elevation of less than 2.5 m above sea level [14], and the city has, therefore, been prone to the flooding that affects the lower Senegal valley in the rainy season (May to October).

In October 2003, to avoid flooding of St. Louis in the wake of a massive rise in the water discharge of the Senegal River, an artificial breach was hastily cut through the Langue de Barbarie, generating rapid reworking of the spit. In the present paper, we describe the recent dynamics of the spit within the framework of development of the Senegal delta and specifically aim at disentangling processes of natural forcing from those of the impact of this breach. Two approaches are used in the study: (1) clearly define the wave climate and longshore sediment transport potential along the Langue de Barbarie; and (2) compare spit behaviour patterns prior to, and following the October 2003 artificial breach. Both of these approaches are important in understanding the current dynamics prevailing along this deltaic coast. They should also be of use in planning of future shoreline management and decision-making regarding the articulations between coastal protection offered by the wave-built spit and flooding of the lower delta plain of the Senegal River.

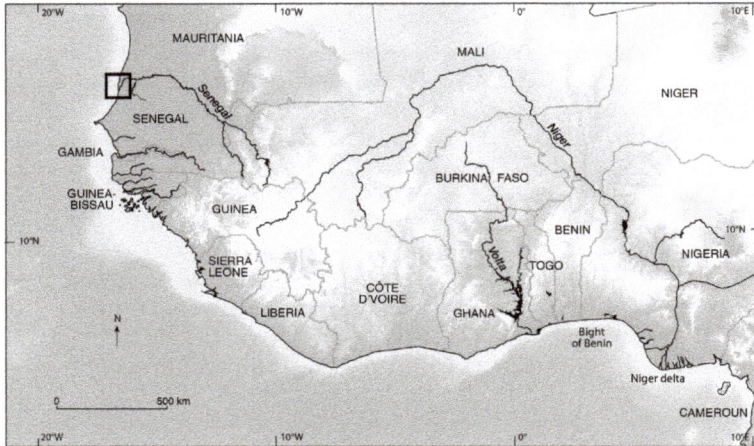

Figure 1. The coast of West Africa, showing the Senegal (box) and other major river deltas. Much of this coast is wave-dominated, and is characterised by beach-ridge sand barriers and spits.

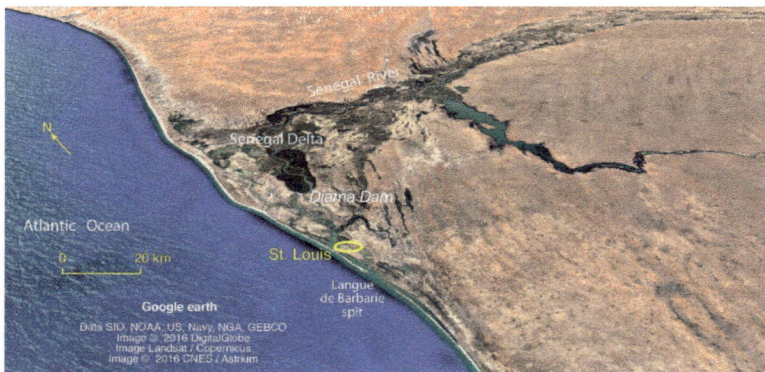

Figure 2. The Senegal River delta, a fine example of a wave-dominated delta characterised by the Langue de Barbarie spit and a river-mouth system subject to strong north-south longshore drift.

2. The Senegal River and Delta

The Senegal River is about 1800 km long, and is the second longest river in West Africa after the Niger. The catchment size has been estimated at 345,000 km^2 [15], much of it covering the arid western Sahel. The river's discharge has been particularly affected by Sahelian droughts since the 1970s [16]. The mean annual water discharge at Bakel, the reference station of the Senegal River, situated 557 km upstream of St. Louis, is 676 m^3/s, and varies from a mean low dry season value of 10 m^3/s in May, to a mean maximum flood value of 3320 m^3/s in September at the height of the rainy season [17]. The interannual variability is extremely high, with a mean annual discharge ranging from 250 to

1400 m^3/s. Little is known of the solid discharge of the Senegal. This solid load has been estimated at 0.9 to 1×10^6 tonnes a year [18], a rather low figure when viewed against the size of the river's catchment and when compared to other tropical rivers. The solid discharge is largely dominated by suspended load transport [19]. The lower Senegal delta is characterised by high biological productivity and by rich agricultural and fishing sectors. In November 1985, the Diama dam (Figure 2) was built in the lower river valley 23 km upstream from St. Louis. The dam was commissioned with the twin aims of preventing saltwater intrusion, which, hitherto, penetrated up to 350 km upstream in the lower Senegal valley, and regulating the river's rainy season discharge in order to improve irrigation of agricultural lands [15]. The delta plain provides 8% of the arable land of Senegal [20].

The Senegal delta coast is fronted by a relatively narrow continental shelf only 15–20 km wide. The dominant waves are from the northwest, and this direction is especially prevalent during the dry season from November to June. One of the objectives of this study is to highlight the salient characteristics of this "West Coast" wave setting (see Results). The tidal regime along the Langue de Barbarie is semi-diurnal and the range microtidal, comprised between 0.5 m at neap tides and 1.6 m at spring tides. The relatively moderate river discharge, including during the flood season, the permanence of moderately energetic waves propagating across a relatively narrow shelf, and the microtidal regime, are three conditions that have been forwarded to explain the wave-dominated character of the Senegal River delta [10].

The stratigraphy and patterns of Holocene geomorphic development of the Senegal delta have been highlighted from borehole data, limited radiocarbon dating, and analysis of plan-view sand barrier and longshore drift patterns in relation to the courses of the river [21,22]. The delta plain prograded as a bayhead delta within a confined setting rich in Late Pleistocene aeolian deposits (Ogolian dunes) that extended as subaerial forms over the then exposed shelf during the last lowstand that peaked at 19,000 year B.P. [21,22]. Mud supplied by the river and fine sand derived from reworking of dunes inland by river-channel meandering have generated up to 8.5 km of essentially fine-grained delta-plain progradation within this bayhead setting. Although the delta plain does not protrude significantly into the Atlantic Ocean (Figure 2), probably because of the combination of this embayed setting and the relatively steep narrow shelf, the Senegal has, nevertheless, formed quite a large delta with an area of about 4254 km^2, much of which is subaerial, the ratio of the subaerial to subaqueous delta being 2:1 [9]. This mud-rich delta plain is bound by massive sandy barriers [21] built by waves propagating over loose aeolian deposits on the submerged narrow shelf. These coarse-grained barriers are separated by swales comprising abandoned river courses. Efficient trapping of river-borne sediments by the aggrading delta plain behind these wave-built sand barriers probably explains the high subaerial-subaqueous ratio of this delta, which is also consistent with the limited delta bulge compared to the more cuspate form commonly evinced by wave-dominated deltas. Remnants of these degraded barriers with beach ridges are discernible within the outer margins of the delta plain south of St. Louis. These spits are ancestral to the present Langue de Barbarie spit. Michel [21] dated the formation of these barriers at between 4000 and 1900 B.P. In essence, therefore, much of the Holocene development of the Senegal delta has consisted in embayment infilling behind the protection of these sand barriers, thus potentially giving rise to two distinct facies arrangements: wave-built sand bodies and back-barrier embayment facies represented by infilling fluvial deposits, including fine sands reworked from the Ogolian dunes by river meandering.

We used European Re-Analysis (ERA) hindcast wave data from 1984 to 2015 generated by the ECMWF Wave Atmospheric Model to characterise the wave climate affecting the Senegal River delta and to calculate longshore sediment transport rates along the spit. We then analysed changes in the position of the river mouth, rates of spit migration and spit width from aerial photographs and satellite images between 1954 and 2015 in order to characterise the shoreline morphodynamic context of the delta (see Materials and Methods).

3. Results

3.1. Wave Climate and Alongshore Sediment Transport

The wave climate of the Senegal delta shoreline is characterised by two components with strongly contrasting behaviour: wind waves generated locally and a dominant component of long swell waves from mid- to high latitudes (Figure 3). The region is not directly affected by major storms or cyclones but the influence of these distant high-energy events in the North Atlantic is materialised in the wave climate. Averaging over the 1984–2015 period gives annual significant swell and wind wave heights respectively of $H_s = 1.52$ m and 0.53, and peak swell and wind wave periods of $T_p = 9.23$ s and 3.06 s. The dominant swell waves originate from WNW to N and have a mean direction of 325°. The direction graph (bottom, Figure 3) shows a brief August swing dominated by swell waves from the south. Wind waves show a much wider directional window and a mean of 295°. There is a clear seasonal modulation, swell activity peaking during the northern hemisphere winter with strong storm activity at mid to high latitudes. Wind waves also show larger day-to-day and monthly variability. Contrary to swell waves, these wind waves are driven by local tropical winds and show peaks in spring and autumn that correspond to the passages of the Intertropical Convergence Zone over Senegal.

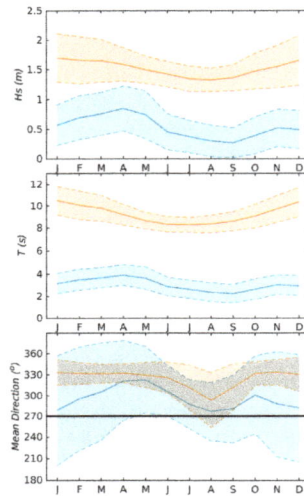

Figure 3. Mean wave characteristics (significant wave height (H_s), peak wave period (T), and incident direction (°)) along the Senegal River delta coast from 1984 to 2015 ERA hindcast data. Orange: swell waves, blue: wind waves.

As both swell and wind waves originate dominantly from W to N, this results in an oblique approach to the coastline that generates a large longshore sediment transport (LST) towards the south. Figure 4 depicts the annual LST along the Senegal delta coast for swell waves and wind waves computed using the formula of Kaczmarek et al. [23] as described in the Methods Section. The mean annual net transport induced by swell waves over the 32-year period of the ERA dataset is of the order of 669×10^3 m^3/year, i.e., ~89% of the total transport, the total wind-wave-induced LST amounting to only 80×10^3 m^3/year. LST is very largely dominated by southwards swell-induced drift which amounts to an annual mean of 611×10^3 m^3/year, while net wind-wave-induced transport in the same direction is only 59×10^3 m^3/year. Counter LST towards the north is nearly an order of magnitude less: 58×10^3 m^3/year for swell waves and 21×10^3 m^3/year for wind waves, i.e., only ~14% of the total LST. These computed sediment transport volumes are remarkably similar to those provided by the French engineering firm [24] SOGREAH (1994) who calculated a drift volume that decreases from north to south along the spit from 700 to 600 $\times$ 10^3 m^3/year.

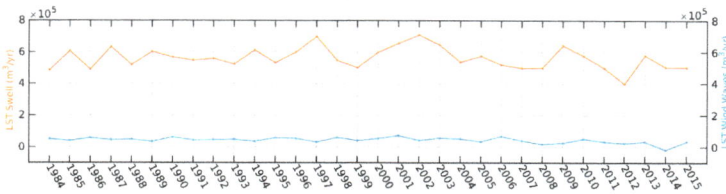

Figure 4. Gross annual longshore sediment transport (LST) along the Senegal River delta coast from 1984 to 2015. Orange: swell waves, blue: wind waves. Note the significant drop in swell-induced LST between 2009 and 2012, corresponding to a decrease of >35%, and the sharp rise the following year.

3.2. LST and Growth Dynamics of the Langue de Barbarie Spit

The Langue de Barbarie spit is a product of the strong wave action and high LST that have controlled the morphosedimentary development of the seaward fringe of the Senegal River delta (Figure 2). These observations and the satellite data also provide insight on the sand sourcing the seaward face of the Langue de Barbarie, which is derived from the coast and shoreface of Mauritania updrift of the historic mouth of the Senegal (Figure 1), in agreement with a conclusion also reached by Barusseau et al. [25]. The satellite data show that the Langue de Barbarie spit is a 100–400 m-wide feature. The spit is capped by aeolian dunes 5–10 m high. Widening of the spit and dune accretion occur through abstraction of the large alongshore sediment supply, especially in the distal section where bare, unvegetated dunes prevail, as well as through distal spit extension [26]. In contrast, the proximal sector, near St. Louis has been characterised by a much more stabilised dune system. Since 1900, a major coastal management preoccupation in the lower Senegal delta has been that of preventing natural breaches in the Langue de Barbarie in the vicinity of St. Louis, as this posed a threat for developing tourist facilities and infrastructure on the spit downdrift of every breach. Spit protection was achieved through the fixing and consolidation of the aeolian dunes via plantations of Filao (*Casuarina equisetifolia*) [27]. The alongshore transport volume would appear to undergo increasingly larger aeolian dune trapping of sand in the relatively poorly vegetated distal zone, compared to the relatively more urbanised and vegetated proximal sector of St. Louis. The former zone also represents one of active remigration following past natural breaches. The longshore gradient in sediment transport highlighted by SOGREAH [24] would appear to correspond to these morphological variations as one goes from the proximal to the distal sector of the spit.

The successive locations of the mouth of the Senegal River have been controlled by spit breaching followed by downdrift spit elongation. Spit breaching has generally been caused by increases in river water level, especially over the narrowest and lowest parts of the spit [26]. Once breaching occurs, the new breach is exploited by river discharge, tidal ingression, and waves, and forms a new river mouth. This leads to the older mouth becoming underfit and sealed by distal spit attachment to the shore. Natural breaching is attended by spit elongation through the classic formation of dune-capped beach ridges at the distal end, and this process has undoubtedly been favoured by the shallow overall depths of the mouth (2.5–3.5 m according to Bâ et al. [28]). The mouth is characterised by bars and spit recurves, remnants of which are identified in updrift locations on the spit. The mouth bars apparently serve as platforms for spit extension and eventual river-mouth diversion southwards.

3.3. Historical and Recent Changes of the Langue de Barbarie Spit Prior to the 2003 Artificial Breach

Joiré [29] and Tricart [30] situated the mouth of the river in the vicinity of St. Louis at about the mid-17th century, while a historical analysis of spit mobility and of the associated locations of mouth openings documented even earlier mouth scars north of St. Louis [27]. The Langue de Barbarie lengthened by 11 km between 1850 and 1900 (about 220 m a year), with a distal tip located 15 km south of St. Louis at the turn of the 20th century, and the spit was affected over this 50-year period by seven breaches [27]. Between 1900 and 1973, 13 other breaches occurred across the Langue de Barbarie [27], thus suggesting a breaching timescale (see Nienhuis et al. [31]) of ~6 years. There were no breaches between 1973 and 2003.

Following the 1973 breach, the Langue de Barbarie lengthened by 12.5 km (at a mean rate of ~400 m/year) before the spit was artificially breached in 2003. Spit elongation calculated from satellite images, aerial photographs and field measurements has, however, fluctuated widely from low values of nearly nil to <170 m/year (1985–1986, 1990–1991) to >1200 m/year (1987–1989, 2000–2002) (Figure 5). Gac et al. [27] showed that the farthest downdrift position of the mouth of the river, which corresponds to the maximal distal spit extension, did not exceed 30 km over the 80-year period covered by their observations, which is close to a value of 28 km reported in an earlier study [32]. The successive locations of the mouth of the Senegal River since 1973, which also correspond to those of the distal tip of the southward-extending spit, are shown in Figure 5, alongside the migration rates. The migration between 1973 and 2003 brought the distal tip of the spit close to the maximum spit length. The data from satellite images show a relatively narrow mouth (0.25–<1 km-wide) with the exception of the years 1968–1973 and 1988–1989 when the width exceeded 1.5 km (Figure 6).

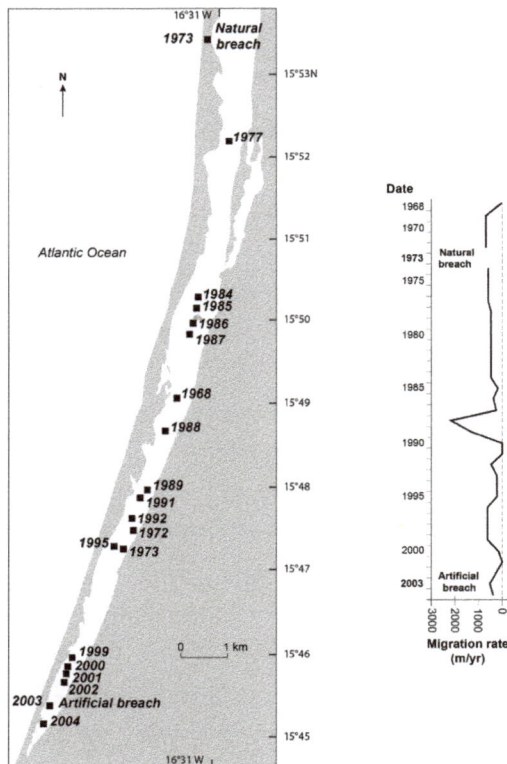

Figure 5. Successive dated locations of the mouth of the Senegal River delta materialised by the distal tip of the Langue de Barbarie spit (**left**); and spit migration rates in m/year from 1968 to 2004 (**right**).

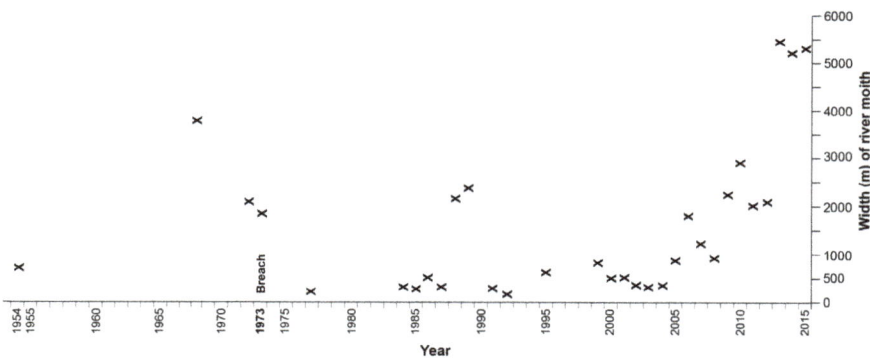

Figure 6. Width of the mouth of the Senegal River delta between 1954 and 2015. Except for the years 1968–1973 and 1987–1988, the width did not exceed 1 km, prior to the 2003 artificial breach. Following this breach, the width of the mouth fluctuated to attain ~1 km in 2008, which corresponds to the average width of the "fluvial" river mouth. A further rapid increase, not related to river-mouth hydraulics (see Discussion), occurred thereafter, peaking in 2013.

3.4. The Artificial Breach in 2003 and Post-Breach Spit and River-Mouth Evolution

An emergency water level in St. Louis prompted artificial breaching, on the night of 3 October 2003, of the Langue de Barbarie in the vicinity of the city to alleviate flooding. This high flood level had been preceded by several other episodes in the 1990s. One function of the Diama dam was to alleviate floods in the lower valley, notably in the deltaic sector. Mietton et al. [33] highlighted the rather mixed results from the flood-control function of the dam since the 1990s, and reported repeated episodes of severe flooding in St. Louis in 1994 (1.26 m above IGN datum), 1995 (1.21 m), 1997 (1.28 m), 1998 (1.43 m), 1999 (1.47 m), 2001 (1.2 m) and 2003 (1.38 m). The latter events preceding the artificial breach are depicted in Figure 7. The water level of 1.47 m above IGN datum attained at the height of the 1999 high-flow season exceeded the 1.2 m flooding threshold for 12 days, and the concern voiced by the population of St. Louis regarding this flooding progressively brought pressure to bear on the administrative authorities in their recourse to artificial breaching [34]. A 4 m-long and 1.5 m-deep trench was cut across a relatively narrow (100 m-wide) portion of the spit about 7 km south of St. Louis by engineers in the night of 3 October 2003. This induced a rapid overnight drop in water level of up to 1 m (Figure 7) that prevented further flooding [34]. Following this opening, the trench widened rapidly (Figure 8) and became the new river mouth, a case of inadvertent delta-mouth diversion generated by humans. The width of this artificial breach grew to 250 m 3–4 days after the opening. The depth of the breach increased to 6 m by 2007 [28], while the width increased to nearly 2 km in October 2006, three years after the breach (Figure 6), before decreasing once more to ~1 km in early 2008. Channelling of the Senegal River flow in the new enlarged mouth led to closure of the former natural mouth located further downdrift. An accelerated phase of widening ensued afterwards, peaking to nearly 5.5 km between October 2012 and June 2013 (Figure 6). Figure 9 summarises the dynamics of the spit and river mouth since the 2003 artificial breach. The rapid widening was related to an additional natural breach created in October 2012 by overwash 500 m south of the new mouth. Much of the remaining spit between this new opening and the mouth was eroded through several other washovers that tended to coalesce, widening the mouth and sea-intrusion pathways, as sand was transported southward by longshore drift.

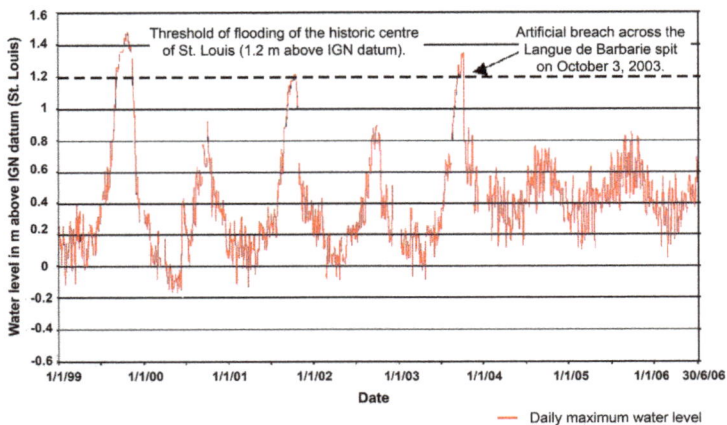

Figure 7. Maximum water levels in the Senegal River channel at St. Louis from 1999 to 2006. Adapted from [34].

Figure 8. Ground photographs showing the initial trench (4 October 2003), dug on the night of 3 October 2003, across the Langue de Barbarie to alleviate flooding of parts of St. Louis. The 5 October 2003 photograph shows the trench considerably widened by river and tidal flow (Photo credit: Service régional de l'Hydraulique, St. Louis du Sénégal).

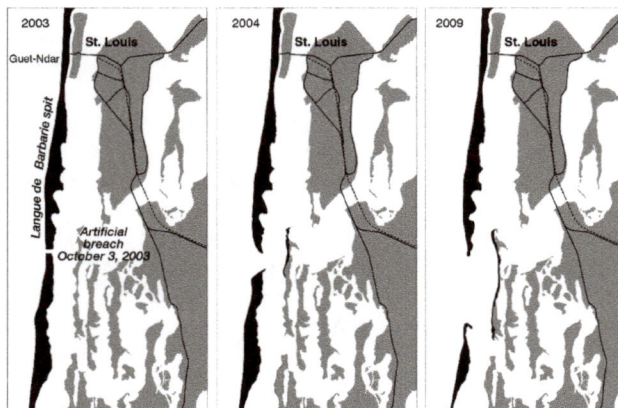

Figure 9. *Cont.*

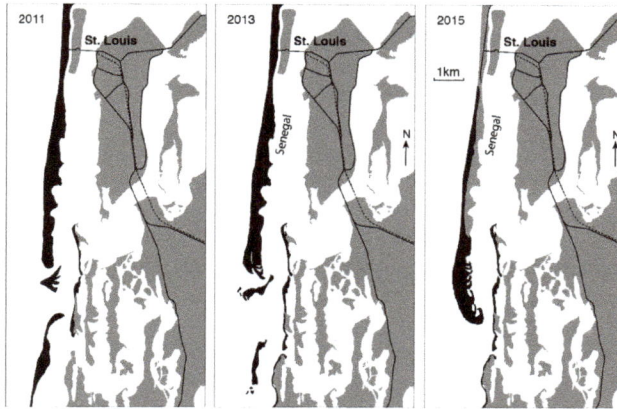

Figure 9. Assemblage from Google Earth images showing changes in the Langue de Barbarie spit and Senegal River mouth between March 2003, prior to the October 2003 artificial breach, and 2015. Black: Langue de Barbarie spit and beach sand; dark grey: subaerial lower delta plain potentially subject to river flooding (including St. Louis); light grey: delta plain seasonally flooded by the Senegal River. From 2012 to 2013, rapid wave-induced erosion of the residual spit downdrift of the mouth led to considerable mouth widening, an increase in tidal influence within the lower Senegal delta, and direct wave attack of parts of the delta plain hitherto protected by the residual spit.

4. Discussion

The shoreline of the Senegal delta offers an interesting example of strong wave influence on delta evolution. Two clear manifestations of this strong influence are the absence of a notable classic deltaic "bulge", and the presence of a persistent sand spit, the Langue de Barbarie, an extremely mobile feature that generates river-mouth diversion. This spit has been subject to repeated past breaches, and delta-mouth migration over a total distance of 28–30 km at least since the mid-17th century. The dominant natural mode of behaviour of the Senegal delta shoreline is thus one imprinted by strong longshore transport of sand generated by Atlantic waves from NW to N. The Senegal River mouth is thus a fine example of a wave-influenced delta illustrating the relationship between river-mouth migration, spit elongation and spit breaching by the river mouth [31], although a simple relationship between these processes cannot be expected because of the influence of fluctuations in river discharge and river-mouth bar dynamics [11]. Whereas high river discharge and the formation of river-mouth bars can lead to reduced sediment bypassing, which affects in turn the river-mouth migration rate and the size of the river-mouth spit [31], reduced discharge at the river mouth, tantamount to a decrease in hydraulic efficiency, can lead to bypassing of sediment around the mouth, thus reducing migration [31,35]. Natural breaches of spits barring river mouths and tidal inlets are a commonly cyclic process determined by a combination of spit lengthening, river discharge and river hydraulic efficiency, and also in many cases, storm wave action [31,36].

The absence of breaching between 1973 and 2003 associated with the lengthening of the Langue de Barbarie spit over this period constitutes a much longer timescale than past breaching timescales [27]. The reasons for this are not clear. They are unlikely to be related to the wave climate, which is devoid of storms, whereas breaching tends to be initiated by high river discharge during the flood season. The longshore transport volumes, of which the period 1984–2003 may be considered as representative, fluctuated but presumably were high enough to ensure spit elongation, without natural breaching updrift that could have been caused by a decrease in the alongshore budget. Spit morphometry (width, depth and migration range) as a criterion for determining the fraction of the LST sequestered in the spit, yields a value of 54%. This value is moderate relative to the relatively high hindcast and predicted values of the sediment bypassing fraction, β [31] (respectively, 0.83 and 0.74, 1 representing 100% bypassing) for the Senegal River mouth. These rates are, however, quite similar to those (0.8–0.9) calculated from our data on spit morphometry and LST using the sediment bypassing fraction equation and 50% of the river mouth depth as an estimate of the "updrift sediment spit depth" (see Materials and Methods). The mouth of the Senegal has thus been characterised by moderate to high bypassing that assured a degree of growth of the Langue de Barbarie but also the stability of the barrier and coast downdrift of the 2003 artificial breach. The absence of breaching over this long phase has been attributed to a decrease in river discharge [37]. Unfortunately, there are no available data on river water discharge to enable us to tie up natural breaches with the hydraulic efficiency of the river mouth. Mietton et al. [33] noted a total absence of critical floods between 1974 and 1993 associated with the Sahelian drought. This period also incorporates the construction of the Diama dam in 1986.

While the breaching timescale since 1973 appears exceptional compared to the pre-1973 conditions, the breaching length is also an important parameter in the onset of breaching [31]. The elevation of the water surface at the upstream boundary of a river channel is directly related to the channel length, such that an increase in the latter, as the river mouth migrates, results in a constant water surface slope, with the eventuality of breaching when a critical channel length is attained [31]. Guilcher [32] and Gac et al. [27] reported that the Langue de Barbarie spit generally did not exceed a maximum length of 28–30 km, beyond which breaching tended to occur. This length probably corresponds to the breaching length defined by Nienhuis et al. [31]. There is a probability, therefore, that a natural breach could have been imminent close to the time frame of the 2003 artificial breach. A reason for advancing this hypothesis is the increase in flooding (Figure 7), which suggests increasing impoundment of flood waters over the lower delta plain and decreasing hydraulic efficiency of the mouth. Whereas natural breaching has been a characteristic of the spit, spit instability since 2003 reflects, in part, the consequences of hasty artificial breaching to solve an impending flooding problem facing St. Louis. By protecting St. Louis and numerous smaller settlements and agricultural land within the delta plain from waves and marine influence, the spit is a major feature of the dynamics and management of the Senegal delta shoreline. Paradoxically, by impounding flood waters of the Senegal River, the spit also contributes to a flood risk that has grown apace with the urban extension of St. Louis. The long phase of absence of breaching between 1973 and 2003 coincided with a period of rapid tourism development in the Senegal delta associated with the emplacement of tourist infrastructure on the rectilinear spit that provided sandy grounds well above flood level. Although much of the lower delta is characterised by a population density of only about ten inhabitants/km^2, there are zones of very high population concentrations, as in St. Louis and certain areas of the Langue de Barbarie such as Guet-Ndar (Figure 9) where the 2013 census shows densities exceeding 80,000 inhabitants/km^2 [6]. The artificial breach annihilated the risk of flooding of St. Louis in 2003 and in the following years by enabling more rapid seaward drainage of river water during the high-flow season [34].

As in the pre-2003 period, the sediment bypassing fraction, β [31], across the mouth of the Senegal River has been quite high (0.8–0.9), although balancing spit morphometry against LST over the same period suggests up to 40% of sand locked up in spit growth, a value lower, however, than that of the pre-2003 breach. There have been marked fluctuations in spit growth, however, with even spit erosion in 2005–2006, 2008–2010 and 2012–2013. Under conditions of spit growth, sand has been incorporated in new recurves that mark the current form of elongation of the residual updrift spit sector, which is also characterised by an enlarged distal tip (Figure 9). The reasons for alternations between spit growth (including widening) and spit erosion are not clear. They may be related to variations in higher-energy waves, and potentially varying LST, as shown by the drop in the number of days with high-energy waves in 2012 (Figure 10) and the correlative drop in LST (Figure 4), but they could also be an outcome of variability in river discharge and sediment bypassing.

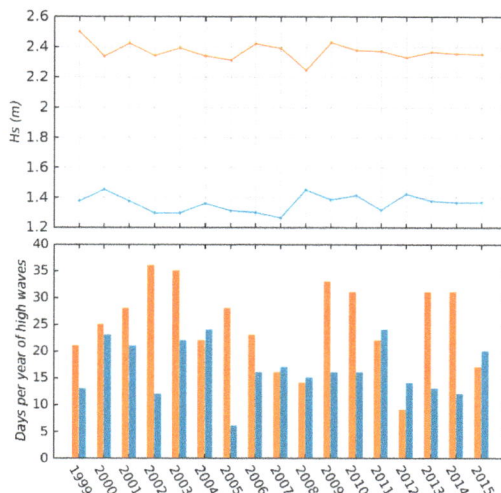

Figure 10. Significant heights (*Hs*) of high-energy waves (±1.6 standard deviations around mean *Hs*) from 1999 to 2015 (**top**); and number of days per year with high-energy waves along the Senegal River delta coast, derived from ERA hindcast data (**bottom**). Orange: swell waves (*Hs* ≥ 2.37 m), blue: wind waves (*Hs* ≥ 1.36 m). Note the significant drop in high-energy swell waves in 2012 (see also Figure 4).

Over this post-2003 period, fluctuations of the width of the river mouth (Figure 6) are presumably a function of the balance between the river's hydraulic efficiency, including the tidal discharge, and incident wave energy and sediment bypassing [11]. The width of the "fluvial" mouth of the river is very likely in the range of ~0.5–1 km, which is the "usual" mouth width (Figure 6) and the stabilised width attained shortly after the artificial breach. The rapid widening between October 2012 and June 2013 occurred following wave overwash and erosion of the remaining spit downdrift of the mouth. This rapid erosion would appear to result from a combination of the most significant drop in LST recorded (2010–2012) over the period 1984–2015 (Figure 4), with a lag effect in time, and possible sequestering of sand in the river mouth. Lower bypassing (due to higher river discharge?) and a sharp increase in LST from 2012 to 2013 (an increase of about ~45% relative to the 2010–2012 LST (Figure 4)) could explain the ensuing exceptionally rapid elongation of the Langue de Barbarie spit between June 2013 and May 2015 (~2 km) (Figure 9). A review of conceptual advances in wave-river-mouth interactions [11] and modelling of alongshore sediment bypassing at river mouths [31] have shown that waves refracting over the river-mouth bar create a zone of low alongshore sediment transport updrift which reduces sediment bypassing. These observations imply that the LST potential south of the new mouth is being assured by a degree of "cannibalisation" of the rest of the spit, as sand transported from the north has been increasingly trapped updrift of the wider mouth, presumably leading to lower bypassing. Except for 2007–2009, and 2010–2011, this sector has been in erosion. This demise of the spit downdrift of the new mouth has led to the destruction of villages, campsites and other tourist structures. The delta plain in this eroding sector is now directly exposed to ocean waves and erosion that are threatening numerous villages.

Much of the lower delta plain and the main river channel are now situated over 20 km upstream of the former mouth, between the new mouth and the anti-salt intrusion Diama dam that confines the tidal prism to the lower delta plain. In consequence, the much wider mouth appears to have become favourable to a larger tidal prism, manifested by an increase in the tidal range in St. Louis, and confirmed by recent studies [33,34]. Durand et al. [34] showed that the maximum semi-diurnal tidal range downstream of the Diama dam has increased three-fold, from a mean of 0.30 m in 2001–2002 to

0.93 m in 2004–2005, whereas the mean maximum spring tide range attained 1.18 m, for a predicted value of 1.29 m, along the Langue de Barbarie spit. These authors have also noted that the semi-diurnal tidal effects are now more clearly expressed even during the high river flood waters. The impacts of these changes are still to be studied, but it may be expected that they are leading to increasing soil salinization in the lower delta plain, to the extension of bare saline flats, and to modifications in biodiversity.

The extent to which accelerated subsidence, one of the two major causes of delta vulnerability (together with rapid and chronic erosion), affects the delta is not known, although it may be inferred that a decreasing sediment load and damming may be contributing to more exacerbated flooding in the delta plain. However, the problem seems to have more to do with accelerated urbanisation of St. Louis over the last few decades, bringing new populations to encroach on areas of the delta that are susceptible to flooding during exceptionally wet years. Durand et al. [34] have highlighted the potential vulnerability of the city and the surrounding low delta plain to sea-level rise. Their model simulating flood propagation in the city, and based on various sea-level scenarios, shows the susceptibility of St. Louis to flooding during the highest annual water levels in the course of the 21st century.

5. Materials and Methods

5.1. Waves and Wave-Induced Longshore Transport

In order to estimate the wave-induced alongshore transport on the Langue de Barbarie, we extracted bulk wave parameters (significant height H_s, peak period T_p and direction of both swell and wind waves) from hindcast data in the Atlantic Ocean between 1984 and 2015, generated by the ECMWF Wave Atmospheric Model (WAM) model [38]. The wave data are part of the ERA-Interim dataset, which involves a reanalysis of global meteorological variables [39,40]. Wave data were extracted from the ECMWF data server on a $0.5° \times 0.5°$ grid, with a 6-h temporal resolution and covering the sector $16.5°$ N/$17°$ W. The ERA-40 and the following ERA-Interim reanalysis are the first in which an ocean wind–wave model is coupled to the atmosphere, and the quality of the wave data has been extensively validated against buoy and altimeter data. Sterl and Caires (2005) [40] demonstrated a very good correlation between the ERA-40 data and these sources, except for high waves ($H_s > 5$ m) and low waves ($H_s < 1$ m), which tend, respectively, to be under- and over-estimated [41]. These critical wave conditions are not typical of the relatively constant wave regime affecting the Senegal delta coast, and extreme wave condition issues reported for ERA-40 are partially resolved for higher resolution ERA-Interim. However, the Senegal coast has scarce observations, and this affects the hindcast quality. ERA-40 and -Interim results in this region should be taken with caution.

Several alongshore sediment transport formulae exist and are widely applied by coastal engineers and dynamicists. However, there is still an important research effort on the improvement of alongshore sediment transport parameters and no large consensus on the choice of a formulation, as dispersion between predictors is often substantial [42], and validation dataset at the regional scale scarce. Here, we chose the formula of Kaczmarek et al. [23] because of its straightforward implementation for remote sites such as the Langue de Barbarie where only limited observations exist and because it has been applied to similar environments [43,44]. The amount of sediment drifting alongshore was computed as follows:

$$Q = 0.023 \left(H_b^2 V \right) \quad if \left(d_b^2 V \right) < 0.15 \tag{1}$$

$$Q = 0.00225 + 0.008 \left(H_b^2 V \right) \quad if \left(d_b^2 V \right) > 0.15 \tag{2}$$

where H_b is the breaking wave height and V an estimation of the alongshore current within the surf zone derived from the commonly used formula of Longuet-Higgins [45]:

$$V = 0.25 k_v \sqrt{\gamma g d_b} \, sin2\alpha_b \tag{3}$$

where α_b is the local breaking wave angle, $\gamma = H_b/d_b = 0.78$ is the breaker parameter constant [46], g the gravitational acceleration (m/s^2), H_b the breaking wave height, d_b the local water depth and k_v an empirical constant. Here, we used $k_v = 2.9$ based on the values of Bertin et al. [43] for wave-dominated environments with similar grain-size characteristics. A separate computation for sediment transport induced, respectively, by wind waves and swell waves was conducted.

Alongshore sediment transport formulae necessitate breaking wave parameters as inputs, but global wave hindcast only provide deepwater characteristics. While a nested model (e.g., SWAN or WW3) to propagate waves from deepwater to the breakpoint would be ideal for a short-term study, the present analysis focuses on seasonal to inter-annual wave variations covering a long period of 32 years. We chose therefore to use the direct breaking wave predictor proposed by Larson et al. [47]. This formula provides breaking wave height H_b and angle α_b from deepwater wave height H_o, period T and incidence angle α_0:

$$H_b = \lambda C^2/g \tag{4}$$

$$\alpha_b = \mathrm{asin}\left(\sin(\alpha_0)\sqrt{\lambda}\right) \tag{5}$$

with a correction factor λ computed as:

$$\lambda = \Delta\lambda_a \tag{6}$$

considering

$$\Delta = 1 + 0.1649\,\xi + 0.5948\,\xi^2 - 1.6787\xi^3 + 2.8573\,\xi^4 \tag{7}$$

$$\xi = \lambda_a \sin\theta_0{}^2, \; \lambda_a = [\cos(\alpha_0)/\theta]^{2/5}, \theta = \left(\frac{C}{\sqrt{gH}}\right)^4\left(\frac{C}{C_g}\right)\gamma^2 \tag{8}$$

where deep water phase celerity is given by $C = 1.56T$, wavelength $L = 1.56T^2$, and group celerity $C_g = C/2$.

5.2. Shoreline Change and Spit and River-Mouth Dynamics

In order to highlight recent deltaic shoreline changes, we resorted to available aerial photographs (1954), a CORONA satellite image (1968) and LANDSAT (1984–1988, 1992, 1999–2004, 2006–2011, 2013, 2015–2016) and SPOT satellite images (2005) with moderate pixel size resolution (30 to 60 m) made available by the USGS and the French IGN. The main items analysed were spit length and corresponding migration rates, spit width, and river-mouth width and the underlying dynamics. The spatial data were chosen to cover the entire "delta-influenced" shoreline for each year of analysis and with a cloud cover not exceeding 10%. We limited our choice to images taken at low tide and systematically in January of every year to minimise seasonal and tidal distortions (tides induce very little variability in the microtidal context of the Senegal River delta). The results on shoreline change were completed by a literature review on the past dynamics of the Langue de Barbarie and by field observations of this spit conducted in 2005, 2007 and 2016.

Based on data from the satellite images and aerial photographs on spit and river-mouth characteristics, the fraction of sediment bypassing the mouth, β, assuming conservation of mass, was inferred from the following relationship [31]:

$$v = Qs(1 - \beta)/Ab \tag{9}$$

where v is the migration rate of the mouth (m·s^{-1}), Qs is the volumetric alongshore sediment transport rate (m^3·s^{-1}), and $Ab = Ws \cdot Ds$ which is the cross-sectional area of the river mouth spit (m^2) composed of blocked littoral sediment from the updrift coast, Ws the width of the spit, and Ds spit updrift sediment depth.

Acknowledgments: This work benefited from the ECMWF ERA Interim dataset (www.ECMWF.Int/research/Era), and from Landsat satellite images provided by the United States Geological Survey, and Spot satellite images provided by Centre National d'Etudes Spatiales. We acknowledge funding from the Belmont Forum Project: *BF-Deltas: Catalyzing Action Towards Sustainability of Deltaic Systems with an Integrated Modeling Framework for Risk Assessment*. Mamadou Sadio benefited from a partial PhD grant provided by the Embassy of France in Senegal. We thank the anonymous reviewers for their salient suggestions for improvement.

Author Contributions: Mamadou Sadio, Edward J. Anthony, Amadou Tahirou Diaw and Alioune Kane designed the project. Mamadou Sadio, Amadou Tahirou Diaw, Philippe Dussouillez and Jules T. Fleury analysed the satellite images. Mamadou Sadio., Edward J. Anthony, Amadou Tahirou Diaw, and Alioune Kane conducted field reconnaissance. Rafael Almar and Elodie Kestenare analysed the E.R.A. data and longshore transport products. All authors wrote the paper.

Conflicts of Interest: We declare no conflicts of interests.

1. Crossland, C.J.; Kremer, H.H.; Lindeboom, H.J.; Marshall Crossland, J.I.; Le Tissier, M.D.A. *Coastal Fluxes in the Anthropocene*; Springer: Berlin, Germany, 2007; p. 231.

2. Van Rijn, L.C. Coastal erosion and control. *Ocean Coast. Manag.* **2011**, *54*, 867–887. [CrossRef]

3. Davies, J.L. *Geographical Variation in Coastal Development*, 2nd ed.; Longman: London, UK, 1980; p. 212.

4. Anthony, E.J. Coastal progradation in response to variations in sediment supply, wave energy and tidal range: Examples from Sierra Leone, West Africa. *Géodynamique* **1991**, *6*, 57–70.

5. Anthony, E.J. The muddy tropical coast of West Africa from Sierra Leone to Guinea-Bissau: Geological heritage, geomorphology and sediment dynamics. *Afr. Geosci. Rev.* **2006**, *13*, 227–237.

6. Diatta, I. L'ouverture d'une Brèche à Travers la Langue de Barbarie (Saint-Louis du Sénégal). Les Autorités Publiques et les Conséquences de la Rupture. Master's Thesis, Université Gaston Berger, St Louis, Senegal, 2004.; p. 116.

7. Sy, B.A. L'ouverture de la brèche sur la Langue de Barbarie et ses conséquences. Approche géomorphologique. *Revue de Géographie de Saint-Louis* **2004**, *4*, 50–60. (In French)

8. Bhattacharya, J.P.; Giosan, L. Wave-influenced deltas: Geomorphological implications for facies reconstruction. *Sedimentology* **2003**, *50*, 187–210. [CrossRef]

9. Coleman, J.M.; Huh, O.K. *Major Deltas of the World: A Perspective from Space*; Coastal Studies Institute, Louisiana State University: Baton Rouge, LA, USA, 2004.

10. Anthony, E.J. Patterns of sand spit development and their management implications on deltaic, drift-aligned coasts: The cases of the Senegal and Volta River delta spits, West Africa. In *Sand and Gravel Spits*; Randazzo, G., Cooper, J.A.G., Eds.; Springer: Berlin, Germany, 2015; Volume 12, pp. 21–36.

11. Anthony, E.J. Wave influence in the construction, shaping and destruction of river deltas: A review. *Mar. Geol.* **2015**, *361*, 53–78. [CrossRef]

12. Nienhuis, J.H.; Ashton, A.D.; Giosan, L. What makes a delta wave-dominated? *Geology* **2015**, *43*, 511–514. [CrossRef]

13. Galloway, W.E. Process framework for describing the morphologic and stratigraphic evolution of delta depositional systems. In *Deltas: Models for Exploration*; Broussard, M.L., Ed.; Texas Geological Society: Houston, TX, USA, 1975; pp. 87–98.

14. Sall, M. Crue et Elévation du Niveau Marin à Saint-Louis du Sénégal: Impacts Potentiels et Mesures D'adaptation. Ph.D. Thesis, Université du Maine, Le Mans, France, 2006.

15. Kamara, S.; Martin, Ph.; Coly, A. Organisation traditionnelle du bas delta du Sénégal et nouvelles régulations hydrauliques. Dimension anthropospatiale d'un développement. *Revue Espaces et Sociétés en Mutation* **2015**, *2015*, 127–144. (In French).

16. Mahé, G.; Olivry, J.C. Variations des précipitations et des écoulements en Afrique de l'Ouest et central de 1951 à 1989. *Sécheresse* **1995**, *6*, 109–117. (In French).

17. Kane, A.; Niang-Fall, A. *Hydrologie du Sénégal*; Atlas Jeune Afrique: Dakar, Sénégal, 2007; p. 14. (In French)

18. Ostenfeld, C.; Jonson, N. *Etude de la Navigabilité et des Ports du Fleuve Sénégal; Études Portuaires à Saint-Louis, Kayes et Ambidebi. Vol. 1: Travaux Préliminaires; Vol. 2, Annexe 2: Rapport Sur les Enquêtes Hydrauliques*; Surveyer-Nenninger et Chevenert Inc.: Montréal, QC, Canada, 1972. (In French)

19. Gac, J.Y.; Kane, A. Le fleuve Sénégal. Bilan hydrique et flux continentaux de matières particulaires à l'embouchure. *Sci. Geol.* **1986**, *39*, 99–130 & 151–172. (In French).

20. Food and Agriculture Organization of the United Nations (FAO). *Caractérisation Des Systèmes de Production Agricole au Senegal*; Document de Synthese; FAO: Rome, Italy, 2007. (In French)

21. Michel, P. The southwestern Sahara margin: Sediments and climate change during the recent Quaternary. *Palaeoecol. Afr. Surround. Isl.* **1980**, *12*, 297–306.

22. Monteillet, J. *Environnements Sédimentaires et Paléohcologie du Delta du Sénégal au Quaternaire*; Lmprimerie des Tilleuls: Millau, France, 1986; p. 267. (In French)

23. Kaczmarek, L.M.; Ostrowski, R.; Pruszak, Z.; Rozynski, G. Selected problems of sediment transport and morphodynamics of a multi-bar nearshore zone. *Estuar. Coast. Shelf Sci.* **2005**, *62*, 415–425. [CrossRef]

24. SOGREAH. *Etudes de Faisabilité et D'avant Projet Sommaire de L'émissaire Delta*; Rapport Final: Grenoble, France, 1994; p. 70. (In French)

25. Barusseau, J.P.; Bâ, M.; Descamps, C.; Diop, E.S.; Diouf, B.; Kane, A.; Saos, J.L.; Soumaré, A. Morphological and sedimentological changes in the Senegal River estuary after the constuction of the Diama dam. *J. Afr. Earth Sci.* **1998**, *26*, 317–326. [CrossRef]

26. Sall, M.M. Dynamique et Morphogenèse Actuelles au Sénégal Occidental. Ph.D. Thesis, Université Louis Pasteur-Strasbourg I, Strasbourg, France, 1982.

27. Gac, J.Y.; Kane, A.; Monteillet, J. Migrations de l'embouchure du fleuve Sénégal depuis 1850. *Cahiers ORSTOM Série Géologie* **1982**, *12*, 73–76. (In French).

28. Bâ, K.; Wade, S.; Niang, I.; Trébossen, H.; Rudant, J.P. Cartographie radar en zone côtière à l'aide d'images multidates RSO d'Ers-2: Application au suivi environnemental de la Langue de Barbarie et de l'estuaire du fleuve Sénégal. *Télédétection* **2007**, *7*, 129–141. (In French).

29. Joiré, J. Amas de coquillages du littoral sénégalais dans la banlieu de Saint-Louis. *Bulletin de l'Institut Français de l'Afrique Noire* **1947**, *9*, 170–340. (In French).

30. Tricart, J. *Notice Explicative de la Carte Géomorphologique du Delta du Sénégal*; Mémoires, B.R.G.M., Ed.; Bureau de Recherches Geologiques et Minieres: Orléans, France, 1961; Volume 8, p. 137. (In French)

31. Nienhuis, J.H.; Ashton, A.D.; Nardin, W.; Fagherazzi, S.; Giosan, L. Alongshore sediment bypassing as a control on river mouth morphodynamics. *J. Geophys. Res. Earth Surf.* **2016**, *121*, 664–683. [CrossRef]

32. Guilcher, A.; Nicholas, J.P. Observation sur la Langue de Barbarie et les bras du Sénégal aux environs de Saint-Louis. *Bulletin d'Information du Comité Océanographique pour les Etudes Côtières* **1954**, *6*, 227–242. (In French)

33. Mietton, M.; Dumas, D.; Hamerlynck, O.; Kane, A.; Coly, A.; Duvail, S.; Baba, M.L.O.; Daddah, M. Le delta du fleuve Sénégal. Une gestion de l'eau dans l'incertitude chronique. In *Incertitudes et Environnement—Mesures, Modèles, Gestion*; d'Allard, P., Denis, F., Picon, B., Eds.; Ecologie Humaine/Edisud: Arles, France, 2006; pp. 321–336. (In French)

34. Durand, P.; Anselme, B.; Thomas, Y.F. L'impact de l'ouverture de la brèche dans la langue de Barbarie à Saint-Louis du Sénégal en 2003: Un changement de nature de l'aléa inondation? *Cybergeo* **2010**, *496*. (In French) [CrossRef]

35. Balouin, Y.; Ciavola, P.; Michel, D. Support of subtidal tracer studies to quantify the complex morphodynamics of a river outlet: The Bevano, NE Italy. *J. Coast. Res.* **2006**, *39*, 602–606.

36. Cooper, J.A.G. Ephemeral stream-mouth bars at flood-breach river mouths on a wave dominated coast: Comparison with ebb-tidal deltas at barrier inlets. *Mar. Geol.* **1990**, *95*, 57–70.

37. Niang, A.J. Les Processus Morphodynamiques, Indicateurs de L'état de la Désertification Dans le Sud-Ouest de la MAURITANIE. Approche Par Analyse Multisource. Ph.D. Thesis, Université de Liège, Liège, Belgium, 2008.

38. The Wamdi Group. The WAM model—A third generation ocean wave prediction model. *J. Phys. Oceanogr.* **1988**, *18*, 1775–1810.

39. Dee, D.P.; Uppala, S.M.; Simmons, A.J.; Berrisford, P.; Poli, P.; Kobayashi, S.; Andrae, U.; Balmaseda, M.A.; Balsamo, G.; Bauer, P.; et al. The ERA-interim reanalysis: Configuration and performance of the data assimilation system. *Q. J. R. Meteorol. Soc. Bull.* **2011**, *137*, 553–597. [CrossRef]

40. Sterl, A.; Caires, S. Climatology, variability and extrema of ocean waves—The web-based KNMI/ERA-40 Wave Atlas. *Int. J. Climatol.* **2005**, *25*, 963–977. [CrossRef]

41. Caires, S.; Swail, V.R.; Wang, X.L. Projection and analysis of extreme wave climate. *J. Clim.* **2006**, *19*, 5581–5605. [CrossRef]
42. Pinto, L.; Fortunato, A.B.; Freire, P. Sensitivity analysis of non-cohesive sediment transport formulae. *Cont. Shelf Res.* **2006**, *26*, 1826–1839. [CrossRef]
43. Bertin, X.; Castelle, B.; Chaumillon, E.; Butel, R.; Quique, R. Alongshore drift estimation and inter-annual variability at a high-energy dissipative beach: St. Trojan Beach, SW Oleron Island, France. *Cont. Shelf Res.* **2008**, *28*, 1316–1332. [CrossRef]
44. Almar, R.; Kestenare, E.; Reyns, J.; Jouanno, J.; Anthony, E.J.; Laibi, R.; Hemer, M.; Du Penhoat, Y.; Ranasinghe, R. Part 1. Wave climate variability and trends in the Gulf of Guinea, West Africa, and consequences for longshore sediment transport. *Cont. Shelf Res.* **2015**, *110*, 48–59. [CrossRef]
45. Longuet-Higgins, M.S. Alongshore currents generated by obliquely incident sea waves. *J. Geophys. Res.* **1970**, *75*, 6788–6801.
46. Battjes, J.A.; Janssen, J.P.F.M. Energy loss and setup due to breaking of random waves. In Proceedings of the ASCE International Conference on Coastal Engineering, Hamburg, Germany, 27 August–3 September 1978; pp. 569–587.
47. Larson, M.; Hoan, L.X.; Hanson, H. A direct formula to compute wave properties at incipient breaking. *J. Waterw. Port Coast. Ocean Eng.* **2010**, *136*, 119–122. [CrossRef]

water

MDPI

Article

Beach Response to Wave Forcing from Event to Inter-Annual Time Scales at Grand Popo, Benin (Gulf of Guinea)

Grégoire Abessolo Ondoa [1,2,*]**, Frédéric Bonou** [2,3,4]**, Folly Serge Tomety** [2,3,4]**,**
Yves du Penhoat [2,3,4]**, Clément Perret** [2,3,4]**, Cossi Georges Epiphane Degbe** [3] **and Rafael Almar** [2]

[1] Fishery Resources Laboratory, University of Douala, BP 2701 Douala, Cameroon
[2] LEGOS (Université Paul Sabatier de Toulouse/CNRS/CNES/IRD), 31400 Toulouse, France;
    fredericbonou@yahoo.fr (F.B.); sertom1@hotmail.fr (F.S.T.); yves.du-penhoat@ird.fr (Y.d.P.);
    clement.perret@outlook.com (C.P.); rafael.almar@ird.fr (R.A.)
[3] Institut de Recherches Halieutiques et Océanologiques du Bénin, 03 BP 1665 Cotonou, Benin;
    gdegbe@yahoo.fr
[4] International Chair in Mathematical Physics and Applications/Unesco Chair, University of Abomey-Calavi,
    01 BP 526 Cotonou, Benin
* Correspondence: gregsolo55@yahoo.fr; Tel.: +237-699-76-73-34

Received: 31 March 2017; Accepted: 16 June 2017; Published: 21 June 2017

Abstract: This paper assesses the morphological storm-event impact, seasonal cycles, trends of wave forcing, and beach's response at the coastal area of Grand Popo, Benin. Three and a half years' worth of data were collected from 2013 to 2016, using a video system calibrated with field data collected during a 10 day experiment. A comparison was carried out with Wavewatch III IOWAGA wave hindcast data. The along-shore-averaged shoreline position exhibited a seasonal pattern, which was related more to the average wave height than the average storm intensity. Storms occur in austral winter (June, July, August, and September). Based on 12 storms, the results revealed that the average storm duration was 1.6 days, with a mean erosion of 3.1 m. The average post-storm beach recovery duration was 15 days, and the average recovery rate was 0.4 m/day. The impact of storms was more or less amplified depending on the eroding and accreting periods of the wave climate. There was an inter-annual eroding trend of about −1.6 m/year, but the causes of this trend could not be explained.

Keywords: shoreline; waves forcing; storms; resilience; post-storm recovery; Bight of Benin; seasonal cycle; trend

1. Introduction

The coastal zone of West Africa is under increasing pressure of overpopulation, as it is a zone of economic interest. Human settlements and livelihood activities have been developing on the shores of the Atlantic Ocean, where the beach evolution varies according to a wide range of different temporal and spatial scales. In this region, beaches are microtidal and swell-dominated environments, where waves and tides are the main drivers of nearshore dynamics [1]. Several findings suggest that along wave-dominated coastlines, regionally-varying wave climates will have an increasing impact on the shoreline in the coming decades, and cannot be ignored in forecasting shoreline variability [2–4]. This highly dynamic behaviour is essentially due to the fact that sandy coasts can undergo adjustments in form and processes, which can change rapidly. Periods of accretion and erosion are generally associated with low- and high-energy wave conditions, respectively, but they also exhibit strong site-specific variations [5]. For many coastal regions, both sea-level rise and changes in the storm-wave climate would result in coastal erosion and an increased frequency with a high intensity of coastal flooding. Storm-events represent a major factor of modulating short- and medium-term morphological evolutions of many sandy shorelines. In the event of changing storm regimes associated with climate change [6], it is important to understand the potential effects of storms on beaches, and how they recover after these high-energy events. Many studies have been carried out on assessing the impact of storms, beaches' responses, and post-storm morphological adjustments in storm-dominated coastlines [6–12]. Managing erosion-induced problems will depend on the resilience of the beach to extreme events, and universal threshold conditions are not likely to be found [5]. Establishing storm thresholds is difficult, especially because they are generally site-specific [11,13]; this is as well as the beach recovery period [7], which has not yet been clearly addressed in the literature. Available literature regarding beaches' responses to storms on tropical microtidal coastlines and the potential impacts of climate change remains scarce.

Along the wave-dominated coastlines of the Gulf of Guinea, the influence of South Atlantic high-energy swells drives strong, eastward longshore sediment transport [14,15]. This littoral drift in the Bight of Benin is one of the largest in the world, with estimations of approximatively 400,000–1,000,000 m^3/year [16]. This transport is mostly driven by swell waves due to the Southern Annular Mode (SAM), rather than wind waves due to the Inter-Tropical Convergence Zone (ITCZ) [15]. This equatorial fluctuation presents a large seasonal and inter-annual variability as well as wave climate [14]. This implies a high seasonal variability in the beaches' responses to equatorial Atlantic forcing, given that seasonal processes dominate the shoreline changes due to seasonal variations of the wave height at several specific sites [5,17]. These findings need to be confirmed, but few measurements are available on the high-frequency evolution of shoreline and beach states in the West and Central African regions.

This study assesses different time scales of beach responses to wave forcing at Grand Popo Coast, Benin, using video-derived shoreline and wave evolution data over a 3.5 year observation period (February 2013 to August 2016). We first investigate an average beach response during and after storm-events. Secondly, we evaluate the impact of storm durations and recurrence with seasonal variability. And finally, we estimate the inter-annual trends.

2. Data and Methods

2.1. Study Area

Located in the Gulf of Guinea, Benin, near the border with Togo, Grand Popo Beach is an ocean-open, sandy stretch of coast facing the South Atlantic Ocean [18]. The beach is far enough from the influence of the major cities: Cotonou (80 km away), and Lome (60 km away). In the last three years, some fields of groynes have been constructed near the town of Anèho, 20 km updrift [19] (see Figure 1b).

The beach dynamics are dominated by the influence of oblique waves (South/Southwest) of moderate energy (mean significant wave height Hs = 1.36 m; mean peak period Tp = 9.4 s). The longshore sediment transport is primarily driven by swell waves (South/Southwest and South/Southeast) generated in the Southern Hemisphere trade-wind region (30–35° S and 45–60° S), rather than wind waves (Southwest) generated locally in the Gulf of Guinea [14–16,20], as shown in Figure 1a. Tides are semi-diurnal with a microtidal range from 0.8 to 1.8 m for neap and spring tides, respectively. The sediment size is medium-to-coarse: 0.4 to 1 mm (median grain size $D50$ = 0.6 mm). Grand Popo Beach is an intermediate low tide terrace (LTT) to reflective beach [14,18,20], according to the classification proposed in [21].

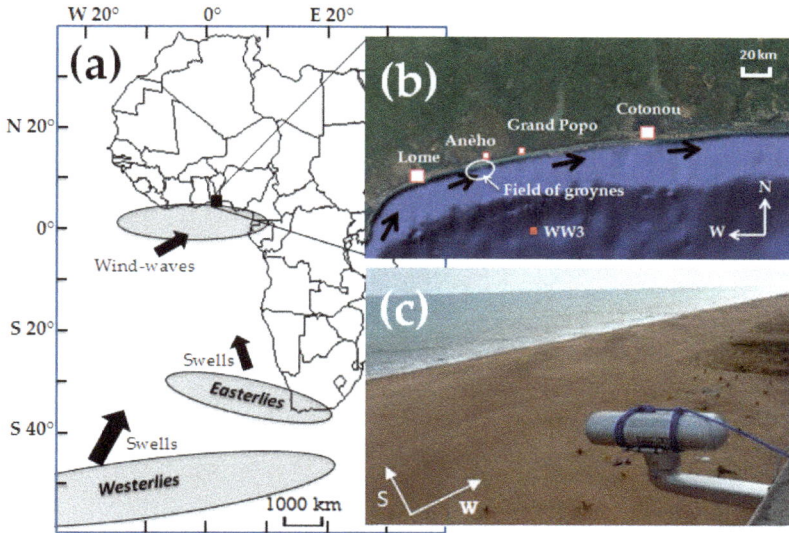

Figure 1. Study site: (**a**) The regions of major generations of swells and wind waves impinging on the Gulf of Guinea are indicated by grey ovals, and their major directions of propagation by the black arrows. (**b**) Focus in the Bight of Benin, with major cities of Cotonou and Lomé. Black arrows represent the longshore sediment transport directions. The red point (2.0° E, 5.5° N) gives the location of the WW3 model output. (**c**) Permanent video camera on a 15 m high semaphore at Grand Popo Beach, Benin.

2.2. Video System and Data

In February 2013, a low-cost video system was installed on the top of a tower of the Navy Forces of Benin in Grand Popo, about 70 m from the shoreline [22]. The system was composed of a VIVOTEK IP 7361 camera (1600 × 728 pixels), which collected data continuously at 2 Hz. An on-site computer processed the raw images and stored three types of secondary images every 15 min: snapshots, cross-shore time-stacks, and 15 min time exposure (or timex) images (Figure 2) [22]. Twenty ground control points were taken with GPS to process image geo-rectification [23], by applying the method of direct linear transformation [24]. This consists of a matrix, which gives the relation between image pixels and control points by taking into account the position of the camera and the correction due to the camera lens distortion [25].

Several recent methods were used to extract hydrodynamic and morphologic parameters. *Hs* video estimations were obtained from the time-stack images [26]. Following the wave signature induced by breaking, wave heights were detected from the pixel intensity threshold *Ipix* = 40. The intensity of breaking pixels was significantly larger (*Ipix* > 80) than that of non-breaking pixels (*Ipix* ~ 10). The pixel intensity peak, which appeared at the wave crests, was calculated by the standard deviation δ of the pixel intensity of each time series. The width of the peak of δ marked the horizontal projection of the wave face covered by the roller (*L*), which was subsequently projected into the vertical direction for a simple rough estimation: $H_b = L \cdot \tan(\beta)$, where β was the camera view angle. The fact that the wave-front slope (α_b) at breaking differed significantly from the vertical direction was taken into account. The common value $\alpha_b \approx 30°$ is used as a breaking criterion in numerical breaking parameterizations, according to [26]. The wave height could therefore be estimated from the equation [26]:

$$H_b = (L - Cor) \tan(\beta) \qquad (1)$$

Cor being a geometrical correction defined as [26]:

$$Cor = \frac{L}{\tan(\alpha_b)} \tan(\beta) \tag{2}$$

The mean wave period T_m was computed from the offshore pixel intensity time series using the mean zero-crossing method on the time-stack images [27]. Wave direction was estimated from the snapshots and 15 min averaged images. The technique consisted of, firstly, subtracting the average image from the snapshot (removing the background, which does not move); secondly, highlighting the wave crests (the time-varying part); then rectifying on a regular grid; and finally, recovering the angle of the crest of the waves by the Radon transform [28].

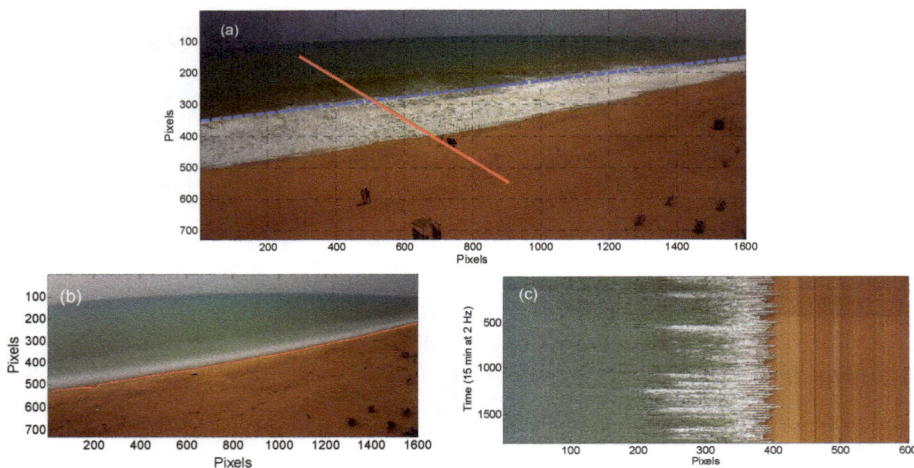

Figure 2. Video image types: (**a**) Snapshot with cross-shore time-stack profile as a red line, and along-shore time-stack profile as a blue dashed line. (**b**) 15 min averaged image; shoreline is shown as a red line. (**c**) Cross-shore time-stacked image (vertical is time: 15 min at 2 Hz).

A 10 day field experiment was conducted at Grand Popo Beach, Benin, in March 2014 [22]. The measurements included both topographic and bathymetric morphological surveys with differential GPS (DGPS) and bathymetric sonar, while offshore forcing (waves and tide) was characterized using an Acoustic Doppler Current Profiler (ADCP) moored at a 10 m depth. *Hs* video data were regressed with 10 days of ACDP field data, with an acceptable root-mean-square error (RMSE) of 0.14 m [18].

Due to technical malfunctions in the video acquisition system, missing video data could be estimated with linear regression between existing video data and Wavewatch III (WW3), model version 4.10 (IOWAGA wave hindcast database [29]) output at the nearest available point, 2.0° E, 5.5° N (Figure 1b), propagated to breakpoint using an empirical predictor formula [30]. This formula was used in a recent study [15] focusing on wave climate variability in the Bight of Benin. This formula directly provides the breaking wave height h_b and angle θ_b, given deep water wave height h_0, period T, and direction θ_0:

$$h_b = \lambda \cdot C^2 / g \tag{3}$$

$$\theta_b = a\sin\left(\sin(\theta_0) \cdot \sqrt{\lambda}\right) \tag{4}$$

with a correction factor λ computed by:

$$\lambda = \Delta \cdot \lambda_a \tag{5}$$

where:

$$\Delta = 1 + 0.1649\zeta + 0.5948\zeta^2 - 1.6787\zeta^3 + 2.8573\zeta^4 \tag{6}$$

$$\zeta = \lambda_a \cdot (\sin\theta_0)^2 \tag{7}$$

$$\lambda_a = [\cos(\theta_0)/\varphi]^{\frac{2}{5}} \tag{8}$$

$$\varphi = \left(\frac{C}{\sqrt{gh_b}}\right)^4 \left(\frac{C}{C_g}\right)\gamma^2 \tag{9}$$

where the phase celerity is given by $C = 1.6T$, group celerity $C_g = \frac{C}{2}$ and breaker depth index $\gamma = 0.78$.

The shoreline location was calculated from timex images as the maximum gradient in the ratio red/green-blue [31]. Beach pixels display high red-channel values and low green values, whereas water pixels exhibit strong green-channel values and low red values [25,32]. The ratio red/green-blue was computed for all pixels and its local minimum stood for the transition between water and beach, namely, the shoreline. The overall error in video detection of the shoreline location owing to water level uncertainties, due to wave breaking or atmospheric pressure variations, or incorrect shoreline detection, was about 0.5 m [12]. Shoreline migration was estimated via the along-shore-averaged location <X> to reduce error data due to the along-shore-digitized shoreline. Determination of the intertidal beach profile and beach slope involved the delineation of the shoreline at different tidal levels [32], and interpolation between daily low and high tides. Tidal levels for the study period 2013–2016 were extracted from the WXTide32 model, version 4.7. As there was no tide gauge at the study site, the tidal subordinate at Lomé, Togo ($1°14'$ E, $6°07'$ N) was referred to the nearest tide gauge station, Takoradi, Ghana (~350 km). The root-mean-square and mean errors in the intertidal profile computed between 7 day DGPS (Grand Popo experiment, 2014) and video data were 0.28 and 0.23 m, respectively [18].

2.3. Event Scale: Storms

The definition of storm-event is site-specific [5,9], and the *Hs* threshold used to define storm conditions or extreme events is selected to produce clear and identifiable storm-events. Three-hourly *Hs* time series were used, and the 5% exceedance probability of the wave height time series over the study period ($Hs_{5\%} = 1.85\ m$) was considered as the threshold for storm-events. A single storm is defined as a continuous period of *Hs* exceeding this threshold and lasting at least one tidal cycle (12 h), following [5,12]. The overall impact of storm-events is assessed through the daily-averaging maximum shoreline moving during the storm. Storm intensity I ($m^2 \cdot h$) is computed as the integration of time-varying *Hs* over the storm duration:

$$I = \int_{t_1}^{t_2} H_S(t)^2 \cdot dt \tag{10}$$

where t_1 and t_2 are times corresponding to the beginning and the end of the storm [12].

There are several ways to define the recovery duration after each storm. It can be defined as the time taken by the nearshore morphology to evolve from a post-storm state (e.g., dissipative/longshore bar and trough) to its modal state (i.e., the most frequently occurring beach state, e.g., rhythmic bar and beach or transverse bar and rip) [9–11]. In this study, the time duration taken to reach the first maximum recovery value of the along-shore-averaged shoreline location <X> after each storm was accepted as the recovery duration [10,12]. This duration referred to the post-storm period of continuous accretion towards its equilibrium pre-storm state (*Tr*), and did not depend on any forcing parameter. The overall recovery duration for the study period was computed as the time for daily-averaging post-storm evolution of continuous accretion [5].

2.4. Seasonal Signal and Trends

To obtain the seasonal signal, monthly nearshore estimations were computed. The test of Mann Kendall was used to check if the time series showed substantial trends [33]. The null hypothesis of trend absence in a time series was tested, against the alternative of having a trend. Each time series was reorganized as a matrix M_{ij} where $1 \leq i \leq N_y$ with N_y representing the number of years of video observation, and $1 \leq j \leq 12$. The seasonal signal S_j was obtained as follows:

$$S_j = \frac{1}{N_y} \cdot \sum_{i=1}^{N_y} M_{ij} \qquad (11)$$

The monthly residual or anomaly signal R_{ij} of each parameter was estimated by removing the seasonal monthly value from each monthly-averaged value computed over the three and a half years. The annual anomaly or trend R_a was computed by averaging monthly anomaly values R_{ij}.

3. Results

3.1. Hydrodynamic and Morphological Variability

Figure 3a–c provides an overview of monthly video and WW3 data [29] over the study period. The same *Hs* seasonality for the two sets of data was observed (Figure 3a), with more energetic waves during the April–October period and less energetic waves during the November–March period. These observations are consistent with the wave climate of the area because of northward migration, by a few degrees, of the wave-generating zone in the high latitudes of the South Atlantic (~40° to 60° S) during the summer period [14,15]. Table 1 gives correlations and errors between the WW3 video and model data on the time series of *Hs*, *Tm*, and the wave direction. There is strong correlation between the two time series of *Hs* ($R^2 = 0.80$). Direction and *Tm* are less correlated ($R^2 = 0.44$ and $R^2 = 0.19$; Figure 3b,c), with a low observed variability.

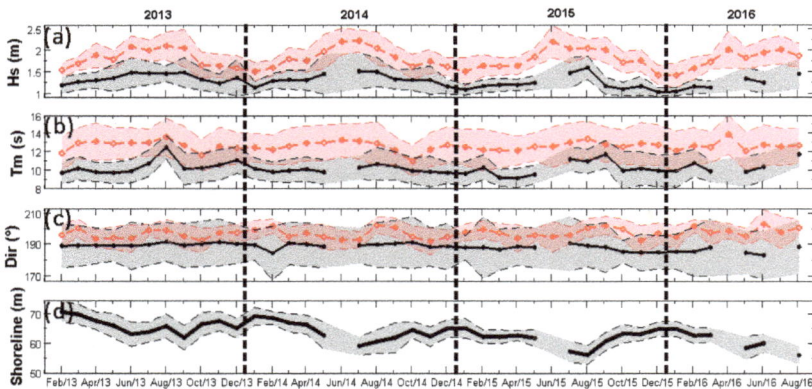

Figure 3. Monthly-averaged video estimates (black) and Wavewatch III data (red): (**a**) wave significant height *Hs*; (**b**) wave mean period *Tm*; (**c**) wave direction *Dir*; and (**d**) shoreline location. Shaded zones stand for day-to-day dispersion (standard deviation).

It should be noted here that the IOWAGA WW3 ocean wave hindcast had not been assimilated with any observations (satellite or buoy). For the previous versions of WW3, random averaged errors between *Hs* model outputs (WW3) and satellite data ranged from 0.3 to 0.4 m for low wave heights (<2 m), and 0.15 m for higher waves [34]. An improvement was made on the recent version used in this work, integrating new parameterizations for the wind–sea and wave dissipation. However, the normalized RMSE for *Hs* remained ~20%, compared to the satellite data in our study area, with biases of more than 0.1 m [29]. This was consistent with the overestimation observed in the model outputs (Figure 3), which were likely amplified due to local unresolved effects of bathymetry for the wave propagation from deep water to breaking point. However, WW3 results were improved even more from linear regression correction than from bias correction, as shown in [35]. Gaps in *Hs* video data could therefore be estimated using a linear regression between the two sets of data.

Table 1. Comparison of daily hydrodynamic video data and WW3 model outputs. WW3 data were propagated from deep water to breakpoint using an empirical direct formula [30]. The root-mean-square error (RMSE) and the mean error (ME) were computed between the two sets of data.

Video–WW3	*Hs* (m)	*Tm* (s)	*Dir* (°)
Correlation	0.8	0.4	0.2
RMSE	0.3	2.4	9.4
ME	0.3	2.3	8.5

3.2. Storms and Morphological Impact

3.2.1. Detection and Statistics of Individual Storms

Thirty-two storms were identified over the study period (Figure 4a). The mean peak storm wave height was 2.05 m (standard deviation $\sigma = 0.05$ m) and the mean wave height throughout the storms was 1.99 m ($\sigma = 0.12$ m). The average duration of a storm was 1.8 days and storms were recorded from April to September, corresponding to austral winter. With the threshold value of $H_{S_{5\%}} = 1.85$ m, no storm with a duration longer than 12 h was recorded from the November to March period, due to less energetic wave conditions. Only 12 storms (average duration of 1.6 days) were further considered for analyses of shoreline responses due to gaps in shoreline data. Individual storms resulted in a wide range of shoreline impacts (Figure 4b), from no change on 2 July 2013 (1 day storm), to significant erosion (−8.7 m) during a 4 day storm (21–25 September 2013). The maximum number of storm-events was recorded in July. In 2014 for example, 12 storms were counted, with four in July. The strongest storm impacts were recorded at the end of austral fall (in May) with an average onshore migration of −3.7 m during a storm of less than 1 day. This impact decreased with the increase of storm numbers until the beginning of austral winter (June and July).

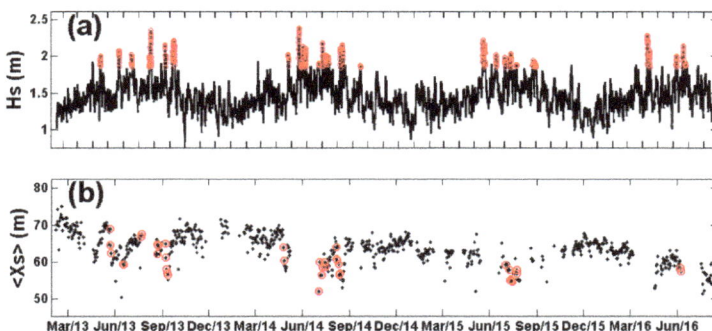

Figure 4. Time series of: (**a**) significant wave height, and (**b**) along-shore-averaged location <*Xs*> from the tower of camera location. Storm periods are marked in red.

3.2.2. Beach Response to Storms and Resilience

Figure 5 shows an ensemble-averaged analysis of the shoreline evolution during the storm and post-storm recovery period, with this period referring to the post-storm period of continuous accretion, at the end of which the beach was assumed to be stabilized [5,9,12]. The day "0" stands for the beginning of the storm, according to the 5% exceedance (1.85 m) of *Hs*. The average storm intensity was 155 m²·h and induced an average beach erosion of 3.1 m. After the end of the storm, the beach attempted to recover during a continuous accretive phase: the shoreline moved offshore (0.46 m/day) and the time needed to reach stabilization was ~15 days. This time was considered as the post-storm recovery duration *Tr*.

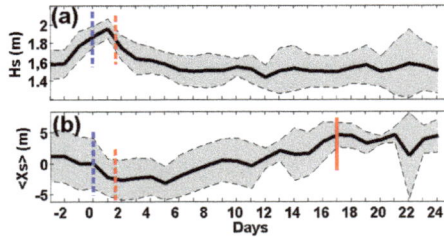

Figure 5. Ensemble-averaged evolution during the storm and post-storm recovery period for: (a) *Hs*, and (b) shoreline location <*Xs*>. Blue dashed lines stand for the beginning of the averaged-storm, red dashed lines for the end of the storm (1.6 day storm duration), and the solid red line stands for the post-storm recovery duration for beach stabilization (15 days).

3.3. Seasonal Cycle

Figures 6 and 7 present the seasonal cycle of several monthly nearshore forcing parameters and beach morphology. The maximum monthly-averaged *Hs* (1.51 m) was obtained in July, corresponding to the maximum monthly-averaged wave flux (15400 J/m·s). The seasonal pattern was highlighted in the along-shore-averaged shoreline position, which was strongly correlated to the monthly *Hs* seasonal cycle ($R^2 = -0.94$), and in contrast, less correlated to the monthly-averaged beach slope ($R^2 = -0.25$). The maximum beach slope was 0.14 rad at the end of April. For the 32 identified storms, the greatest number of storms was recorded in July, but these were shorter (average duration of 1.34 days), leading to a lesser average intensity compared to the other months. Figure 6a,d and Figure 7a show that the beach response (shoreline location) was most related to the monthly-averaged *Hs* rather than the intensity of storm-events.

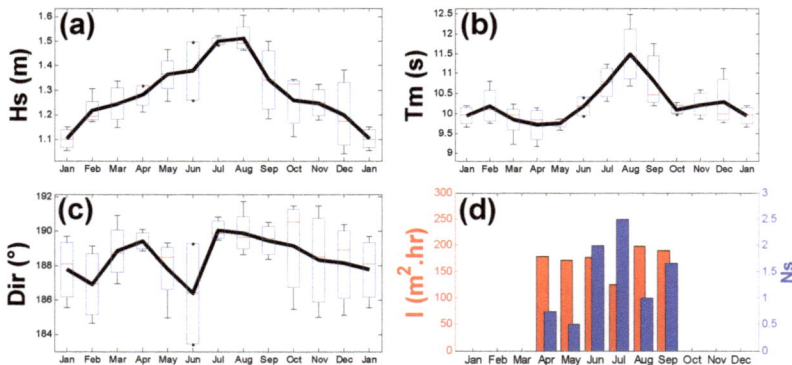

Figure 6. Seasonal variability (monthly average) of: (a) *Hs*, (b) *Tm*, (c) wave direction (*Dir*), and (d) storm intensity in m²·h (red) and storm number *Ns* (blue). For each box, the central mark (red line) is the median, the edges of the box (blue) are the 25th and 75th percentiles, and the whiskers extend to the most extreme data points not considered outliers.

The seasonal cycle of the shoreline presented two main different phases, the erosive phase (January to August) and the accretive phase (August to December), due to the changes of Hs (Figure 6a) and Tm (Figure 6b), which were driven by swell waves. The standard deviation of the beach slope variation was smaller during the eroding period, reflecting the low variability of the beach slope according to low-energy waves.

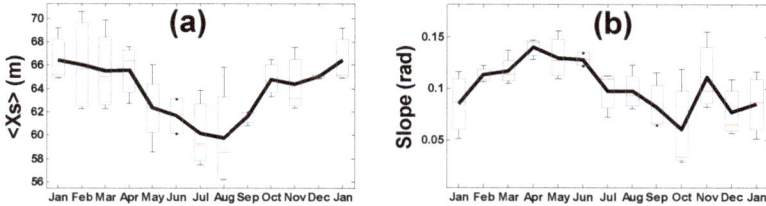

Figure 7. Monthly average seasonal variability: (**a**) mean shoreline location <Xs> from video camera location, and (**b**) beach slope. For each box, the central mark (red line) is the median, the edges of the box (blue) are the 25th and 75th percentiles, and the whiskers extend to the most extreme data points not considered outliers.

Storms occurred at the end of the eroding period (April to July) and at the beginning of the accreting period (August to September). The beach exhibited particular responses to storm-events, depending on the concerned period, as shown in Figure 8, where only 12 storms were considered with their corresponding shoreline data. During this eroding period, the recovery duration seemed to be shorter (~10 days). In contrast, in the accreting period, the time recovery duration seemed to be longer and the storm impact increased with the storm duration, consistent with [17]. The post-storm recovery was more significant in the accreting period than in the eroding period. In August, the beach experienced an onshore migration of 2.4 m during a 1.4 day averaged-duration storm, while in September, a 3.7 day averaged-duration storm caused an erosion of −5.4 m. The beach response was therefore influenced by the erosive or accretive period of the wave climate oscillation.

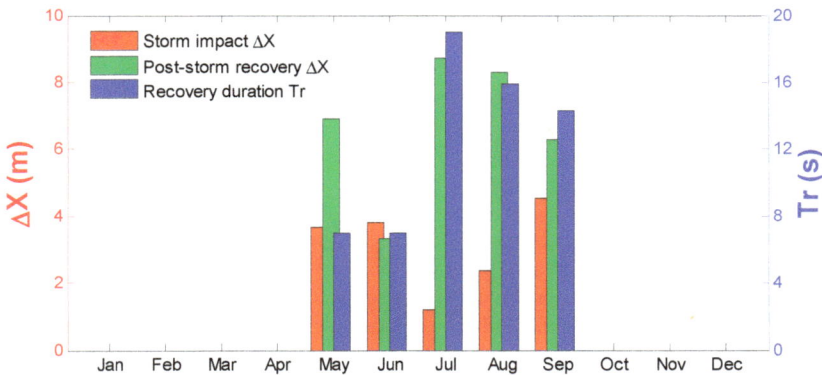

Figure 8. Ensemble-averaged shoreline variation <ΔX_j> during and after storm-events: storm impact (red), beach recovery (green), and recovery duration (blue) per month.

3.4. Trends and Inter-Annual Evolution

Trends of waves and the shoreline were investigated using the test of Mann Kendall [33]. The results showed that the shoreline position Hs and wave direction data presented substantial trends. A failure to reject the null hypothesis (absence of trend) at 95% significance level was obtained for the wave period and energy, and the beach slope. The residual signal reflected trends over the study period and was obtained by removing the seasonal cycle from the daily data. Table 2 shows the annual average and residuals of each studied parameter. During the 3.5 year study period, the shoreline migrated 6 m onshore. The annual anomaly or residual of the shoreline decreased from +2.8 m in 2013 to −2.9 m in 2016, while the annual residual of the beach slope did not change. A decrease of 3° in the wave direction was observed during the study period.

Table 2. Annual averaged values and anomalies (R_a) for the study period.

Study Period	2013		2014		2015		2016 (January–August)	
	Mean	R_a	Mean	R_a	Mean	R_a	Mean	R_a
Hs (m)	1.35	+0.05	1.34	+0.04	1.23	−0.07	1.28	−0.06
Tm (s)	10.4	+0.2	10.1	−0.2	10.2	−0.1	10.6	+0.3
Dir (°)	188.9	+1.4	188.2	+0.5	186.4	−1.3	185.2	−1.8
Shoreline position (m)	66.4	+2.8	63.9	+0.2	61.8	−1.9	60.1	−2.9
Beach slope (rad)	0.098	−0.006	0.101	0	0.109	+0.008	0.113	−0.04

4. Discussion

Perhaps more than other survey techniques (e.g., DGPS, LIght Detection And Ranging LIDAR) used in coastal science, the video system remains less costly. However, video measurements are subject to large uncertainties [10]. In particular, the shoreline-detection methods are sensitive to waves, lighting conditions, and water levels, which can scale the effects of both the setup and run-up, and reduce the colour signal strength [32]. Previous works on video error detection [18,23,36] compared shorelines measured by video to topographic surveys, and the results suggested a reasonable error (about 0.23 m at Grand Popo, Benin, following [18]). In this work, tidal heights were estimated using the WXTide2 model, to compute the shoreline location. This did not take into account all regional and coastal components of the sea level (including wave-induced setup and run-up) and nearshore bathymetry that were measured by the video [18].

The results presented here are consistent with the wave climate observations in the studied area [14,15]. A seasonal pattern was clearly observed in the shoreline position, beach slope, and wave characteristics. Data exhibited two specific periods: an accreting period (August to December), where *Hs* and *Tm* decreased; and an eroding period, where *Hs* and *Tm* increased (January to July). This was consistent with the oscillation of the Southern Annular Mode (SAM), which has a predominant influence on transport induced by swell waves [15]. The low values of the beach slope standard deviation observed during February to April reflected a stabilization reached during low wave-energy conditions. Computing the beach slope variability could lead to an understanding of the nature of the waves breaking, as the beach slope is connected to the surf-similarity parameter. A recent finding [20] suggested that the wave reflection is mostly governed by swash dynamics, whereby the reflected spectrum essentially depends on the swash slope.

The 3.5 years of video of the shoreline location suggests that beach dynamics observed at Grand Popo Beach are affected by storm-events. This study revealed that the impact on shoreline migration can be significant: −8.7 m on 21 September 2013, during a 4 day sequence of storms. The mean duration of the 12 observed storms was 1.6 days, with an average storm-erosion of −3.1 m. At the end of the storm, the shoreline migrated offshore at an average distance of 6 m within 15 days. However, the storm impact and post-storm recovery were very dependent on the observed seasonal pattern, consistent with [17]. The recovery duration in the eroding period was shorter (~10 days) than in the accreting period (>15 days). The end of the period of stabilization observed between February and April was marked by short storms of an average duration of less than 1 day, which caused significant erosion on the beach, averaging 3.7 m. During the accreting period, it took longer storms to observe a significant impact (mean of 3.7 days for −5.4 m). The beach's response was therefore mostly related to the energy of average wave conditions, rather than to the energy of extreme wave conditions.

The study of inter-annual trends in this work demonstrated a gradual decline of the shoreline cross-shore location during the period 2013–2016. However, the length of the data set was not enough to assess the inter-annual variability. A field of nine groins of 100 m lengths and 20 m widths was constructed between 2012 and 2014 over a distance of 3.5 km, near the city of Anèho (Togo), about 20 km from the video system at Grand Popo Beach. This field could reduce the sediment supply of the eastward coastal drift in the direction of Grand Popo. A recent study [19] investigated the coastline evolution between Grand Popo (Benin) and Anèho (Togo) from 1984 to 2011, using Landsat TM (Thematic Mapper) and ETM+ (Enhanced Thematic Mapper Plus) images. This study showed that the locality of Grand Popo was in dynamic equilibrium from 1984 to 2000, while a significant accretion occurred with +0.4 m/year between 2000 and 2011. The impact of the field of groins was also investigated, but no clear conclusions were drawn on its impacts on the coastal area of Grand Popo; although immediately downstream from the groynes, the region is experiencing significant visual erosion.

The trend of oceanic forcing (wave direction of −3° for the period 2013–2015) is another possible factor of the observed erosion. The diminution of the wave direction from 188.9° to 185.2° during the study period increased the importance of cross-shore processes compared to along-shore processes, resulting in a decrease in the along-shore sediment transport (~5% per year), computed with an empirical formula [30] presented in [15], in our study area. The longshore sediment transport is very dependent on the shore's normal wave direction, and the resulting littoral drift in the Bight of Benin is one of the largest in the world, following [14,15]. Nevertheless, the video estimates of the wave direction had uncertainties. A comparison of video and WW3 wave direction outputs showed that the video data presented a larger standard deviation than the model output, respectively 13.4° (video) and 6.4° (WW3), and the video wave direction trend was not observed in the WW3 data. A previous study estimated the RMSE with field ADCP measurements to be about 9.25°, and the mean error (ME) to be 2.25°, within the range of observed variation from 2013 to 2016 (−3°).

5. Conclusions

Three and a half years of video-derived shoreline and wave evolution data at Grand Popo Beach was used to investigate the beach's response to wave forcing from event to inter-annual time scales in the coastal area of Grand Popo Beach, Benin. The beach exhibited a seasonal pattern in wave conditions and the along-shore-averaged shoreline position was most related to the monthly-averaged wave height rather than the average storm intensity. The seasonal pattern of the shoreline indicated an eroding period and an accreting period, corresponding to austral and winter periods, respectively. Thirty-two storms were identified for the period 2013–2016. The mean storm duration was 1.6 days for the 12 observed storms due to gaps in video data, and the average storm erosion was −3.1 m. Ensemble-averaged storm recovery conditions showed that the beach recovered within 15 days, and the average recovery rate was 0.4m/day. This study underlines that the impact of storms is more or less amplified depending on the eroding and accreting periods of the wave climate: (i) the recovery duration is longer in the accreting period than in the eroding period, (ii) the storm-impact is more significant in the eroding period than in the accreting period, (iii) storms are longer in the accreting period than in the eroding period, and (iv) storm numbers are high during the transition from the eroding to accretive phases. A trend on the along-shore-averaged shoreline location was observed (−1.6 m/year), however our data were not enough to draw conclusions at inter-annual time scales.

Acknowledgments: This publication was made possible through support provided by the IRD. We acknowledge the use of the WW3 dataset (IOWAGA wave hindcast database; http://tds1.ifremer.fr/thredds/IOWAGA-WW3-HINDCAST). We would like to express our gratitude to IRD/JEAI-RELIFOME (Jeune Equipe Associée à l'IRD) for its financial support and to support the three months stay at Gao in Benin. Thanks are due to Gaël ALory for his technical support in maintaining the video system. The Grand Popo experiment was supported by the French INSU/CNRS EC2CO-LEFE/IRD, UNESCO Co-Chair ICMPA/UAC. We are indebted to the "Forces Navales" of Benin at Grand Popo for their logistic support during field experiments and for allowing the installation of the permanent video system on the semaphore. This work was supported by French ANR project COASTVAR (ANR-14-ASTR-0019).

Author Contributions: All authors were implicated in maintaining the video system at Grand Popo and analysis of data.

Conflicts of Interest: The authors declare no conflict of interest.

References

1. Stive, M.J.F.; Aarninkhof, S.G.J.; Hamm, L.; Hanson, H.; Larson, M.; Wijnberg, K.M.; Nicholls, R.J.; Capobianco, M. Variability of shore and shoreline evolution. *Coast. Eng.* **2002**, *47*, 211–235. [CrossRef]
2. Ranasinghe, R.; Callaghan, D.; Stive, M.J.F. Estimating coastal recession due to sea level rise: Beyond the Bruun rule. *Clim. Chang.* **2012**, *110*, 561–574. [CrossRef]
3. Ruggiero, P.; Cote, J.; Kaminsky, G.; Gelfenbaum, G. Scales of variability along the Columbia River littoral cell. In Proceedings of the Coastal Sediments '99: The 4th International Symposium on Coastal Engineering and Science of Coastal Sediment Processes, Hauppauge, NY, USA, 21–23 June 1999; pp. 1692–1707.
4. Brunel, C.; Sabatier, F. Potential influence of sea-level rise in controlling shoreline position on the French Mediterranean Coast. *Geomorphology* **2009**, *107*, 47–57. [CrossRef]
5. Senechal, N.; Coco, G.; Castelle, B.; Marieu, V. Storm impact on the seasonal shoreline dynamics of a meso-to macrotidal open sandy beach (Biscarrosse, France). *Geomorphology* **2015**, *228*, 448–461. [CrossRef]
6. Zhang, K.; Douglas, B.; Leatherman, S. Do storms cause long-term beach erosion along the U.S. East Barrier Coast? *J. Geol.* **2002**, *110*, 493–502. [CrossRef]
7. Morton, R.A.; Paine, J.G.; Gibeaut, J.G. Stages and durations of post-storm beach recovery, southeastern Texas coast, USA. *J. Coast. Res.* **1994**, *10*, 884–908.
8. Morton, R.A.; Gibeaut, J.C.; Paine, J.G. Meso-scale transfer of sand during and after storms: Implications for prediction of shoreline movement. *Mar. Geol.* **1995**, *126*, 161–179. [CrossRef]
9. Castelle, B.; Marieu, V.; Bujan, S.; Splinter, K.D.; Robinet, A.; Senechal, N.J.; Ferreira, S. Impact of the winter 2013–2014 series of severe Western Europe storms on a double-barred sandy coast: Beach and dune erosion and megacusp embayments. *Geomorphology* **2015**, *238*, 135–148. [CrossRef]
10. Masselink, G.; Scott, T.; Russel, P.; Davidson, M.A.; Conley, D.C. The extreme 2013/2014 winter storms: Hydrodynamic forcing and coastal response along the southwest coast of England. *Earth Surf. Process. Landf.* **2015**, *41*, 378–391. [CrossRef]
11. Almeida, L.P.; Vousdoukas, M.V.; Ferreira, Ó.; Rodrigues, B.A.; Matias, A. Thresholds for storm impacts on an exposed sandy coastal area in southern Portugal. *Geomorphology* **2012**, *143*, 3–12. [CrossRef]
12. Angnuureng, D.B.; Almar, R.; Senechal, N.; Castelle, B.; Addo, K.A.; Marieu, V.; Ranasinghe, R. Shoreline resilience to individual storms and storm clusters on a meso-macrotidal barred beach. *Geomorphology* **2017**, *290*, 265–276. [CrossRef]
13. Ba, A.; Senechal, N. Extreme winter storm versus summer storm: Morphological impact on a sandy beach. *J. Coast. Res.* **2013**, *1*, 648–653. [CrossRef]
14. Laibi, R.; Anthony, E.; Almar, R.; Castelle, B.; Senechal, N.; Kestenare, E. Longshore drift cell development on the human-impacted Bight of Benin sand barrier coast, West Africa. *J. Coast. Res.* **2014**, *70*, 78–83. [CrossRef]
15. Almar, R.; Kestenare, E.; Reyns, J.; Jouanno, J.; Anthony, E.J.; Laibi, R.; Hemer, M.; Du Penhoat, Y.; Ranasinghe, R. Response of the Bight of Benin (Gulf of Guinea, West Africa) coastline to anthropogenic and natural forcing, Part1: Wave climate variability and impacts on the longshore sediment transport. *Cont. Shelf Res.* **2015**, *110*, 48–59. [CrossRef]
16. Anthony, E.J.; Blivi, A.B. Morphosedimentary evolution of a delta-sourced, drift-aligned sand barrier–lagoon complex, western Bight of Benin western Bight of Benin. *Mar. Geol.* **1999**, *158*, 161–176. [CrossRef]
17. Yates, M.L.; Guza, R.T.; O'Reilly, W.C. Equilibrium shoreline response: Observations and modeling. *Geophys. Res.* **2009**, *114*. [CrossRef]
18. Abessolo, O.G.; Almar, R.; Kestenare, E.; Bahini, A.; Houngue, G.H.; Jouanno, J.; Du Penhoat, Y.; Castelle, B.; Melet, A.; Meyssignac, B.; et al. Potential of video cameras in assessing event and seasonal coastline behaviour: Grand Popo, Benin (Gulf of Guinea). *J. Coast. Res.* **2016**, 442–446. [CrossRef]

19. Degbe, C.G.E.; Laibi, R.; Sohou, Z.; Oyede, M.L.; Du Penhoat, Y.; Djara, M.B. Diachronic analysis of coastline evolution between Grand-Popo and Hillacondji (Benin), from 1984 to 2011. *Water* **2016**, submitted.

20. Almar, R.; Ibaceta, R.; Blenkinsopp, C.; Catalan, P.; Cienfuegos, R.; Viet, N.T.; Duong Hai, T.; Uu, D.V.; Lefebvre, J.P.; Laryea, W.S.; et al. Swash-based wave energy reflection on natural Beaches. In Proceedings of the Coastal Sediments 2015, San Diego, CA, USA, 11–15 May 2015.

21. Wright, L.D.; Short, A.D. Morphodynamic variability of surf zones and beaches: A synthesis. *Mar. Geol.* **1984**, *56*, 93–118. [CrossRef]

22. Almar, R.; Honkonnou, N.; Anthony, E.J.; Castelle, B.; Senechal, N.; Laibi, R.; Mensah-Senoo, T.; Degbe, G.; Quenum, M.; Dorel, M.; et al. The Grand Popo beach 2013 experiment, Benin, West Africa: From short timescale processes to their integrated impact over long-term coastal evolution. *J. Coast. Res.* **2014**, 651–656. [CrossRef]

23. Angnuureng, D.B.; Almar, R.; Addo, K.A.; Castelle, B.; Senechal, N.; Laryea, S.W.; Wiafe, G. Video observation of waves and shoreline change on the Microtidal James Town Beach in Ghana. *J. Coast. Res.* **2016**, 1022–1026. [CrossRef]

24. Holland, K.T.; Holman, R.A.; Lippmann, T.C. Practical use of video imagery in near-shore oceanographic field studies. *IEEE J. Ocean. Eng.* **1997**, *22*, 81–92. [CrossRef]

25. Almar, R.; Ranasinghe, R.; Senechal, N.; Bonneton, P.; Roelvink, D.; Bryan, K.; Marieu, V.; Parisot, J.P. Video-based detection of shorelines at Complex Meso–Macro Tidal Beaches. *J. Coast. Res.* **2012**, *28*, 1040–1048. [CrossRef]

26. Almar, R.; Cienfuegos, R.; Catalán, P.A.; Michallet, H.; Castelle, B.; Bonneton, P.; Marieu, V. A new breaking wave height direct estimator from video imagery. *Coast. Eng.* **2012**, *61*, 42–48. [CrossRef]

27. Almar, R.; Senechal, N.; Bonneton, P.; Roelvink, D. Wave celerity from video imaging: A new method. Proceedings of Coastal Engineering 2008, Hamburg, Germany, 31 August–5 September 2008; pp. 661–673.

28. Almar, R.; Michallet, H.; Cienfuegos, R.; Bonneton, P.; Ruessink, B.G.; Tissier, M. On the use of the radon transform in studying nearshore wave dynamics. *Coast. Eng.* **2014**, *92*, 24–30. [CrossRef]

29. Rascle, N.; Ardhuin, F. Global wave parameter data base for geophysical applications. Part II: Model validation with improves source term parameterization. *Ocean Model.* **2013**, *70*, 145–151. [CrossRef]

30. Larson, M.; Hoan, L.X.; Hanson, H. Direct formula to compute wave height and angle at incipient breaking. *J. Waterw. Port Coast. Ocean Eng.* **2010**, *136*, 119–122. [CrossRef]

31. Boak, E.H.; Turner, I.L. Shoreline definition and detection: A review. *J. Coast. Res.* **2005**, *21*, 688–703. [CrossRef]

32. Aarninkhof, S.G.J.; Turner, I.L.; Dronkers, D.T.; Caljouw, M.; Nipius, L. A video-based technique for mapping intertidal beach bathymetry. *Coast. Eng.* **2003**, *49*, 275–289. [CrossRef]

33. Hamed, K.H.; Rao, A.R. A modified Mann-Kendall trend test for autocorrelated data. *J. Hydrol.* **1998**, *204*, 182–196. [CrossRef]

34. Tolman, H.L. Limiters in third-generation wind wave models. *Glob. Atmos. Ocean Syst.* **2002**, *8*, 67–83. [CrossRef]

35. Woodcock, F.; Greenslade, D.J.M. Consensus of numerical model forecasts of significant wave heights. *Weather Forecast.* **2007**, *22*, 792–803. [CrossRef]

36. Ranasinghe, R.; Holman, R.; de Schipper, M.A.; Lippmann, T.; Wehof, J.; Minh Duong, T.; Roelvink, D.; Stive, M.J.F. Quantification of nearshore morphological recovery time scales using Argus video imaging: Palm Beach, Sydney and Duck, NC. *Coast. Eng. Proc.* **2012**, *1*, 24.

water

MDPI

Article

The June 2016 Australian East Coast Low: Importance of Wave Direction for Coastal Erosion Assessment

Thomas R. Mortlock [1,*], Ian D. Goodwin [2], John K. McAneney [1] and Kevin Roche [1]

[1] Risk Frontiers, Macquarie University, North Ryde, NSW 2109, Australia;
     john.mcaneney@mq.edu.au (J.K.M.); kevin.roche@mq.edu.au (K.R.)
[2] Marine Climate Risk Group, Department of Environmental Sciences, Macquarie University,
     North Ryde, NSW 2109, Australia; ian.goodwin@mq.edu.au
*   Correspondence: thomas.mortlock@mq.edu.au

Academic Editor: Sylvain Ouillon
Received: 5 December 2016; Accepted: 6 February 2017; Published: 14 February 2017

Abstract: In June 2016, an unusual East Coast Low storm affected some 2000 km of the eastern seaboard of Australia bringing heavy rain, strong winds and powerful wave conditions. While wave heights offshore of Sydney were not exceptional, nearshore wave conditions were such that beaches experienced some of the worst erosion in 40 years. Hydrodynamic modelling of wave and current behaviour as well as contemporaneous sand transport shows the east to north-east storm wave direction to be the major determinant of erosion magnitude. This arises because of reduced energy attenuation across the continental shelf and the focussing of wave energy on coastal sections not equilibrated with such wave exposure under the prevailing south-easterly wave climate. Narrabeen–Collaroy, a well-known erosion hot spot on Sydney's Northern Beaches, is shown to be particularly vulnerable to storms from this direction because the destructive erosion potential is amplified by the influence of the local embayment geometry. We demonstrate the magnified erosion response that occurs when there is bi-directionality between an extreme wave event and preceding modal conditions and the importance of considering wave direction in extreme value analyses.

Keywords: East Coast Low; nearshore processes; coastal erosion; coastal management; climate change; numerical modelling; Southeast Australia

1. Introduction

East Coast Low (ECL) storms bring heavy rain, strong winds and powerful coastal wave conditions to the Southeast Australian coast. They are often responsible for significant beach erosion and lowland inundation and pose a threat to coastal infrastructure and public safety. The 'Pasha Bulker' storm in June 2007, for example, was responsible for normalised insurance losses of AUD$1.97 billion [1,2]; economic losses arising from that event are likely to have been at least double this figure.

ECLs are a common feature across the Southern Hemisphere extra-tropics [3,4]. In Australia, they typically form in the late Austral autumn to early winter and are a regular feature of the winter climate [5]. They typically bring storm wave conditions over a 3-day period before decaying eastwards into the Tasman Sea [6]. Their steep build up to peak storm wave conditions makes them one of the more dangerous weather systems affecting the New South Wales coast (NSW) and poses difficulties for forecasting.

The weather pattern of the June 2016 event was unusual in the context of the last few decades and is referred to as a "Black Nor'easter" because the black skies and north-easterly winds recorded by mariners in the late 1800s. The heavy rainfall and unusual wave direction coincided with some of the highest tides of the year further amplifying impacts at the coast. During most ECLs, the rotation of the low-pressure cell (clockwise flow in the Southern Hemisphere) and extra-tropical origin usually produce a south to south-easterly wave direction. The June 2016 event, however, produced east to north-east storm waves along the entire east coast (~2000 km of shoreline length) (Figure 1). Wave heights increased from north to south and the Eden buoy (far south coast of NSW) recorded a maximum individual wave height of 17.7 m—the largest wave ever recorded along the NSW coast [7]. Peak storm significant wave heights offshore of Sydney were considerably smaller (1-hourly H_s 6.4 m) yet still led to significant erosion at the coast.

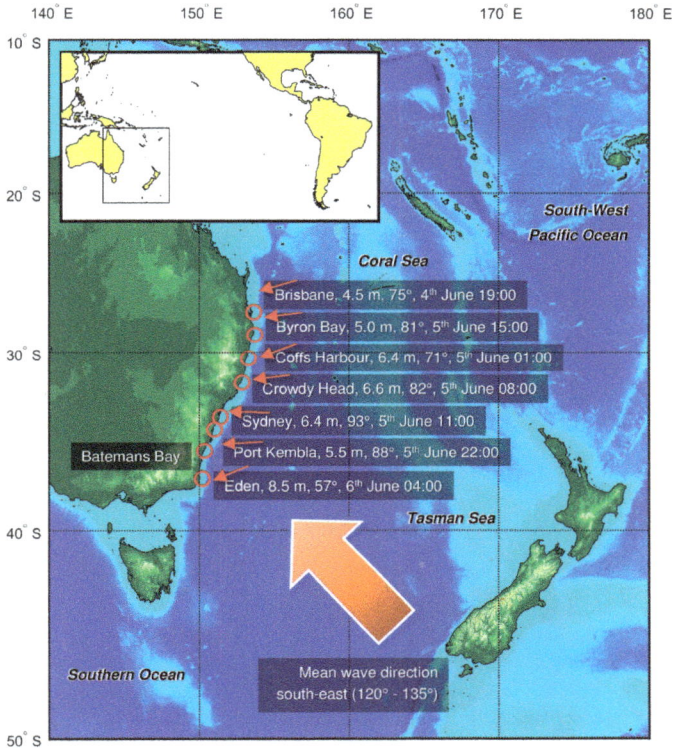

Figure 1. Map of the Tasman and Coral Seas region with locations of waverider buoys in Southeast Australia (red circles). The Tasman Sea borders the east coast of Australia, and extends west to New Zealand. It extends north to where it meets the Coral Sea at approximately 30° S [8]. The peak storm hourly significant wave height, mean wave direction, and time of peak storm wave conditions during the June 2016 event, are given for each buoy location. The Batemans Bay buoy data transmission failed during the event [7]. Red arrows illustrate peak storm wave direction. Mean wave direction (range of averages for all buoys over portions of records having directional observations) is also shown. Inset shows the study region in relation to the Pacific Basin.

The north-easterly wave direction was the result of a hybrid Anticyclonic Intensification (AI)/Easterly Trough Low (ETL) synoptic weather pattern as described in [5]. The low-pressure cell developed over the Coral Sea and northern Tasman Sea and was directed down the east coast by a strong blocking anticyclone which intensified over the South Island of New Zealand (Figure 2). The long fetch of approximately 1500 km produced sustained long wave periods (in excess of 14 s) that are unusual for the Tasman Sea from this direction and further contributed to powerful wave conditions at the coast.

Storm wave events from the north-east are uncommon, having occurred on average twice per decade at Sydney over the past 40 years [9]. The long-term mean annual wave direction is 135° (south-easterly), and waves approaching clockwise of east constitute almost 70% of observations. The south-easterly wave climate is moderately oblique to the shoreline and leads to increasing south-to-north alongshore wave energy and littoral transport [10]. The beach morphology is equilibrated to this wave energy gradient [11] and provides a natural buffer to erosion when the storm wave direction is similar to the modal (non-storm) wave direction. A more extreme erosion response occurs when the storm and modal wave directions are different, as occurred in June 2016. Even a relatively subtle change in wave obliquity can make a big difference to the magnitude and direction of littoral transport, as demonstrated empirically [12], and can drive a significant shoreline response on wave-dominated sandy coasts [13].

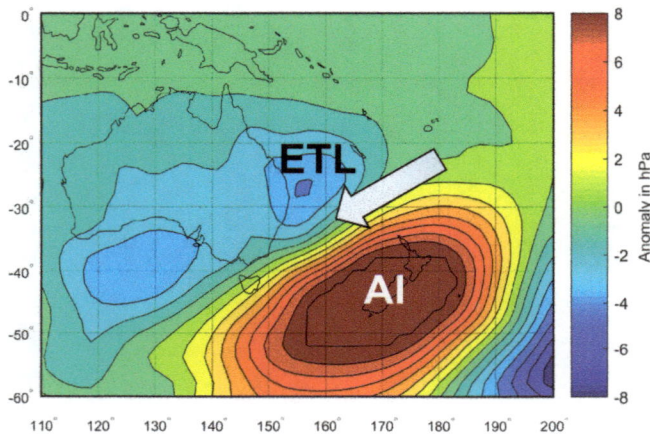

Figure 2. Composite sea level pressure (SLP) anomaly (in hectopascals, hPa) for all east-northeast Anticyclonic Intensification (AI)/Easterly Trough Low (ETL) storm wave events since 1974, using events identified from buoy records in Shand et al. [9], and the European Centre for Medium-Range Weather Forecasts (ECMWF) ERA 20th Century reanalysis (ERA-20C) [14]. The composite anomaly was calculated relative to the ERA-20C long-term mean (1900–2010).

One of the worst affected areas was Collaroy, a suburb situated at the south end of the Narrabeen–Collaroy embayment on Sydney's Northern Beaches (Figure 3). Beach erosion and accretion trends have been monitored here since 1976 [15] and have been shown to be synchronous with other beach compartments along the NSW coast [16,17]. Despite being well recognised as an erosion hot spot, the shorefront of Collaroy is characterised by a legacy of inappropriate development into the active beach zone [18]. How to best to manage this legacy poses a significant challenge and fraught policy area for governments [19]. Management decisions are purportedly based on cost-benefit analyses, which are chiefly controlled by the value of the land and property at risk. As population density and demand for coastal property continues to increase, the cost of not providing ongoing protection also rises. Contemporaneously, the hazard threatening coastal communities is also likely to increase with climate change.

The primary focus of our study is to demonstrate the importance of storm wave direction for coastal erosion impacts, using Narrabeen–Collaroy as an example of regional significance. The hydrodynamics controlling the erosion risk are examined, by comparing impacts from the June 2016 event with another ECL event that impacted Sydney in April 2015 but with a different offshore wave direction. A re-assessment of extreme wave conditions at Sydney, combined with nearshore buoy observations, also demonstrates the importance of wave direction considerations when assessing erosion potential in a changing climate.

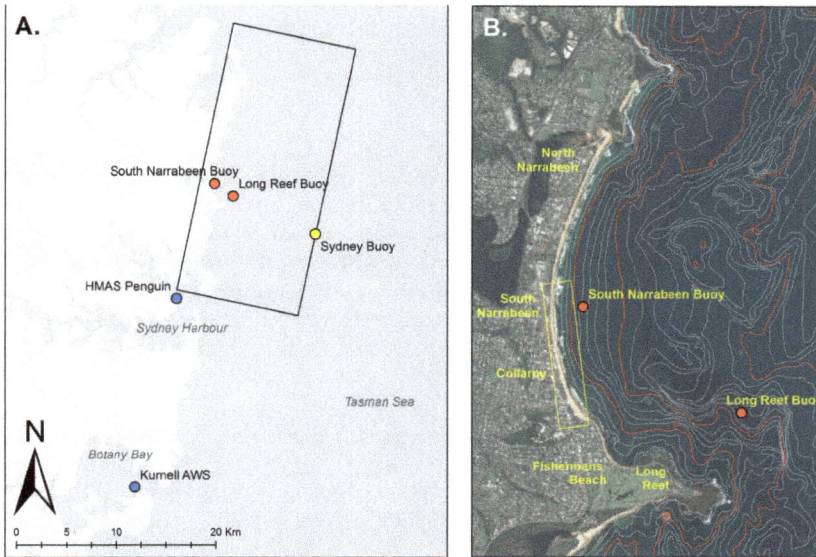

Figure 3. (**A**) Regional map showing approximate extents of numerical model domain (black box); locations of nearshore wave buoys (red); location of Sydney wave buoy (yellow); and locations of tide and wind stations used in this study (blue); and (**B**) local map of Narrabeen–Collaroy beach with bathymetry (5, 15 and 30 m isobaths in red); locations of nearshore wave buoys shown in (**A**); and approximate area of erosion damage at Collaroy during the June 2016 storm (yellow box).

2. Observational Data

2.1. Wave Conditions

Hourly parametric wave data were sourced over the period 3 to 10 June 2016 from observations at the Sydney waverider buoy located approximately 9 km off the Narrabeen–Collaroy embayment in 90 m water depth (Figure 3). The significant wave height, H_s (m), maximum wave height, H_{max} (m), peak wave period at the first and second spectral peak, Tp_1 and Tp_2 (s) and mean wave direction at the first and second spectral peak, MWD_{Tp1} and MWD_{Tp2} (degrees true north coming from) were the wave parameters used. The long-term parametric wave record from the Sydney buoy (non-directional since 1987, directional since 1992) was also used in this study. Observations from two other wave buoys deployed inside the Narrabeen embayment (Figure 3) between August and November 2011 were also used. These buoys were deployed at South Narrabeen (~12 m depth) and at Long Reef (~20 m depth). The South Narrabeen buoy was directional, while the Long Reef buoy was not.

2.2. Tide Conditions

Tide levels over the duration of the storm were taken at 15-min intervals from the gauge at the naval base of Her Majesty's Australian Ship (HMAS) Penguin, Middle Head, Sydney Harbour (Figure 3). This station is considered representative of the open coast tide. Tides are semi diurnal and micro tidal (mean range of 1.3 m) along the NSW coast and approach almost perpendicular to the shelf, so that geographical differences in the timing of high and low tides across the state are negligible. Both forecast and observed tidal data were used to measure the surge component of the water level.

2.3. Wind Conditions

Half-hourly wind observations of velocity (m/s) and direction (degrees true north coming from) were obtained from the Kurnell weather station, 10 m above sea level, in Botany Bay (Figure 3). In the absence of over-sea wind measurements, Wood et al. [20] found the wind climate at Kurnell to best represent coastal ocean winds for the Sydney region.

2.4. Coastal Erosion

Pre- and post-storm beach surveys (above the water line) were undertaken at Narrabeen–Collaroy by University of New South Wales as part of a long-term beach monitoring program [15]. Shore-normal transects were surveyed at five locations along the beach using GPS, each extending from the dune behind the beach down to the water line. Pre- and post-storm hydrographic surveys (below the water line) were also undertaken at Narrabeen–Collaroy by NSW Office of Environment and Heritage [21]. Both surveys used a jet ski bottom-mounted single-beam echo sounder, narrow-beam transducer and GPS system and covered the shoreface from approximately 2 m to 15 m water depth with 50 m-spaced shore-normal survey lines.

3. Numerical Modelling

Wave, water level and wind observations over the duration of the storm were input to a coupled wave, flow and morphological model of the Sydney region to investigate coastal processes associated with the June 2016 event. A MIKE 21/3 Coupled Model developed by the Danish Hydraulic Institute (DHI) was used [22]. In this instance, the modelling system coupled a spectral wave model with a hydrodynamic flow model and a sediment transport model to simulate waves, currents, water levels and bed elevation change during the storm.

The wave model (MIKE 21 Spectral Wave, SW) simulates the growth, decay and transformation of wind-generated waves and swell in coastal locations. Directionally-decoupled and quasi-stationary formulations were used. Wave- and wind-driven currents and water level variations drive the hydrostatic flow model (MIKE 21 Flow Model), which is based on the two-dimensional Reynolds-averaged Navier–Stokes equations. Hydrodynamic conditions from the flow model and wave radiation stress terms from the wave model were used as input to the sand transport model (STPQ3D) to simulate morphological changes during the storm. The effect of bed ripples, bed slope, cross-current transport, rips and undertow were all included in the calculations. All model components were dynamically coupled, in other words a full feedback between the morphology, waves and currents took place at each time step. For a more detailed description of coupled model physics, the reader is referred to [22].

3.1. Model Bathymetry

The model domain covered the Sydney Northern Beaches area from North Head, Sydney Harbour in the south to Palm Beach in the north (Figure 3) and extended offshore to the 90-m isobath where the Sydney wave buoy is located. A mosaic of best-available bathymetries was used to generate a seabed topography in the model, including the pre-storm hydrographic survey undertaken at Narrabeen–Collaroy [21]. Bathymetric data also included a series of Single-Beam Echosounder (SBES) surveys of the upper shoreface of the Sydney Northern Beaches area undertaken by the Office of Environment and Heritage (OEH) between 2011 and 2016 (average 50 m line spacing); a Multi-Beam Echosounder (MBES) survey of the lower shoreface area offshore of the Northern Beaches undertaken by OEH in 2014 (5 m point density); and a series of SBES survey lines of the inner shelf offshore of the Northern Beaches area undertaken by the Royal Australian Navy (RAN) in the 1970s (average 200 m line spacing). The bathymetric data was mapped onto a flexible computational mesh, composed of a series of irregular Delaney triangles (elements), using a natural neighbor interpolation method. The interpolated bathymetric mesh was then refined based on a depth/gradient ratio to improve the computational resolution for shallow water and areas of complex subaqueous reef. This led to a computational mesh with an average element length of 200 m offshore (in approximately 90 m water depth), grading to 20 m at the shoreline. An approximate 20 m resolution at the shoreline was considered sufficient to resolve cross- and along-shore transport processes and ensured all available bathymetry soundings were used.

The coastal planform and elevation of the subaerial beach was idealized using the pre-storm cross-shore profile data from the Narrabeen beach monitoring program [15], and recent aerial photography. The dry beach was included in the model to provide a sediment source for storm cut and for flooding and drying at the shoreline.

3.2. Model Boundary Conditions

Boundary wind, wave and tide data were derived from point-source measurements close to or within the model domain, as described in Section 2. These time-series data (hourly waves, half-hourly winds, and 15-min tides) were applied at all three open boundaries of the model (south, east and north), apart from the waves which were only applied on the east (ocean) boundary. A depth-integrated approximation based on linear wave theory [22] was used for wave forcing on both lateral boundaries (north and south) to estimate waves entering the two sides of the model domain. Sensitivity testing ensured that the model domain was sufficiently wide to avoid potential errors propagating from the lateral boundaries into the area of interest. A computational time step of 15 min was used to resolve all boundary conditions, and a 12-h model spin up period was imposed.

An inherent assumption of the model was that the measured waves, tides and winds were representative of, and constant along, the length of the model boundaries on which they were applied. The east (ocean) boundary ran through the moored location of the Sydney waverider buoy and followed the same isobath (90 m). Wind data was sourced from the most representative source and was chosen over other wind hindcasts as the latter can under-estimate near-coast wind fields in this area [20,23]. Because an atmospheric model was not used, the wind field was spatially constant throughout the model domain for each time step, so localized variations were not captured. Tide measurements were applied as variations in the surface water elevation at each boundary. This included a barometric surge element as measured on top of the astronomical tide. Local wind and wave set-up were generated by the model.

Parametric wave data from the Sydney buoy (H_s, Tp_1, MWD_{Tp1}) were used assuming a JONSWAP spectrum as full spectral conditions from the buoy during the storm were unavailable at the time of writing. Thus, any bi-modality during the storm was not replicated in the model and is likely to have contributed to residual errors after calibration (Section 3.5). Directional spreading of the parametric wave data was estimated for each time step based on wave-age curves generated using the method of [24]. The entire Sydney buoy record (1987–2016) was used to approximate wave steepness limits of wind-sea, intermediate sea-swell, and swell, based on H_s and Tp_1 (Figure A2). All wave events classed as wind-sea were assigned a spreading of ~25°, intermediate sea-swell ~20°, and swell ~15°, after [22]. From Figure A2, it is interesting to note that most wave conditions during this event were theoretically swell waves because of the unusually long wave periods.

3.3. Model Bed Characteristics

Bed resistance, bottom friction and bed layer thickness varied spatially within the model domain in accord with a comprehensive set of observations of subaqueous reef areas and sediment samples collated in [25]. In this way, the influence of the roughness and non-erodibility of rock reefs was accounted for in the model. An estimate of the physical roughness height of rock reef was derived from Light Detection And Ranging (LiDAR) imagery of low-tide exposed rock platforms around Long Reef headland. Transects were taken through the reef sections, following the method of [26]. Results ranged from 0.08 m to 0.33 m with a mean of 0.16 m. These values are similar to those reported for coral reef platforms (e.g., 0.16 m [27]) with the upper values reflecting the greater rugosity of bare rock reef than that of coral. The mean value of 0.16 m was used to calculate bed resistance effects for all areas of rock reef within the model domain. A value of 0.05 m was used elsewhere to account for the effect of small sand ripples on flow resistance [22,28].

To describe the bottom frictional effects of reef, the physical roughness height, k_w, was converted to a hydraulic (Nikuradse) roughness value, k_n, using the approximation $k_n \approx 2.k_w$ [29], giving a k_n of 0.32 m. Using similar logic, Huang et al. [30] arrived at a comparable value for k_n of 0.27 m for coral reef. For non-reef areas, a k_n of 0.04 m was used to approximate the effect of small sand ripples on the frictional dissipation of wave energy [31].

Where reef areas existed, the bed thickness was set to zero to ensure that no erosion could take place. Elsewhere, an infinite thickness was used. At the land boundary, areas of reef headland were assigned a reflection coefficient of 0.8 (1.0 being fully reflective) to describe the intensity of a reflected wave heights relative to incident wave heights against a headland [22]. Full wave reflection effects (such as standing waves) were not accounted for. Beach sections were fully absorptive.

A cross-shore gradation in the median grain diameter, d_{50}, was used based on observations in [25]. d_{50} varied from 0.33 mm on the sub-aerial beach, to 0.22 mm in the surf zone, coarsening thereafter to 0.48 mm on the lower shoreface. Observations were not of sufficient density to derive any along-shore variation in grain size. Sediment samples from various sources collated in [25] suggest the grain size distribution around d_{50} did not vary significantly in a cross-shore direction, thus, a single grading coefficient of 1.34 was used. A porosity of 0.4 was used consistent with that of quartz sand. Carbonate material present in beach sediments along the Northern Beaches is likely to vary the porosity (and specific density) of sand in a cross-shore direction, but is not included here.

3.4. Model Calibration

The sensitivity of the model to varying mesh resolution, seabed substrate type, wave breaking, directional resolution and wave theory were examined. A Brier Skill Score (BSS) was used to assess the model's morphological performance, and by inference the nearshore flow field, using observed bathymetric change from pre- and post-storm hydrographic surveys undertaken at Narrabeen–Collaroy [21]. A skill score was calculated for each mesh element within the survey area so the spatial variance in model error could be assessed. The scoring method and classification system described in Sutherland et al. [32] was used.

The most significant improvement in model skill was obtained by increasing the resolution of the computational mesh from the 30-m isobath to the shoreline. Changes to the mesh resolution seaward of this had negligible effects on model performance. A coarse resolution mesh grading from an approximate element length of 250 m along the 30-m isobath to 50 m at the shoreline, returned a median BSS across the survey area of −5 ('Bad'). A more highly-resolved mesh covering the whole of the Narrabeen–Collaroy embayment (from 30-m isobath to shoreline) at a 20-m resolution, dramatically improved the median BSS to 0.22 ('Good').

The cost of not including areas of rock reef in the model, and their effect on flow resistance, bottom friction and erodible layer thickness, was a reduction in model skill from 0.22 ('Good') to 0.12 ('Fair'). A wave breaking parameter, γ, of 0.9, best replicated the observed volume in the surf zone bar, as has been also found by [33] in a similar setting. Increasing the directional resolution of the wave forcing (from 10° to 5°) made no significant improvement, perhaps because the coarser binning in part compensated for deficiencies in the directionality of boundary wave forcing (Section 3.2).

The model sensitivity to two wave theories was tested; the first using a combination of Stokes and Cnoidal fifth-order classic wave theories, and another combining the semi-empirical wave theories of Doering and Bowen [34] and Isobe and Horikawa [35]. The Stokes/Cnoidal theories led to a wider area of small-scale bed lowering seaward of the surf zone bar but did not replicate the accretion volume in the bar as well as the Doering and Bowen/Isobe and Horikawa formulation. Thus, the latter was used here.

3.5. Model Verification

The wave component of the model was verified against the four months of directional, hourly wave buoy observations collected from the South Narrabeen and Long Reef nearshore wave buoys (described in Section 2.1, locations shown in Figure 3). The model showed good predictive skill for significant wave height (R^2 = 0.9, slope = 0.9, at both buoy locations) and mean wave direction (R^2 = 0.8, slope = 0.9) but exhibited a small positive bias in the mean wave period (1–2 s) at both buoy locations.

After calibration, the morphological model attained a median skill score across the surveyed area of 0.35, a figure considered 'Good' [32]. In describing the location and geometry of the surf zone storm bar, in most places the skill score exceeded 0.7, a rating considered 'Excellent' (Figure 4). However, morphological change at the southern end of the embayment was not well replicated. This may have been because data transmission issues at the buoy during the storm [7] meant the wave direction and period around the storm peak are interpolated values. Moreover, there was no spectral wave data available at the time of writing thus wave bi-modality was not accounted for in the model. These factors collectively may have led to an under-representation of north-easterly wave energy responsible for bar build-up at the south end of the embayment.

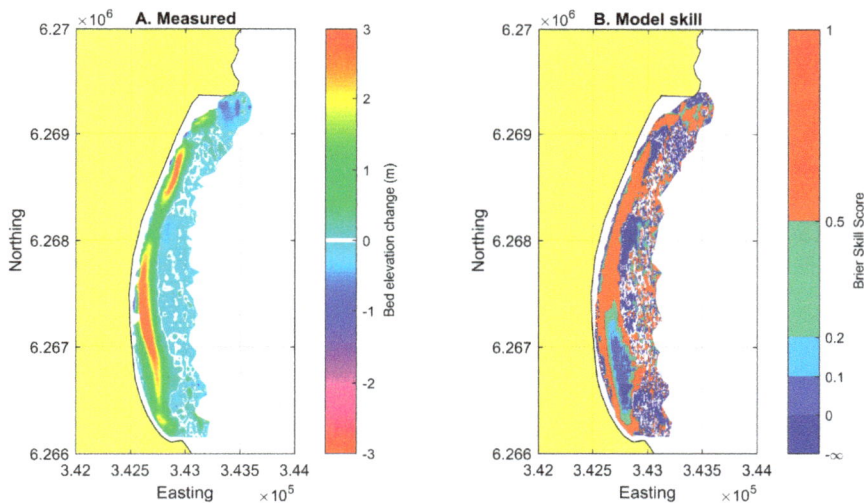

Figure 4. Measured morphological change (in metres) during the June 2016 storm event (**A**) and model skill in replicating observed change (**B**). The pre- and post-storm surveys in (**A**) extend from ~2 m to ~15 m water depth. The colour scheme in (**B**) reflects the Brier skill assessment of [32]. Morphological change within the vertical error of the surveyed data (±0.03, [21]) was omitted from both plots.

4. Coastal Conditions at Sydney

4.1. Peak Storm Conditions

Storm peak wave conditions coincided with the winter solstice spring tide, a period of strong onshore winds and heavy rainfall. Figure 5 shows the wave heights and water levels observed at Sydney during the storm. Wave period and direction and wind speed and direction observations are given in the Appendix A (Figure A1).

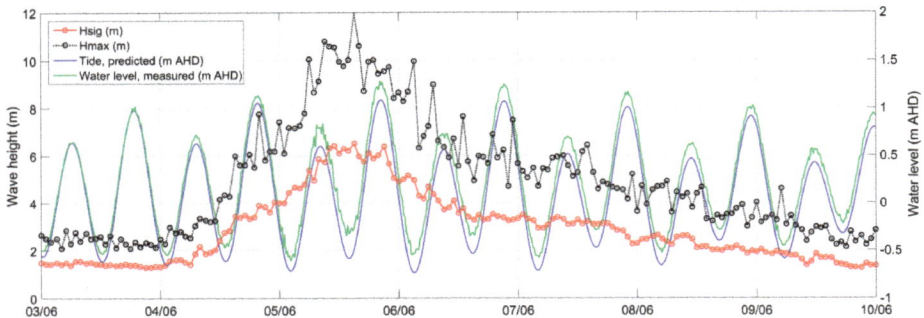

Figure 5. Observations of significant and maximum hourly wave heights and 15-min water levels observed and predicted at the Sydney wave buoy and the tide gauge at Her Majesty's Australian Ship (HMAS) Penguin, Sydney Harbour, from 3 to 10 June 2016. Water levels are given relative to Australian Height Datum (AHD) which approximates to the mean sea level.

A maximum water level of 1.29 m Australian Height Datum (AHD) was recorded at the tide gauge at HMAS Penguin around 20:00 on 5 June, which represents an Average Recurrence Interval (ARI) of approximately 2.5 years [7]. This included a positive tidal residual of up to 0.34 m (storm surge). Water levels on the open coast would have variably been higher than this because of wave and wind set up. Modelling here suggests that maximum water levels may have exceeded 1.7 m AHD around 21:00 at Long Reef headland and North Narrabeen (comprising a ~0.6 m surge). This does not include wave run up which can easily be an order of magnitude greater than the surge component [36]. Onshore winds observed at the Kurnell weather station averaging 15 m·s^{-1} (~30 knots) were sustained for almost 20 h leading up to and during the storm peak, with gusts of up to 24 m·s^{-1} (~60 knots).

Storm wave conditions at Sydney (defined as the hourly H_s exceeding 2 m for a period of 72 h or more, after [9]) began around midday on Friday 4 June and lasted until late evening on Wednesday 8 June (106 h). A peak storm wave height (H_s) of 6.4 m was recorded at 11:00 on Sunday 5 June. H_s then remained around 6 m for the next 10 h (until 21:00), after which it began to gradually decline. Maximum wave heights (H_{max}) exceeded 8 m between 06:00 on the Sunday to 03:00 the following morning peaking at 12 m at 15:00 on Sunday afternoon.

Some wave period and direction information was missing due to transmission problems at the buoy over the peak of the storm [7], but interpolated values suggest a storm peak direction (MWD_{Tp1}) of around 93°, and period (Tp_1) of around 11.5 s. While the exact variation of MWD_{Tp1} and Tp_1 over the period of missing data is unknown, the interpolated values are in broad agreement with the peak direction and period of other mid-shelf wave buoys along the coast (Figure 1). As the storm intensified, the wave period at Sydney lengthened to around 14 s and the wave direction simultaneously rotated clockwise from north-east to east. Over this period, the synoptic weather pattern suggests that the deep-water wind and wave field was still north-east (Figure 6C,D).

The more easterly direction recorded at the buoy may be a result of the longer-period waves having already refracted around towards shore-normal by the time they reached the buoy location. Wave base—the depth at which surface waves begin to be influenced by the sea bed—was around 150 m, suggesting that waves had already refracted on the edge of the continental shelf far seawards of the buoy location (on the 90-m isobath).

Synoptic storm pattern
3ʳᵈ - 9ᵗʰ June 2016 (AEST)

Sea Level Pressure (Pascals)

-3000.0   -1800.0   -600.0   600.0   1800.0   3000.0

A. FRI 3/6 10:00   **B.** SAT 4/6 10:00   **C.** SUN 5/6 10:00   **D.** MON 6/6 10:00

E. TUE 7/6 10:00   **F.** WED 8/6 10:00   **G.** THU 9/6 10:00

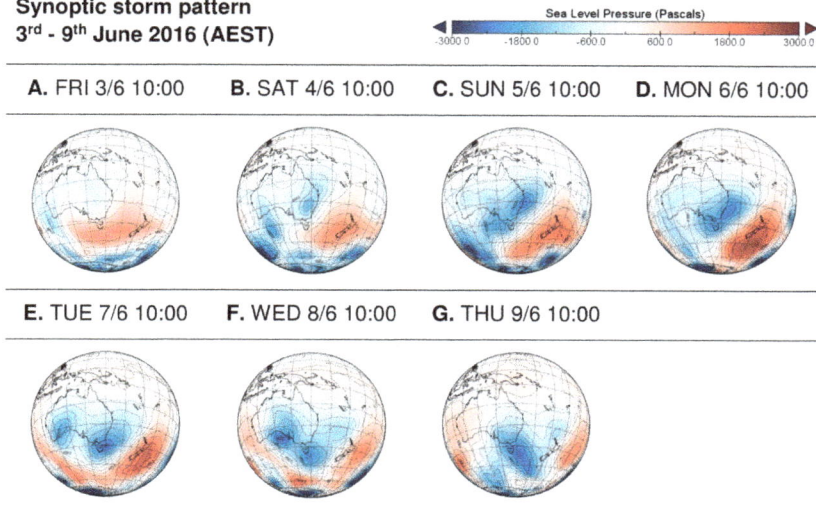

Figure 6. Sea level pressure patterns over the Tasman Sea every 24 h from 3 to 9 June 2016 from the NCEP/NCAR reanalysis at a 2.5° resolution [37]. High pressure (red) is associated with anti-cyclonic winds and low pressure (blue) is associated with cyclonic winds. The low-pressure trough that formed in (**A,B**) was directed down the length of the New South Wales coast in (**C,D**) and into Tasmania in (**E**) by the anticyclone that intensified over New Zealand. Only when the anticyclone migrated eastwards into the South-West Pacific in (**F**) could the low-pressure cell move off the coast into the South Tasman Sea (**G**).

4.2. Post Storm Conditions

As the low-pressure trough tracked south down the coast, the mean wave direction at Sydney rotated clockwise from around 70° (East–North–East) at the start of the storm to around 110° (East–South–East) by the early morning of Monday 6 June. Up until this point, buoy observations showed that wave conditions were largely uni-modal, meaning most wave energy was travelling at the same speed from the same direction. This suggests that wave conditions during the peak of the storm were generated from a single source (the low-pressure system). This can be seen in Figure A1 where wave periods and directions at the first and second spectral peaks have similar values. As the storm moved away from Sydney during Monday, the parametric wave data suggests that conditions became bi-modal as the amount of wave energy generated by the low-pressure system was replaced by longer period (Tp_1–14 s) swell from the east. Figure 6D suggests that this long-period easterly swell was generated off the northern limb of the anticyclone situated over New Zealand. Long wave periods between 12 and 14 s were sustained for the following 96 h as wave heights decreased. At the same time, very oblique (southerly) waves were still being produced by the low-pressure system as it tracked south down the east coast (MWD_{Tp2}, Figure A1A).

5. The Importance of Wave Direction for Coastal Risk

5.1. Wave Direction Control on Nearshore Wave Heights

Traditional assessments of coastal risk relate the offshore wave height to beach erosion and inundation with secondary, or no, consideration of wave direction. The June 2016 event highlighted the importance of storm wave direction for coastal impacts at Sydney. The peak storm offshore wave height (6.4 m) was unremarkable yet wave energy conditions at the coast were extra-ordinary (Figure 7).

Figure 7. Nearshore wave conditions at South Narrabeen on Monday 6 June 2016. Authors' and photographer's visual estimates of breaking wave heights from these images are between 4 and 5 m (~13 to 16 ft.). Reproduced with permission from Mark Onorati [38]. The images in (**A,C**) are of the beach section behind Narrabeen Fire Station (red brick building in foreground in (**C**)) approximately half the way along the Narrabeen-Collaroy embayment; the image in (**B**) is 100 m north of (**A,C**) adjacent to Robertson Street.

The high-energy coastal wave conditions were a result of the unusual offshore wave direction rather than large offshore wave heights. Concurrent wave data previously recorded at the Sydney (mid-shelf) buoy and the nearshore buoy at South Narrabeen (over the period August to November 2011—not during the June 2016 storm), suggest that waves from the north-east to east undergo less energy dissipation (reduction in wave height) and refraction (change in wave direction) when travelling across the continental shelf towards the coast than do waves approaching from the south-east to south (Figure 8). The dissipation coefficients in Figure 8A were calculated as the nearshore H_s divided by the offshore H_s. Likewise, the refraction coefficients in Figure 8B were calculated as nearshore MWD_{Tp1} divided by the offshore MWD_{Tp1}. These values were then normalized between 0 and 1 to make rates of dissipation and refraction comparable across wave directions.

Nearshore wave heights (seaward of shoaling in shallow water) are usually smaller than offshore wave heights because wave energy is dissipated across the shelf and shoreface, principally because of friction with the seabed. The South Narrabeen buoy was moored in ~12 m water depth, meaning most waves recorded at this location were only weakly shoaled and unbroken. Thus, most nearshore wave heights recorded over the observation period were lower than the offshore (mid-shelf) wave height because of energy dissipation across the shelf.

However, the rate of dissipation is often a function of wave direction. Figure 8A indicates that energy dissipation is lowest for waves from the north-east to east, and increases as the wave direction rotates towards the south. Likewise, Figure 8B shows that waves from the south-east to south undergo most refraction to reach the Sydney coast.

The dissipation and refraction coefficients (red lines Figure 8A) were applied to the entire directional wave record at the mid-shelf buoy (1992–2016) to obtain the likely long-term wave height distribution (Figure 8C) and probability of occurrence of wave directions (Figure 8D) at Narrabeen, compared to the offshore wave data. Figure 8C shows that the highest waves recorded offshore come from 160° to 190°, while the highest waves at Narrabeen (centre of embayment, ~12 m water depth) are from 70° to 90°. Similarly, Figure 8D shows that the most frequent waves offshore are from 160° to 180° while the most frequent waves at Narrabeen are from 90° to 110°.

This highlights two important points: first, that the offshore wave observations are not a good representation of the nearshore wave climate at Sydney. Second, that there is a strong directional control on coastal wave conditions at Sydney, which explains how extreme surf zone waves, as shown in Figure 7, can occur for modest offshore storm wave heights when approaching from the north-east to east.

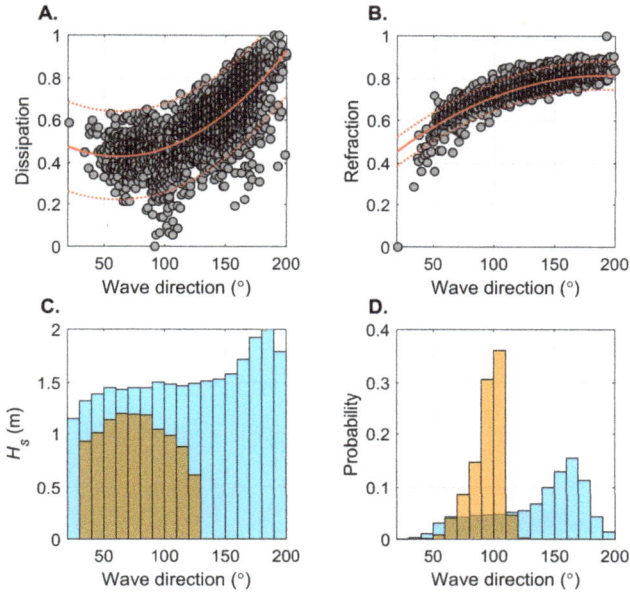

Figure 8. Rates of energy dissipation (**A**) and refraction (**B**) for waves approaching Narrabeen between August and November 2011. Wave direction on the *x*-axis relates to the offshore (mid-shelf) wave direction as measured at the Sydney buoy. Polynomial functions (solid red lines) describe the dissipation/refraction curves with 95% confidence intervals (dashed red lines). The long-term wave height distribution (**C**) and occurrence probability of wave directions (**D**) are shown for the Sydney offshore (blue, observed) and Narrabeen nearshore (orange, modelled) for the period 1992–2016. The Narrabeen nearshore data was obtained by applying the dissipation and refraction coefficients in (**A,B**) to the offshore data.

5.2. Wave Direction Control on Nearshore Hydrodynamics

As shown above, the Sydney coast is most exposed to north-east to easterly wave energy, despite south-east to southerly waves being more powerful and frequent offshore. Modelling here shows that the geometry and aspect of the Narrabeen–Collaroy compartment further increases the vulnerability of the Collaroy beachfront to north-east to easterly storm waves because of the effect of wave direction on the nearshore hydrodynamics.

To demonstrate this, the model was run for the June 2016 storm and for another ECL event that impacted Sydney between 19 and 23 April 2015. The April 2015 event had a significantly larger peak storm wave height (H_s 8.0 m) and some of the strongest-ever winds recorded at the Kurnell weather station with gusts exceeding 130 km·h^{-1}. It also produced a more 'regular' south-easterly storm wave direction (145° at the storm peak) than in June 2016. The wave refraction pattern, nearshore current field and sediment transport patterns at Narrabeen–Collaroy at the peak of the storm in April 2015 and June 2016 are shown in Figures 9A–C and 10A–C, respectively. The modelled bathymetric change after each storm event is shown in Figures 9D and 10D.

April 2015 storm   *peak H_{sig} 8.0 m, deepwater wave direction SE*

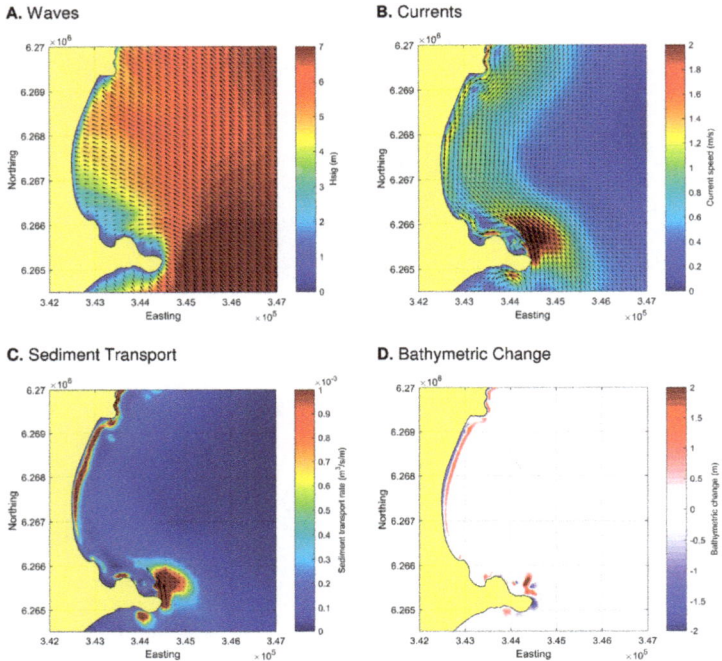

Figure 9. Modelling results for (**A**) significant wave height and mean wave direction; (**B**) current velocity and direction (**C**) sediment transport rate and direction; and (**D**) bathymetric change at Narrabeen–Collaroy during the April 2015 storm event. Results shown in (**A**–**C**) are at the storm peak; while (**D**) is the cumulative change post-event. Bathymetric change in (**D**) less than ±0.03 m was omitted, consistent with limits of model calibration (Section 3).

The modelled flow and transport pattern in June 2016 was very different. The easterly nearshore wave direction produced a southward, rather than northward, alongshore current for almost all beaches on Sydney's Northern Beaches. The only exception to this regional pattern was the Long Reef to Collaroy coastal section where a strong northward littoral current exceeding 2 ms^{-1} at the peak of the storm occurred (Figure 10B). This current met a powerful southward current around Collaroy/South Narrabeen, which the model suggests produced an offshore-directed mega-rip extending out to ~20 m water depth offshore of Collaroy. This current then deflected south and re-entered the surf zone at Long Reef forming a large rip cell.

The positioning of rip currents is well known to correspond to 'erosion hot spots', as the offshore-directed flow scours a trough, lowering the beach level and leaving adjacent dunes (and property) more exposed to storm wave erosion. Figure 11 illustrates how the positioning of the modelled mega-rip current during the peak of the June storm aligns with locations of observed erosion damage. The northward-directed flow between Long Reef and Collaroy was unique within the Northern Beaches region because of the eastward extension of Long Reef headland, which bifurcated the wave-driven current south towards the neighbouring embayment, DeeWhy, and north towards Collaroy.

June 2016 storm — *peak H_sig 6.5 m, deepwater wave direction NE*

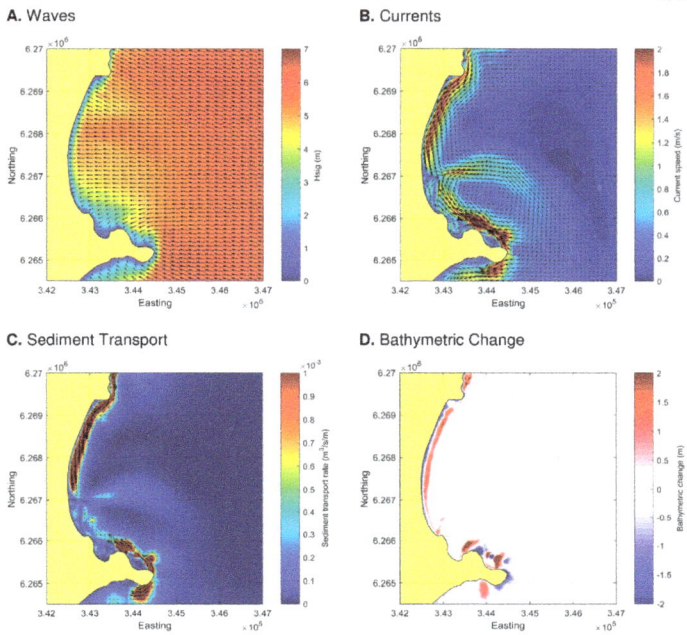

Figure 10. As for Figure 9 but for the June 2016 storm event.

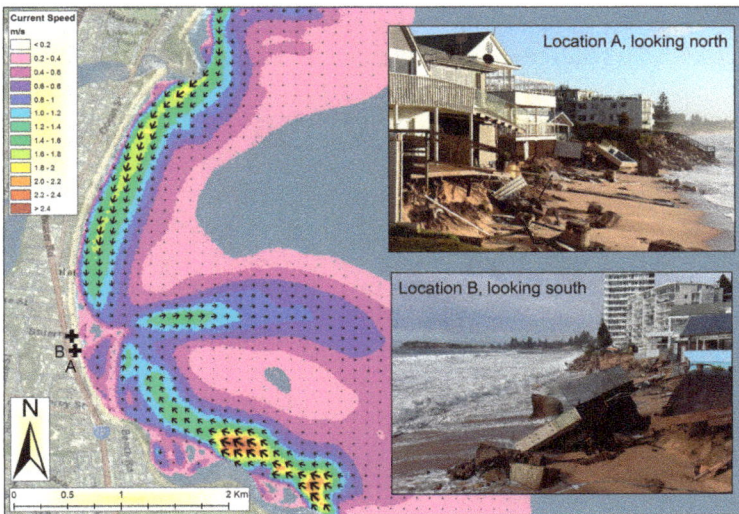

Figure 11. Modelled nearshore current pattern (m/s) at Narrabeen–Collaroy during the peak of the June 2016 storm (as shown in Figure 10B), and locations of observed severe erosion damage at Collaroy. Area of most extreme erosion corresponds to the location of the modelled storm rip current. Source of images Australian Associated Press (rights for reproduction purchased) [39].

The modelled littoral flow velocities, sediment transport rates and volumes of offshore sand movement at Narrabeen–Collaroy were all greater in June 2016 than April 2015, despite significantly lower offshore wave heights in 2016. Long-term beach monitoring [15] also shows that the 2016 storm led to the largest eroded subaerial beach volume at Narrabeen–Collaroy since recording began in 1976. Erosion of the subaerial beach was not explicitly modelled here because of insufficient information on the pre-storm beach condition. On the other hand, the amount of sand moved below the water line within the surf zone is a robust indicator of erosion to the subaerial beach, since the surf zone and beach morphologies at Narrabeen are closely coupled [40]. The same starting bathymetry was used for both storm events, meaning any differences in post-storm surf zone bathymetry are primarily a result of nearshore wave conditions during the two storms.

Our modelling suggests that in April 2015 the surf-zone bar accreted by approximately 90,500 m^3 (an average of ~25 m^3 per m of alongshore length) compared to an accretion of approximately 205,000 m^3 (~60 m^3 per m of alongshore length) in June 2016. The 'bar' was defined as any positive post-storm cumulative bed elevation change over 0.3 m on the upper shoreface. These volumes may not reflect the actual change because of several factors not included in the model as discussed in Section 3. Moreover, not all sand that forms the storm bar comes from the beach. Some sand is also moved from depths seaward of the surf zone because of wave asymmetry leading to a net shoreward transport in depths where there are no breaking waves or undertow [41–43]. The ratio between the offshore/onshore sources depends on the dominance of alongshore or cross-shore transport, which in turn depends on the storm wave direction. Our modelling indicates that during April 2015 over 95% of sand that formed the storm bar originated from the beach, whereas in June 2016 around 85% came from the beach and the remainder from onshore transport outside the surf zone.

The beach monitoring program at Narrabeen shows a slightly lower but comparable ~40% difference between the amount of subaerial beach erosion in June 2016 and April 2015. This is based on the Subaerial Beach Volume Index (SVI), a ratio of the dry beach volume relative to the long-term mean [15]. Prior to the April 2015 event, the beach was close to the mean state (SVI + 2.6), but after the storm it had been reduced to −24.9 (a net change of 27.5). Prior to the June 2016 event, the early winter was characterised by a period of quiescent, southerly waves that had acted to considerably build up the beach (+13.6). After the storm, the beach was down to its lowest SVI value recorded (−34.0)—a net change of 47.6 (thus, a ~40% difference between the April 2015 and June 2016 storms). While this is not directly comparable to the ~50% difference derived from modelling in this study (the monitoring is of the subaerial beach while the modelling is of the subaqueous shoreface), it serves to highlight the magnitude differences in the erosion impacts between the events and is broadly in agreement with the modelled shoreface change.

Bathymetric surveys at Narrabeen (Section 3) and well as the modelling here indicate that sand removed from the subaerial beach during storms does not leave the embayment. Narrabeen and other embayments along the central and south coast of NSW are known to be essentially 'closed' each with a finite sediment budget [44]. Most sand during a storm is transferred down the upper shoreface to around 4–5 m water depth, after which it is progressively reworked shoreward during subsequent non-storm wave conditions. This loss/recovery cycle can be seen in the long-term beach monitoring record at Narrabeen, which shows no significant net change in the subaerial beach volume over the past 40 years [15]. Indeed, previous analyses of the Narrabeen record highlight the importance of time intervals between successive storms in allowing the post-storm recovery of the beach [45,46]. While the beach may recover after a storm, erosion to the foredune and damage sustained to property is lasting.

5.3. Wave Direction Control on Recurrence Estimates

Estimates of the return period, or Average Recurrence Interval (ARI), of storm wave conditions are usually derived from an extreme value analysis of storm peak wave heights from a long-term set of observations. Since wave height is the most important design parameter for coastal engineering, ARI estimates are traditionally non-directional; i.e., one extreme value curve is used for storm wave conditions from all directions. Per this approach, the ARI estimate for the June 2016 event is only ~2 years (Figure 12A). However, this estimate is misleading because the powerful inshore wave conditions generated by this event were far greater than that seen on average every two years.

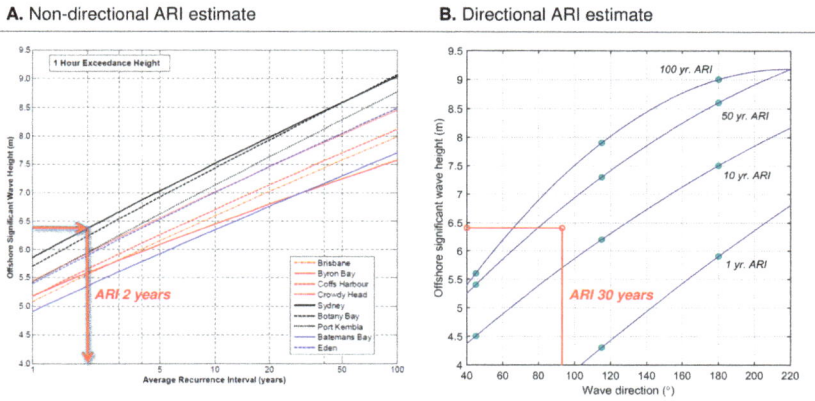

Figure 12. (A) Average Recurrence Interval (ARI) estimates for storm wave heights not accounting for wave direction. The peak storm wave height and ARI estimate for the June 2016 event are highlighted (red) for the Sydney buoy (black solid line). Original figure reproduced from [47]; (B) ARI estimates for storm wave heights at Sydney accounting for wave direction. Original data from [47].

A more relevant ARI of the June 2016 storm needs to consider wave direction. Shand et al. [47] re-examined ARI estimates at Sydney (up to 2009) for storm waves coming from three directional quadrants; <90° (north-east to east), 90°–135° (south-east), and >135° (south-east to south). We interpolated these results across directional bins (using a second-order polynomial) to obtain an indication of the ARI of peak storm wave heights across all wave directions (Figure 12B). Indicative results here suggest that the June 2016 peak storm conditions (6.4 m H_s from 93°) have an ARI of approximately 30 years when wave direction is considered. It is important to note that this estimate is derived from only 17 years of directional wave data (1992–2009, yet to be updated), which is not considered sufficient for reliable extreme value analysis on the NSW coast [7]. Nonetheless, it serves to demonstrate the importance of accounting for wave direction when estimating the likelihood of recurrence of storm wave conditions. Incorporating wave directional effects into extreme value analyses can significantly affect the design of coastal structures [48,49], especially when projecting future extreme conditions where there may be shifts in the directional wind and wave climate [50,51].

6. Implications for Coastal Management in a Changing Climate

The June 2016 East Coast Low was an unusual storm, both in terms of the synoptic configuration and wave conditions. However, storm wave events from this direction are projected to become more common in Southeast Australia in the future with tropical expansion [6]. One of the most robust signatures of present and near-future climate warming is a widening of the tropics, with a continued poleward expansion of ~1° to 2° projected for later this century [52]. This tropical expansion may lead to an increased frequency of easterly and north-easterly waves along the Southeast Australian coast [51] and, as a result, the regional wave climate is predicted to rotate anticlockwise [50].

It is unclear, however, whether this will be manifest in both the modal and storm wave climate, or whether tropical expansion will lead to an increase in bi-directionality between extreme wave events and the mean state. Currently, there is a ~10° difference between the long-term (past 40 years) modal and storm wave directions along the NSW coast, meaning beach systems are largely equilibrated with the distribution of wave energy during and between storms. However, when extreme wave events come from different directions to modal conditions, the erosion response is magnified. The June 2016 storm exemplified this, as the east to north-east wave conditions followed several months of quiescent southerly swell.

At present, the minimum criteria for estimating current and future rates of shoreline recession in coastal zone management plans in NSW include recession due to sediment budget deficits and projected sea level rise [53]. The requirements for beach erosion revolve around a storm bite magnitude arising from an event with an ARI of ~100 years [53]. The requirements for both shoreline recession and beach erosion ignore potential changes in wave direction. An allowance for beach rotation (resulting from inter-annual changes in wave direction associated with El Niño Southern Oscillation) is usually included, but this does not address the control that long-term changes in storm wave direction may have on the propagation of wave energy into the nearshore and, as demonstrated here and in previous studies [54,55], on coastal erosion response. Neither does it address the significant influence that wave direction has on the return period estimates of storm events and their resultant erosion potential.

7. Conclusions

The June 2016 East Coast Low storm demonstrates the importance of wave direction for coastal impacts in Southeast Australia. Our modelling shows that the direction of wave propagation across the shelf was the primary control on the amount of energy in the nearshore, rather than the magnitude of storm wave conditions offshore. Dissipation and refraction coefficients derived from simultaneous offshore and nearshore buoy observations also illustrate the importance of wave direction on the nearshore distribution of wave heights and indicate that the offshore wave record alone is a misleading proxy for erosion risk. Hydrodynamic modelling suggests that the Collaroy to South Narrabeen coastal section is particularly vulnerable to erosion impacts during storms from the east to north-east because of the local embayment geometry. A re-assessment of extreme wave heights at Sydney also demonstrates the importance of considering wave direction when deriving storm recurrence parameters for coastal engineering.

Both observational and climate model based studies suggest an anti-clockwise rotation in the mean wave direction for the South-West Pacific region over the coming decades in association with a poleward expansion of the tropics. It is unclear, however, whether these changes will lead to an increase in bi-directionality between the extreme and mean wave climate. As demonstrated in June 2016, the coastal erosion response is magnified when the modal and storm wave directions are different. Assessments of erosion risk in Southeast Australia do not yet consider impacts of future changes to the directional wave climate. This is symptomatic of the global emphasis on sea level rise over wave climate change, and should be identified as an important knowledge gap for coastal management. Our findings are also of relevance for other Southern Hemisphere east coasts in the sub-tropics, such as the Southern Brazil and Natal to Mozambique regions, where tropical expansion may lead to similar changes in the directional wave climate.

Acknowledgments: This research was part funded by a NSW Office for Environment and Heritage (OEH) grant to CI Goodwin as part of the OEH/Sydney Institute of Marine Science (SIMS), Coastal Processes and Response Research Node (Grant Project 1A Quantification of Sand Supply from the NSW Shoreface). Nearshore wave data from buoys deployed at South Narrabeen and Long Reef were collected as part of an Australian Research Council Linkage Project (grant number LP100200348). Wave data from these deployments are available on request from the authors. Wave data from waverider buoys in New South Wales (NSW) and tidal data from the HMAS Penguin tide gauge are collected as part of the NSW Coastal Data Network Program managed by the Office of Environment and Heritage (OEH) and was sourced from Manly Hydraulic Laboratory. Wave data from the Brisbane waverider buoy is funded by State of Queensland, Department of Science, Information Technology, Innovation and the Arts and is available from https://data.qld.gov.au/dataset/coastal-data-system-waves-brisbane. Wind data from the Kurnell automatic weather station was sourced from the Bureau of Meteorology. Beach profile information from the long-term monitoring program at Narrabeen–Collaroy beach was sourced from http://narrabeen.wrl.unsw.edu.au. Pre- and post-storm hydrographic surveys of Narrabeen–Collaroy beach were undertaken by OEH for the OEH/SIMS Coastal Processes and Response Research Node of the NSW Climate Change Adaptation Hub. All other bathymetric data used for modelling was obtained on request from OEH. A Macquarie University academic licence for MIKE by DHI software was used for modelling. Permission for reproduction of images of wave conditions in Figure 7 was kindly given by the photographer Mark Onorati for use in this publication only. Rights of reproduction of images of coastal damage in Figure 11 was purchased by Risk Frontiers from Australian Associated Press under a Multimedia Commercial Use agreement. The authors also thank the anonymous reviewers of this paper whose comments significantly improved the quality of the manuscript.

Author Contributions: All authors jointly conceived the idea for the paper; Thomas R. Mortlock performed all data analysis and modelling; Thomas R. Mortlock wrote the bulk of the paper; Ian D. Goodwin and Thomas R. Mortlock made observations of the storm meteorology, nearshore wave conditions and coastal impacts; Ian D. Goodwin, Kevin Roche and John K. McAneney made substantial edits to the final draft of the paper.

Conflicts of Interest: The authors declare no conflict of interest.

Appendix A

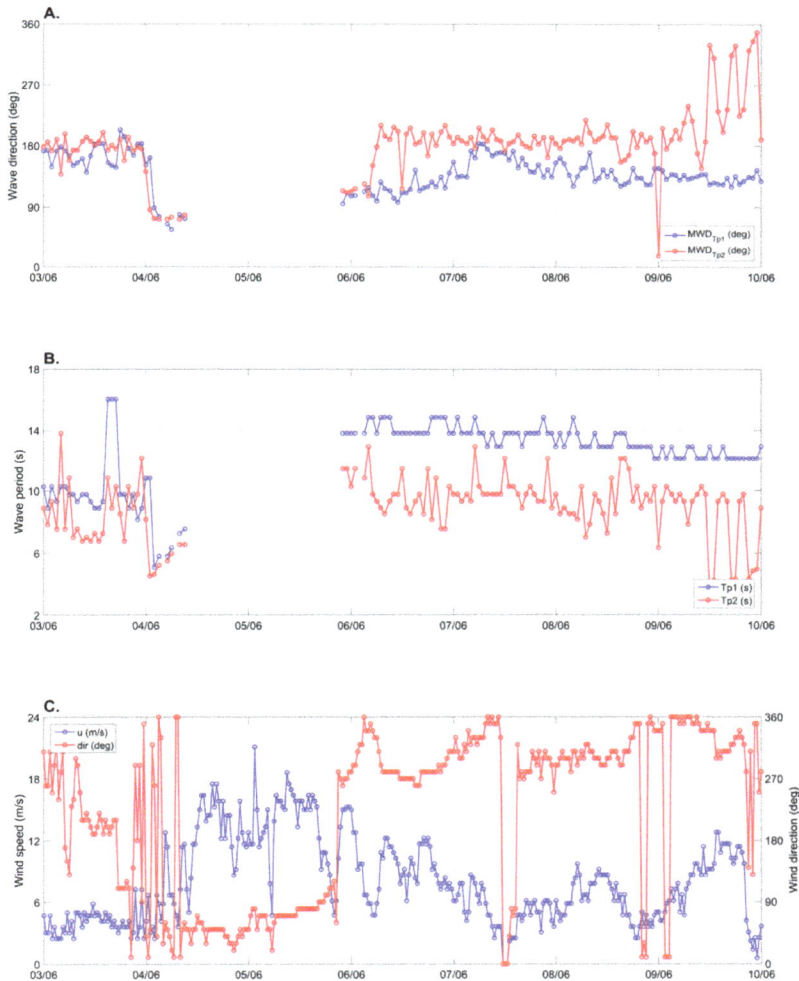

Figure A1. Observations of (**A**) mean wave direction at the first and second spectral peaks (MWD_{Tp1} and MWD_{Tp2}); (**B**) wave period at the first and second spectral peaks (Tp_1 and Tp_2); and (**C**) wind speed (u) and direction (dir) from 3 to 10 June 2016. Wave data was recorded at the Sydney waverider buoy and wind data was recorded at the Kurnell Automatic Weather Station (locations Figure 3).

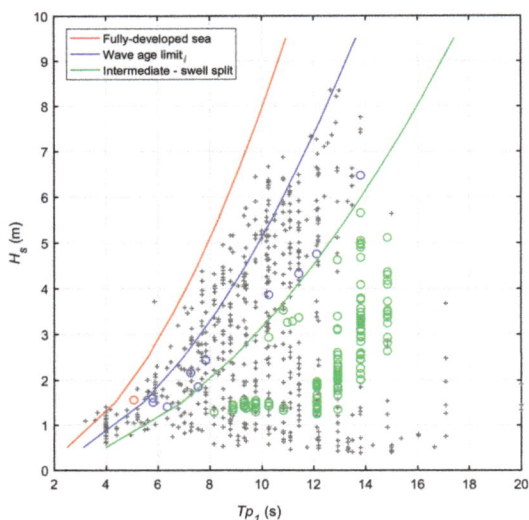

Figure A2. Wave age curves derived from all H_s and T_{p1} values recorded at the Sydney waverider buoy (1987–2016). Grey crosses represent unique wave steepness occurrences at the buoy and circles represent wave conditions as measured during the June 2016 storm. The red line represents the Pierson–Moskowitz limit for fully-developed seas [56]; the green line represents the wave age limit between wind-sea and swell [57]; and the blue line is a line of constant steepness that divides the sea and swell components into two equal parts [24]. All wave events falling between the red and blue lines are 'wind-sea'; between the blue and green lines are 'intermediate sea-swell'; and beyond the green line are 'swell'. In this way, the long-term wave climate defines the divisions between wave types. Values for directional spreading of parametric wave data can then be defined (Section 3.2).

References

1. Crompton, R.P.; McAneney, K.J. Normalised Australian insured losses from meteorological hazards: 1967–2006. *Environ. Sci. Policy* **2008**, *1*, 371–378. [CrossRef]
2. McAneney, J.; McAneney, D.; Musulin, R.; Walker, G.; Crompton, R. Government-sponsored natural disaster insurance pools: A view from down-under. *Int. J. Disaster Risk Reduct.* **2006**, *15*, 1–9. [CrossRef]
3. Taljaard, J.J. *Atmospheric Circulation Systems, Synoptic Climatology and Weather Phenomena of South Africa. Part 2: Atmospheric Circulation Systems in the South African Region;* Department of Environmental Affairs and Tourism, South Africa, Report No. 27-32; South African Weather Bureau: Pretoria, South Africa, 1995.
4. Cuchiara, D.C.; Fernandes, E.H.; Strauch, J.C.; Winterwerp, J.C.; Calliari, L.J. Determination of the wave climate for the southern Brazilian shelf. *Cont. Shelf Res.* **2009**, *29*, 545–555. [CrossRef]
5. Browning, S.; Goodwin, I.D. Large-Scale Influences on the Evolution of Winter Subtropical Maritime Cyclones Affecting Australia's East Coast. *Mon. Weather Rev.* **2013**, *141*, 2416–2431. [CrossRef]
6. Goodwin, I.D.; Mortlock, T.R.; Browning, S. Tropical and Extratropical-Origin Storm Wave Types and Their Contrasting Directional Power Distribution on the Inner Shelf. *J. Geophys. Res. Oceans* **2016**, *121*, 4833–4853. [CrossRef]
7. Louis, S.; Couriel, E.; Lewis, G.; Glatz, M.; Kulmar, M.; Golding, J.; Hanslow, D. NSW East Coast Low Event—3 to 7 June 2016 Weather, Wave and Water Level Matters. In Proceedings of the NSW Coastal Conference, Coffs Harbour, Australia, 9–11 November 2016.
8. International Hydrographic Organization (IHO). *Limits of Oceans and Seas;* International Hydrographic Organization: Bremerhaven, Germany, 1953.

9. Shand, T.D.; Goodwin, I.D.; Mole, M.A.; Carley, J.T.; Browning, S.; Coghlan, I.; Harley, M.D.; Pierson, W.J. *NSW Coastal Inundation Hazards Study: Coastal Storms and Extreme Waves*; Water Research Laboratory (WRL) & Climate Futures at Macquarie University, WRL Technical Report 2010/16; University of New South Wales: Sydney, Australia, 2011.

10. Short, A.D. Role of geological inheritance in Australian beach morphodynamics. *Coast. Eng.* **2010**, *57*, 92–97. [CrossRef]

11. Wright, L.D.; Short, A.D. Morphodynamic Variability of Surf Zones and Beaches: A Synthesis. *Mar. Geol.* **1984**, *56*, 93–118. [CrossRef]

12. Rosati, J.D.; Walton, T.L.; Bodge, K. Chapter III-2 Longshore Sediment Transport. In *Coastal Engineering Manual, Part III*; Vincent, L., Demirbilek, Z., Eds.; U.S. Army Corp of Engineers: Washington, DC, USA, 2002; pp. III-2-1–III-2-109.

13. Splinter, K.D.; Turner, I.L.; Reinhardt, M.; Ruessink, G. Rapid adjustment of shoreline behavior to changing seasonality of storms: Observations and modelling at an open-coast beach. *Earth Surf. Process. Landf.* **2017**. [CrossRef]

14. Poli, P.; Hersbach, H.; Dee, D.P.; Berrisford, P.; Simmons, A.J.; Vitart, F.; Laloyaux, P.; Tan, D.G.H.; Peubey, C.; Thépaut, N.; et al. ERA-20C: An Atmospheric Reanalysis of the Twentieth Century. *J. Clim.* **2016**, *29*, 4083–4097. [CrossRef]

15. Turner, I.L.; Harley, M.D.; Short, A.D.; Simmons, J.A.; Bracs, M.A.; Phillips, M.S.; Splinter, K.D. A multi-decade dataset of monthly beach profile surveys and inshore wave forcing at Narrabeen, Australia. *Sci. Data* **2016**, *3*, 160024. [CrossRef] [PubMed]

16. Short, A.D.; Bracs, M.A.; Turner, I.L. Beach oscillation and rotation: Local and regional response on three beaches in southeast Australia. *J. Coast. Res.* **2014**, *66*, 712–717. [CrossRef]

17. Bracs, M.; Turner, I.L.; Splinter, K.D.; Short, A.D.; Mortlock, T.R. Synchronised patterns of erosion and deposition observed at two beaches. *Mar. Geol.* **2016**, *380*, 196–204. [CrossRef]

18. Nielsen Lord Associates. *Narrabeen-Collaroy Fishermans Beach Coastal Management Strategy, Phase One: Hazard Definition*; Nielsen Lord Associates Report No. 87020.01.003; Prepared for Warringah Shire Council; Wavelength Press: Sydney, Australia, 1988; p. 17.

19. Roche, K.; Goodwin, I.D.; McAneney, J. Management of the coastal zone in Bryon Bay: The neglect of medium-term considerations. *Agenda* **2013**, *20*, 21–39.

20. Wood, J.E.; Roughan, M.; Tate, P.M. Finding a proxy for wind stress over the coastal ocean. *Mar. Freshw. Res.* **2012**, *63*, 528–544. [CrossRef]

21. Holtznagel, S.; Ingleton, T. Field Survey QAQC for Sydney Northern Beaches—Narrabeen Beach Hydro June 2016. In *Single-Beam QAQC Metadata Sheet*; NSW Office of Environment and Heritage: Sydney, Australia, 2016.

22. Danish Hydraulics Institute (DHI). *MIKE21/3 Coupled Model FM User Guide*; MIKE by DHI; Danish Hydraulics Institute: Hørsholm, Denmark, 2016.

23. Cardno. *NSW Coastal Waves: Numerical Modelling Final Report*; Cardno Report No. LJ2949/R2745; Prepared for Office of Environment and Heritage (NSW); Cardno: St Leonards, Australia, 2012; p. 281.

24. Le Cozannet, G.; Lecacheux, S.; Delvallee, E.; Desramaut, N.; Oliveros, C.; Pedreros, R. Teleconnection Pattern Influence on Sea-Wave Climate in the Bay of Biscay. *J. Clim.* **2011**, *24*, 641–652. [CrossRef]

25. Goodwin, I.D.; Mortlock, T.R.; Ribó, M.; O'Brien, P. *Wave Climate and the Distribution of Inner to Middle Shelf Sand Bodies on the South-Eastern Australian Shelf*; Prepared for NSW Office of Environment and Heritage; Marine Climate Risk Group, Macquarie University: Sydney, Australia, 2017.

26. McCormick, M.I. Comparison of field methods for measuring surface topography and their associations with a tropical reef fish assemblage. *Mar. Ecol. Prog. Ser.* **1994**, *112*, 87–94. [CrossRef]

27. Lowe, R.J.; Falter, J.L.; Bandet, M.D.; Pawlak, G.; Atkinson, M.J.; Monismith, S.G.; Koseff, J. Spectral wave dissipation over a barrier reef. *J. Geophys. Res.* **2005**, *110*, C04001. [CrossRef]

28. Thornton, E.B.; Guza, R.T. Transformation of wave height distribution. *J. Geophys. Res.* **1983**, *88*, 5925–5938. [CrossRef]

29. Nielsen, P. *Coastal Bottom Boundary Layers and Sediment Transport*; World Scientific: Singapore, 1992; Volume 4.

30. Huang, Z.-C.; Lenain, L.; Kendall Melville, W.; Middleton, J.H.; Reineman, B.; Statom, N.; McCabe, R.M. Dissipation of wave energy and turbulence in a shallow coral reef lagoon. *J. Geophys. Res.* **2012**, *117*, C03015. [CrossRef]

31. Weber, N. Bottom Friction for Wind Sea and Swell in Extreme Depth-Limited Situations. *J. Phys. Oceanogr.* **1991**, *21*, 149–172. [CrossRef]

32. Sutherland, J.; Peet, A.H.; Soulsby, R.L. Evaluating the performance of morphological models. *Coast. Eng.* **2004**, *51*, 917–939. [CrossRef]

33. Sedigh, M.; Tomlinson, R.; Cartwright, N.; Etemad-Shahidi, A. Numerical modelling of the Gold Coast Seaway area hydrodynamics and littoral drift. *Ocean Eng.* **2016**, *121*, 47–61. [CrossRef]

34. Doering, J.C.; Bowen, A.J. Parameterization of orbital velocity asymmetries of shoaling and breaking waves using bispectral analysis. *Coast. Eng.* **1995**, *26*, 15–33. [CrossRef]

35. Isobe, M.; Horikawa, K. Study on water particle velocities of shoaling and breaking waves. *Coast. Eng. Jpn.* **1982**, *25*, 109–123.

36. Morris, B.; Foulsham, E.; Laine, R.; Wiecek, D.; Hanslow, D. Evaluation of Runup Characteristics on the NSW Coast. *J. Coast. Res.* **2016**, *75*, 1187–1191. [CrossRef]

37. Kalnay, E.; Kanamitsu, M.; Kistler, R.; Collins, W.; Deaven, D.; Gandin, L.; Iredell, M.; Saha, S.; White, G.; Woollen, J.; et al. The NCEP/NCAR 40-year reanalysis project. *Bull. Am. Meteorol. Soc.* **1996**, *77*, 437–471. [CrossRef]

38. Onorati, M. Images of Coastal Wave Conditions at South Narrabeen during the June 2016 Storm. Images taken 06/06/16 Reproduced with Permission from M. Onorati for This Publication Only. Available online: https://www.instagram.com/markonorati/ (accessed on 5 November 2016).

39. Australian Associated Press (AAP). *Images of Coastal Erosion Damage at South Narrabeen During the June 2016 Storm*; Images Taken 07/06/16; Image ID 20160607001262647757 and 20160607001262620563, Purchased under Commercial Agreement by Risk Frontiers 02/12/16; Australian Associated Press: Sydney, Australia, 2016.

40. Harley, M.D.; Turner, I.L.; Short, A.D. New insights into embayed beach rotation: The importance of wave exposure and cross-shore processes. *J. Geophys. Res. Earth Surf.* **2015**, *120*, 1470–1484. [CrossRef]

41. Patterson, D.C.; Nielsen, P. Depth, bed slope and wave climate dependence of long term average sand transport across the lower shoreface. *Coast. Eng.* **2016**, *117*, 113–125. [CrossRef]

42. O'Donoghue, T.; Wright, S. Flow tunnel measurements of velocities and sand flux in oscillatory sheet flow for well-sorted and graded sands. *Coast. Eng.* **2004**, *51*, 1163–1184. [CrossRef]

43. Ruessink, B.G.; Michallet, H.; Abreu, T.; Sancho, F.; Van der A, D.A.; Van der Werf, J.J.; Silva, P.A. Observations of velocities, sand concentrations, and fluxes under velocity-asymmetric oscillatory flows. *J. Geophys. Res. Oceans* **2011**, *116*, 2156–2202. [CrossRef]

44. Short, A.D. *Beaches of the New South Wales Coast: A Guide to Their Nature, Characteristics, Surf and Safety*, 2nd ed.; Sydney University Press: Sydney, Australia, 2007.

45. Callaghan, D.P.; Nielsen, P.; Short, A.; Ranasinghe, R. Statistical simulation of wave climate and extreme beach erosion. *Coast. Eng.* **2008**, *55*, 375–390. [CrossRef]

46. Karunarathna, H.; Pender, D.; Ranasinghe, R.; Short, A.D.; Reeve, D.E. The effects of storm clustering on beach profile variability. *Mar. Geol.* **2014**, *348*, 103–112. [CrossRef]

47. Shand, T.; Mole, M.A.; Carley, J.T.; Peirson, W.L.; Cox, R.J. *Coastal Storm Data Analysis: Provision of Extreme Wave Data for Adaptation Planning*; Water Research Laboratory (WRL) Technical Report 2011/242; University of New South Wales: Sydney, Australia, 2011.

48. Jonathon, P.; Ewans, K. The effect of directionality on extreme wave design criteria. *Ocean Eng.* **2007**, *34*, 1977–1994. [CrossRef]

49. Thompson, P.; Cai, Y.; Reeve, D.; Stander, J. Automated threshold selection methods for extreme wave analysis. *Coast. Eng.* **2009**, *10*, 1013–1021. [CrossRef]

50. Hemer, M.A.; Fan, Y.; Mori, N.; Semedo, A.; Wang, X.L. Projected changes in wave climate from a multi-model ensemble. *Nat. Clim. Chang.* **2013**, *3*, 471–476. [CrossRef]

51. Mortlock, T.R.; Goodwin, I.D. Directional wave climate and power variability along the Southeast Australian shelf. *Cont. Shelf Res.* **2015**, *98*, 36–53. [CrossRef]

52. Lucas, C.; Timbal, B.; Nguyen, H. The expanding tropics: A critical assessment of the observational and modeling studies. *WIREs Clim. Chang.* **2014**, *5*, 89–112. [CrossRef]

53. Office of Environment and Heritage (OEH). *Guidelines for Preparing Coastal Zone Management Plans*; Report No. OEH 2013/0224; Office of Environment and Heritage: Sydney, Australia, 2013; p. 26.

54. Splinter, K.D.; Davidson, M.A.; Golshani, A.; Tomlinson, R. Climate controls on longshore sediment transport. *Cont. Shelf Res.* **2012**, *48*, 146–156. [CrossRef]
55. Zacharioudaki, A.; Reeve, D.E. Shoreline evolution under climate change wave scenarios. *Clim. Chang.* **2011**, *108*, 73–105. [CrossRef]
56. Pierson, W.J.; Moskowitz, L. A proposed spectral form for fully developed wind seas based on the similarity theory of A.A. Kitaigorodskii. *J. Geophys. Res.* **1964**, *69*, 5181–5190. [CrossRef]
57. Carter, D.J.T. Prediction of wave height and period for a constant wind velocity using the JONSWAP results. *Ocean. Eng.* **1982**, *9*, 17–33. [CrossRef]

water MDPI

Article

Shoreline Response to a Sequence of Typhoon and Monsoon Events

Rafael Almar [1,*], **Patrick Marchesiello** [1], **Luis Pedro Almeida** [1], **Duong Hai Thuan** [1,2], **Hitoshi Tanaka** [3] and **Nguyen Trung Viet** [2]

[1]   LEGOS (Université de Toulouse/CNRS/CNES/IRD), 31400 Toulouse, France; Patrick.Marchesiello@legos.obs-mip.fr (P.M.); luis.pedro.almeida@legos.obs-mip.fr (L.P.A.); duonghaithuan@tlu.edu.vn (D.H.T.)

[2]   Faculty of Marine and Coastal Engineering, Thuyloi University, Hanoi, Vietnam; nguyentrungviet@tlu.edu.vn

[3]   Department of Civil and Environmental Engineering, Tohoku University, Sendai 980-8576, Japan; hitoshi.tanaka.b7@tohoku.ac.jp

*   Correspondence: rafael.almar@ird.fr; Tel.: +33-05-6133-3006

Academic Editor: Maurizio Barbieri

Received: 11 April 2017; Accepted: 18 May 2017; Published: 23 May 2017

Abstract: Shoreline continuously adapts to changing multi-scale wave forcing. This study investigates the shoreline evolution of tropical beaches exposed to monsoon events and storms with a case study in Vietnam, facing the South China Sea, over the particularly active 2013–2014 season, including the Cat-5 Haiyan typhoon. Our continuous video observations show for the first time that long-lasting monsoon events have more persistent impact (longer beach recovery phase) than typhoons. Using a shoreline equilibrium model, we estimate that the seasonal shoreline behavior is driven by the envelope of intra-seasonal events rather than monthly-averaged waves. Finally, the study suggests that the interplay between intra-seasonal event intensity and duration on the one hand and recovery conditions on the other might be of key significance. Their evolution in a variable or changing climate should be considered.

Keywords: Vietnam; South China Sea; erosion; recovery; storminess; winter monsoon; typhoons

1. Introduction

It would be a mistake to consider the vulnerability of coastal regions as a simple response to sea level change, assuming static coastal morphology [1,2]. On the contrary, coastal morphology is in a constant process of equilibration at various timescales. It is generally assumed that waves are the main driver of coastal evolution but their role is strongly non-linear, and the coastal response to unsteady forcing is unclear [3].

Beach recovery to extreme events is also still debated as there is not even agreement on their transient or persistent impacts [4,5]. For isolated events, departure from equilibrium is related to the event's intensity and duration [2,6,7]. However, no clear conclusion can be drawn when considering a sequence of events, since both enhanced [8] and weakened effects are observed [9–11], e.g., during the particularly stormy winter of 2013–2014 in Europe, e.g., [12]. The timescales' interplay between recurring events and recovery conditions appears determinant.

Existing shoreline equilibrium models (among others: [2,13,14]) show appreciable skills in predicting shoreline location from wave energy at monthly or longer time-scales for mid-latitude, storm-dominated coasts. However, these skills may be at fault in a so-called low-energy environment as often encountered in the tropics. There, the beach is mostly active during occasional events and is generally found in equilibrium with the preceding energetic event rather than current conditions [15].

Existing equilibrium models might not be able to describe such behavior, in particular when energetic wave events do not occur concomitantly with the seasonal peak of wave energy.

Tropical beaches are exposed to infrequent short (1–3 days) but paroxysmal storms such as cyclones (typhoons in the western Pacific) and can rapidly adapt to these very energetic conditions [16]. They slowly recover under persisting low to moderate waves during the rest of the year. However, all the tropical environments are not strictly low-energy, and this is particularly true in Southeast Asia as it is affected by monsoons [17]. Typical winter monsoon events last from three days to three weeks and can bring strong persistent swells of somewhat lower energy but longer duration than tropical storms. There is substantial literature on the atmospheric cold intrusion affecting the Southeast Asian coastal states every winter, e.g., [18] but the role of these energetic events on shoreline evolution has not been investigated. Clearly, they are active processes for shoreline erosion and must be compared with the effect of short-term storms. Their particularly long duration may be a crucial element of their beach response as the beach may have sufficient time to adjust to the energetic conditions and reach equilibrium.

In this paper, we investigate the video-derived shoreline evolution of Nha Trang beach, Vietnam, over the particularly active 2013–2014 season, with numerous winter monsoon events and storms, including the Cat-5 Haiyan typhoon. We first investigate the role of monsoon events on shoreline evolution compared with storms, and secondly the seasonal behavior of the beach in response to both monthly-averaged wave forcing and wave events using a shoreline equilibrium model [2].

2. Study Site

Nha Trang is an embayed beach located in southeastern Vietnam coast, facing the South China Sea (Figure 1, upper panel). This 6 km bay is oriented north–south and is partially sheltered from waves by a group of islands at its southern end. This medium-sized (D_{50} = 0.4 mm) sandy beach is rather uniform along the shore, and is characterized by a steep (slope ~ 0.1) upper face and a flat low-tide terrace (~40 m wide). The tide is a mix of diurnal and semi-diurnal, with a small tidal range (<1.6 m).

2.1. Typhoons

The Northwest Pacific is the most cyclogenetic region on earth. Of the 16 tropical storms that turn into typhoons (JTWC 2013) annually, about one-third propagate westward to South China Sea [19]. Every year, 4–6 typhoons hit Vietnam [20], typically between August and December, but the risk of landfall varies strongly at seasonal and interannual scales, e.g., [19,21]. The year 2013 came after two years of La Niña conditions, resulting in strong sea surface temperatures, which favored cyclone generation [22]. As a consequence, 2013 was observed to be the most active typhoon season since 2004, and the one with most casualties since 1975. Among the 10 typhoons landing in Vietnam in 2013, Cat. 5 Haiyan in early November turned into one of the world strongest recorded tropical cyclones [23].

2.2. Monsoons

Summer monsoons (May to September) drive relatively weak, short-period southwesterly waves in the South China Sea. The inception of winter monsoon (October to April), caused by high-pressure systems in Siberia, drives strong northeast winds. Because these pressure systems form every three days to three weeks, wind pulses occur at these timescales. As a result, the winter monsoon generates energetic waves larger than 2.5 m off the Vietnamese coast [24], reaching values up to 4 m, which stands for the 10% exceedance level of wave climate [25].

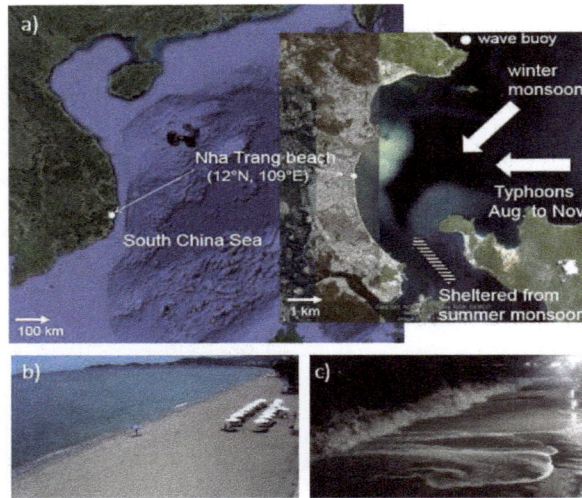

Figure 1. Study site, (**a**) Nha Trang each, Vietnam, facing the South China Sea. Images from the video system during (**b**) calm summer season and (**c**) Cat. 5 Haiyan typhoon.

From its orientation, Nha Trang region is mainly sheltered from summer monsoon which consists in very calm conditions. It can be considered that this stretch of coast is under the influence of winter monsoons and typhoons only [26,27].

3. Methods and Data

A video station was installed in May 2013 [26] in the central part of Nha Trang Bay, is considered far enough from the influence of the edges of the bay and is predominantly influenced by cross-shore rather than by longshore dynamics [27,28]. Hydrodynamic (waves, currents, tides) and morphology (intertidal and submerged bathymetry and shoreline) can be extracted from secondary images, timestack, and average images [27,29]. In this study, the shoreline was extracted manually at a single cross-shore section of the beach. It is estimated as the video-based average between maximum and minimum runup excursions over 15-min images and during daylight and night hours. Hourly tidal modulation of the shoreline location was averaged out using daily means. Wave fields were extracted from ERA-interim global reanalysis provided by the European Centre for Medium-Range Weather Forecasts (resolution of 0.5°, every 6 h [30]) at the closest node off Nha Trang, and validated over a two month period using a local wave gauge. This validation was successful (coefficient of determination $R^2 = 0.87$, RMSE = 0.26 m) down to event scale, which made it possible to extend our study over a full annual period, from 1 August 2013 to 1 August 2014. The period starts from summer monsoon conditions until October, and then winter monsoon lasts until April when a new summer monsoon begins.

Observations of waves and shoreline changes were used to calibrate the parametrical model of equilibrium shoreline position ShoreFor [2,31] accounting for cross-shore transport processes. This model was chosen as it was applied successfully at various sites for predicting the daily to seasonal shoreline response to waves compared to other models more dedicated to long-term interannual evolution [32]. This one-dimensional shoreline prediction model has the form

$$\frac{dx}{dt} = b + c\left(F^+ + rF^-\right) \tag{1}$$

where b and c are calibrated coefficients, r the ratio between erosive and accretive shoreline change, and $F^{\pm}$ is the shoreline forcing, which depends on disequilibrium with anteceding wave conditions

$$F^{\pm} = P^{0.5}(\Omega_{eq} - \Omega) \tag{2}$$

where P is wave power ($\propto Hs^2 Tp$), with Hs and Tp as the deep water significant wave height and peak period, Ω and Ω_{eq} are the instant and time-varying dimensionless fall velocity such as

$$\Omega_{eq} = \left[\sum_{i=1}^{2\phi} \Omega_{eq} 10^{-i/\phi}\right] \tag{3}$$

with Ω defined as $\Omega = H_s / w T_p$, where w is the settling velocity and is a function of the site-specific median grain size (D_{50}). In Equation (3), i is the day prior to present, ϕ the number of past days where the decaying exponential function reaches 10% (more details in [2,14]).

4. Results

The offshore wave forcing for the one year study period is presented in Figure 2a. The southeast coast of Vietnam is partly sheltered from summer monsoon low-energy wind-waves ($Hs < 1$ m, $Tp < 4$s). The winter season presents more energetic swell ($Hs \sim 1.7$ m, $Tp = 7$–8 s) and larger variability, with largest values ($Hs \sim 3$ m, $Tp = 10$ s) during monsoon events. Waves during typhoon events have similar magnitude but shorter duration (<3 days). The shoreline (Figure 2a) is stable or even slightly in accretion till the end of summer monsoon in October, and then begins an erosive phase during winter monsoon until February when it reaches its most landward location, before migrating seaward again.

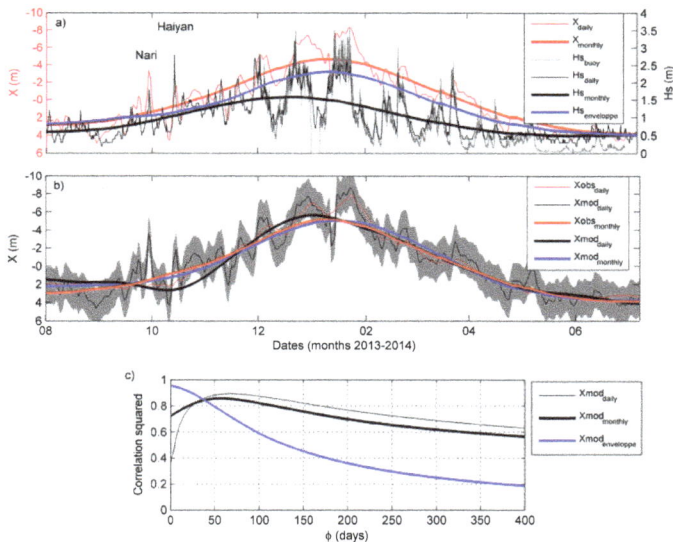

Figure 2. (**a**) Video-derived shoreline location X (red), wave height Hs from Era-Interim (black) and from buoy (gray) off Nha Trang, with the monthly envelope of intra-seasonal Hs (from 3 to 30 days) in blue. (**b**) Results from the shoreline equilibrium model ShoreFor (black, with uncertainty as shaded area), compared with the observed daily shoreline position (red) and in (**c**) the respective correlations squared for different ϕ values. In panels a,b, negative and positive X values stand for erosion and accretion from the mean, respectively. In a–c, thin and thick lines are daily and monthly data respectively.

This seasonal pattern (duration > 30 days) hides important intra-seasonal (or synoptic scale, i.e., winter monsoon events) variability with shoreline variations reaching 10 m during typhoon events and during long-lasting winter-monsoon events with durations from three days to three weeks. Even though the magnitude of shoreline retreat of these events is similar, Figure 2a shows that monsoon events have a more persistent impact with longer recovery (i.e., 10 to 20 days) than typhoons which takes a few days at most (sometimes within less than a day). While typhoon events were frequent at the beginning of the winter season, from October to December 2013, the magnitude of monsoon events increased and peaked in January to February of 2014. This coincided with already decreasing monthly-mean wave height but, surprisingly, the monthly-mean shoreline still eroded.

To investigate this two-month lag between monthly-mean waves and shoreline, the empirical shoreline equilibrium model ShoreFor (Equation (1)) was applied on daily wave and shoreline data. In Figure 2b the model shows good skills (R^2 = 0.8, RMS = 1.2 m) at intra-seasonal (larger than three days) and seasonal scales but miss short-lived storm impacts. The best correlation appears for a lag ϕ of 50 days (Figure 2c), which is much larger than the actual morphological response time of a few days observed at this beach. To investigate this point further, the model was forced with monthly-averaged waves and with the monthly envelop of intra-seasonal events (Figure 2b). The latter is done by means of the Hilbert transform, already applied successfully to study wave groupiness [33,34], $\left| Hs^{hf}(t) + H\left\{ Hs^{hf}(t) \right\} \right|^{lf}$ where H denotes the Hilbert transform operator and $| \; |^{lf}$ a low-pass filter operator, taking 30 days as the cutoff period separating short term (hf) and monthly (lf) timescales. Both monthly estimates well predict the seasonal shoreline behavior (Figure 2b), although that using the envelope gives more accuracy (R^2 = 0.8 and R^2 = 0.9, RMS = 0.5 and RMS = 0.3 m for monthly-averaged and envelope cases, respectively). Interestingly, the main difference rises from ϕ value in Figure 2c, which is similar (52 days) for monthly-averaged and daily data, but 0 for the envelope, indicating that the shoreline is in phase with the monthly envelope of energetic intra-seasonal events. Therefore, the shoreline is in better equilibrium with wave intra-seasonal events than monthly mean wave energy.

5. Discussion

One of the most striking points of this study is the wave energy provided by monsoon events and their dominant role on shoreline evolution. In contrast, while typhoons have large ephemeral impacts, our results show that the shoreline recovers rapidly. There is no evidence for a persistent influence, as suggested for short-lived storms by [35] and [36]. This is confirmed here using the ShoreFor model that presents good skills at predicting intra-seasonal and seasonal scales but poorly describes short-term typhoon-induced dynamics (see Figure 2b, October to November). Recurrence of typhoons is more than 10 days so that they can be considered isolated and without cumulative effect [5]. Winter monsoon events have impacts of similar magnitude to typhoons but with longer recovery, which is close to their observed recurrence period. Therefore, they can be considered as a sequence, as the beach cannot fully recover between events and is constantly moving towards a high energetic equilibrium. Note that the ratio of erosion and accretion rates for typhoons is nearly constant over time, but not so for winter monsoons: while the erosion rate is constant, the accretion rate decreases after each new event. This result highlights the importance of the wave event duration, to the extent that the erosive potential will be satisfied only if the event lasts the necessary time for the beach to establish a new equilibrium [6]. Thus for short-lived typhoons, the full erosion potential is not achieved. Despite the significant shoreline retreat, the beach profile is likely to be far from a new equilibrium. As a consequence, the shoreline recovery to its previous state (and shoreline position) occurs within a few days and does not affect longer term evolution [37]. On the other hand, winter monsoon events have enough duration to achieve their full erosion potential, thus modifying the beach to a fully new equilibrium. The return of the shoreline to its previous position takes longer, leaving the beach more vulnerable and for longer.

It was previously observed that Nha Trang's embayed beach has a seasonal rotation with modulation of waves incidence [27], from northward summer to winter southward transport: the north end of the beach enlarges in summer and erodes in winter, the center of rotation being localized in the central part of the beach [38]. It is noteworthy that, even if summer monsoon has only a weak influence compared to the energetic winter monsoon, locally generated wind-waves induce this northward transport [26]. The shoreline at the central part of the beach close to the video system mainly experiences translation due to cross-shore dynamics and is rather dominated by event scale. This is similar to what has been observed elsewhere [39] where rotation and translation of the shoreline were quantified separately.

A main outcome of this study is the long lag (50–60 days) observed between monthly waves and shoreline location, while the envelope (Hilbert transform) of intra-seasonal monsoon events is in closer phase with the shoreline. This suggests that, contrarily to shoreline equilibrium model paradigms [2,13], the Nha Trang shoreline is actually in equilibrium with energetic wave conditions. This is in line with observations by [15] at low-energy environments where the beach is assumed to be in equilibrium with previous energetic wave events rather than with current conditions. The beach is considered inactive the rest of the time. Here, it is the particularly long duration of winter monsoon events that presumably drives most of the shoreline changes, with very gentle wave conditions in between which limit the recovery potential, as observed elsewhere by [10]. In this sense, the phase-lag observed here between seasonal means and the intra-seasonal envelope is crucial for shoreline equilibrium.

6. Conclusions

In this paper, we addressed the shoreline evolution of the tropical Nha Trang beach, Vietnam, over the particularly active 2013–2014 season. Our results show for the first time that long-lasting (3–10 days) monsoon events have more persistent impact than typhoons (less than 3 days), of similar amplitude but rather transient with fast recovery. The ShoreFor shoreline equilibrium model shows good skills in predicting seasonal shoreline behavior. The seasonal shoreline appears driven by the intra-seasonal event envelope (from Hilbert transform) rather than monthly-averaged waves. Finally, this paper suggests that the interplay between intra-seasonal event intensity and duration, on the one hand, and recovery conditions, on the other, might be of key significance. Therefore, their evolution in a variable or changing climate should be considered.

Acknowledgments: This work was supported by Vietnamese grant (MOST2/216/QD/BKHCN-No2994) and French ANR project COASTVAR (ANR-14-ASTR-0019). We acknowledge use of the ECMWF ERAInterim dataset (www.ECMWF.Int/research/Era). DHT PhD supported by ARTS-IRD program.

Author Contributions: Rafael Almar, Patrick Marchesiello, Nguyen Trung Viet and Luis Pedro Almeida designed the project and conducted analyses. Duong Hai Thuan analyzed the video images and Hitoshi Tanaka provided expertise on shoreline dynamics. All authors wrote the paper.

Conflicts of Interest: The authors declare no conflict of interest.

References

1. Ashton, A.D.; Evans, R.L.; Donnelly, J.P. Discussion of the Potential Impacts of Climate Change on the Shorelines of the Northeastern USA. *Mitig. Adapt. Strateg. Glob. Chang.* **2008**, *13*, 719–743. [CrossRef]
2. Davidson, M.A.; Splinter, K.D.; Turner, I.L. A simple equilibrium model for predicting shoreline change. *Coast. Eng.* **2013**, *73*, 191–202. [CrossRef]
3. Cooper, J.A.G.; Pilkey, O.H. Sea-level rise and shoreline retreat: Time to abandon the Bruun Rule. *Glob. Planet. Chang.* **2004**, *43*, 157–171. [CrossRef]
4. Anderson, T.R.; Frazer, L.N.; Fletcher, C.H. Transient and persistent shoreline change from a storm. *Geophys. Res. Lett.* **2010**, *37*. [CrossRef]
5. Ranasinghe, R.; Holman, R.; de Schipper, M.A.; Lippmann, T.; Wehof, J.; Minh Duong, T.; Roelvink, D.; Stive, M.J.F. Quantification of nearshore morphological recovery time scales using Argus video imaging: Palm Beach, Sydney and Duck, NC. *Coast. Eng. Proc.* **2012**, *1*, 24.

6. Kriebel, D.L.; Dean, R.G. Convolution method for time dependent beach profile response. *J. Waterw. Port. Coast. Ocean Eng.* **1993**, *119*, 204–226. [CrossRef]
7. Frazer, L.N.; Anderson, T.R.; Fletcher, C.H. Modeling storms improves estimates of long-term shoreline change. *Geophys. Res. Lett.* **2009**, *36*. [CrossRef]
8. Ferreira, Ó. Storm groups versus extreme single storms: Predicted erosion and management consequences. *J. Coast. Res.* **2005**, *21*, 221–227.
9. Coco, G.; Senechal, N.; Rejas, A.; Bryan, K.R.; Capo, S.; Parisot, J.P.; Brown, J.A.; MacMahan, J.H.M. Beach response to a sequence of extreme storms. *Geomorphology* **2014**, *204*, 493–501. [CrossRef]
10. Karunarathna, H.; Pender, D.; Ranasinghe, R.; Short, A.D.; Reeve, D.E. The effects of storm clustering on beach profile variability. *Mar. Geo.* **2014**, *348*, 103–112. [CrossRef]
11. Splinter, K.D.; Carley, J.T.; Golshani, A.; Tomlinson, R. A relationship to describe the cumulative impact of storm clusters on beach erosion. *Coast. Eng.* **2014**, *83*, 49–55. [CrossRef]
12. Masselink, G.; Scott, T.; Poate, T.; Russell, P.; Davidson, M.; Conley, D. The extreme 2013/14 winter storms: Hydrodynamic forcing and coastal response along the southwest coast of England. *Earth Surf. Proc. Landf.* **2016**, *41*, 378–391. [CrossRef]
13. Yates, M.L.; Guza, R.T.; O'Reilly, W.C. Equilibrium shoreline response: Observations and modeling. *Geophys. Res.* **2009**, *114*. [CrossRef]
14. Splinter, K.D.; Turner, I.L.; Davidson, M.A.; Barnard, P.; Castelle, B.; Oltman-Shay, J. A generalized equilibrium model for predicting daily to interannual shoreline response. *J. Geophys. Res. Earth Surf.* **2014**, *119*, 1936–1958. [CrossRef]
15. Jackson, N.L.; Nordstrom, K.F.; Eliot, I.; Masselink, G. "Low energy" sandy beaches in marine and estuarine environments: A review. *Geomorphology* **2002**, *48*, 147–162. [CrossRef]
16. Chen, S.S.; Curcic, M. Ocean surface waves in Hurricane Ike (2008) and Superstorm Sandy (2012): Coupled modeling and observations. *Ocean Mod.* **2016**, *103*, 161–176. [CrossRef]
17. Chang, C.-P.; Wang, Z.; Hendon, H. *The Asian Monsoon*; Springer: Heidelberg, Germany, 2006; pp. 89–128.
18. Wu, M.C.; Chan, J.C.L. Surface features of winter monsoon surges over South China. *Mon. Wea. Rev.* **1995**, *123*, 662–680. [CrossRef]
19. Camargo, S.J.; Robertson, A.W.; Gaffney, S.J.; Smyth, P.; Ghil, M. Cluster analysis of typhoon tracks, Part I: General properties. *J. Clim.* **2007**, *20*, 3635–3653. [CrossRef]
20. Nicholls, R.J.; Hoozemans, F.M.; Marchand, M. Increasing flood risk and wetland losses due to global sea-level rise: Regional and global analyses. In *Global Environmental Change*; Elsevier: Amsterdam, The Netherlands, 1999; Volume 9, pp. 69–87.
21. Chan, J.C.L. Tropical cyclone activity in the northwest Pacific in relation to El Niño/Southern Oscillation phenomenon. *Mon. Weather Rev.* **1985**, *113*, 599–606. [CrossRef]
22. Nakamura, T.; Yamazaki, K.; Iwamoto, K.; Honda, M.; Miyoshi, Y.; Ogawa, Y.; Tomikawa, Y.; Ukita, J. The stratospheric pathway for Arctic impacts on midlatitude climate. *Geophys. Res. Lett.* **2016**, *43*, 3494–3501. [CrossRef]
23. Nakamura, R.; Shibayama, T.; Esteban, M.; Iwamoto, T. Future typhoon and storm surges under different global warming scenarios: Case study of typhoon Haiyan (2013). *Nat. Hazards* **2016**, *82*, 1645–1681. [CrossRef]
24. Chu, P.C.; Qi, Y.; Chen, Y.; Shi, P.; Mao, Q. South China Sea wind- wave characteristics. Part I: validation of Wavewatch-III using TOPEX/Poseidon data. *J. Atmos. Ocean Technol.* **2004**, *21*, 1718–1733. [CrossRef]
25. Mirzaei, A.; Tangang, F.; Juneng, L.; Mustapha, M. A.; Husain, M. L.; Akhir, M. F. Wave climate simulation for southern region of the South China Sea. *Ocean Dyn.* **2013**, *63*, 961–977. [CrossRef]
26. Lefebvre, J.-P.; Almar, R.; Viet, N.T.; Uu, D.V.; Thuan, D.H.; Binh, L.T.; Ibaceta, R.; Duc, N.V. Contribution of swash processes generated by low energy wind waves in the recovery of a beach impacted by extreme events: Nha Trang, Vietnam. *J. Coast. Res.* **2014**, *70*, 663–668. [CrossRef]
27. Duong, H.T.; Binh, L.T.; Viet, N.T.; Hanh, K.D.; Almar, R.; Marchesiello, P. Typhoon impact and shoreline recovery from continuous video monitoring: a case study from Nha Trang beach, Vietnam. Proceedings of the 14th International Coastal Symposium (Sydney, Australia). *J. Coast. Res.* **2016**, *75*, 263–267.
28. Almeida, L.P.; Almar, R.; Marchesiello, P.; Blenkinsopp, C.; Martins, K.; Sénéchal, N.; Floc'H, F.; Bergsma, E.; Benshila, R.; Caulet, C.; et al. Swash zone dynamics of a reflective beach with a low tide terrace. *Mar. Geol.* **2017**, submitted for publication.

29. Abessolo, O.G.; Almar, R.; Kestenare, E.; Bahini, A.; Houngue, G.H.; Jouanno, J.; Du, P.Y.; Castelle, B.; Melet, A.; Messignac, B.; et al. Potential of video cameras in assessing event and seasonal coastline behaviour: Grand Popo, Benin (Gulf of Guinea). *J. Coast. Res.* **2016**, *75*, 442–446. [CrossRef]

30. Dee, D.P.; Uppala, S.M.; Simmons, A.J.; Berrisford, P.; Poli, P.; Kobayashi, S.; Andrae, U.; Balsamo, G.; Bauer, P.; Bechtold, P. The ERA-Interim reanalysis: Configuration and performance of the data assimilation system. *Q. J. R. Meteorol. Soc.* **2011**, *137*, 553–597. [CrossRef]

31. Davidson, M.A.; Turner, I.L. A behavioral template beach profile model for predicting seasonal to interannual shoreline. *J. Geophys. Res.* **2009**, *114*, F01020. [CrossRef]

32. Ruggiero, P.; Buijsman, M.; Kaminsky, G.M.; Gelfenbaum, G. Modeling the effects of wave climate and sediment supply variability on large-scale shoreline change. *Mar. Geol.* **2010**, *273*, 127–140.

33. Veltcheva, A.D. Wave and group transformation by a Hilbert spectrum. *Coast. Eng. J.* **2002**, *44*, 283–300. [CrossRef]

34. Ortega, J.; Smith, G.H. Hilbert-Huang transform analysis of storm waves. *Appl. Ocean Res.* **2009**, *31*, 212–219. [CrossRef]

35. Douglas, B.C.; Crowell, M. Long-Term Shoreline Position Pre diction and Error Propagation. *J. Coast. Res.* **2000**, *16*, 145–152.

36. Zhang, K.; Douglas, B.C.; Leatherman, S.P. Do Storms Cause Long-Term Beach Erosion along the U.S. East Barrier Coast? *J. Geol.* **2002**, *110*, 493–502. [CrossRef]

37. Hansen, J.E.; Barnard, P.L. Sub-weekly to interannual variability of a high-energy shoreline. *Coast. Eng.* **2010**, *57*, 959–972. [CrossRef]

38. Thanh, T.M.; Tanaka, H.; Viet, N.T.; Mitobe, Y.; Hoang, V.C. Evaluation of longshore sendiment transport on Nha Trang coast considering influence of Northeast monsoon waves. *J. Jpn. Soc. Civ. Eng.* **2015**, *71*, 1681–1686.

39. Turki, I.; Medina, R.; Gonzalez, M.; Coco, G. Natural variability of shoreline position: Observations at three pocket beaches. *Mar. Geol.* **2013**, *338*, 76–89. [CrossRef]

water

MDPI

Article

Numerical Simulations of Suspended Sediment Dynamics Due to Seasonal Forcing in the Mekong Coastal Area

Vu Duy Vinh [1,*], Sylvain Ouillon [2,3], Nguyen Van Thao [1] and Nguyen Ngoc Tien [4]

[1] Institute of Marine Environment and Resources, VAST, 246 Danang Street, Haiphong City 180000, Vietnam; thaonv@imer.ac.vn

[2] UMR LEGOS, Université de Toulouse, IRD, CNES, CNRS, UPS, 14 avenue Edouard Belin, 31400 Toulouse, France; sylvain.ouillon@legos.obs-mip.fr

[3] Department Water-Environment-Oceanography, University of Science and Technology of Hanoi, 18 Hoang Quoc Viet, Hanoi 100000, Vietnam

[4] Institute of Marine Geology and Geophysics, VAST, 18 Hoang Quoc Viet, Hanoi 100000, Vietnam; nntien@imgg.vast.vn

*   Correspondence: vinhvd@imer.ac.vn; Tel.: +84-912-799-629

Academic Editor: Y. Jun Xu
Received: 27 April 2016; Accepted: 9 June 2016; Published: 16 June 2016

Abstract: The Mekong River is ranked as the 8th in terms of water discharge and as the 10th in terms of sediment load in the world. During the last 4500 years, its delta prograded more than 250 km to the south due to a tremendous amount of sediments deposited, and turned from a "tide-dominated" delta into a "wave-and-tide dominated" delta. This study aims at completing our knowledge on the fate of sediments that may be stored in estuarine or coastal systems, or dispersed over the continental shelf and slope. Sediment transport in the Mekong River Delta (MRD) coastal area was studied by numerical simulations using the Delft3D model. The model configuration was calibrated and validated from data collected *in situ* during 4 periods from 2012 to 2014. Then, 50 scenarios corresponding to different wave conditions (derived from the wave climate) and river discharge values typical of low flow and flood seasons enabled us to quantify the dispersal patterns of fluvial sediments close to the mouths and along the coast. Sediments mostly settled in the estuary and close to the mouths under calm conditions, and suspended sediment with higher concentrations extend further offshore with higher waves. Waves from the Southeast enhanced the concentration all along the MRD coastal zone. Waves from the South and Southwest induced coastal erosion, higher suspended sediment concentrations in front of the southern delta, and a net transport towards the Northeast of the delta. Because of episodes of Southern and Southwestern waves during the low flow season, the net alongshore suspended sediment transport is oriented Northeastward and decreases from the Southwestern part of the coastal zone ($\sim960 \times 10^3$ t yr^{-1}) to the Northeastern part ($\sim650 \times 10^3$ t yr^{-1}).

Keywords: suspended sediment; sediment transport; coastal hydraulics; Mekong; river plume; monsoon; mathematical model

1. Introduction

Rivers originating from the Tibetan plateau and the Himalayas (Ganges, Brahmaputra, Yellow River, Yangtze, Mekong, Irrawady, Red River, Pearl River) are huge providers of sediments to the ocean at global scale [1]. However, recent human activities have severely altered their sediment discharge, mainly as a consequence of artificial impoundments, and also by activities such as groundwater pumping, irrigation, dredging, and deforestation [2]. At a global scale, around 53% of river sediment flux is now potentially trapped in reservoirs [3], and this reduction dramatically affects deltas [4]. For example, in Asia, sediment discharge decreased by 87% (from 1200 to 150 $\times$ 10^6 t yr^{-1}) over a 40 year period in the Yellow River (Huanghe) [5,6] and by 61% (from 119 to 46 $\times$ 10^6 t yr^{-1}) in the Red River after the Hoa Binh dam settlement [7].

Sediment dynamics have considerable impacts, not only in terms of geomorphology but also in terms of geochemical cycles [8,9], biogeochemistry [10], microbiology [11], fate of metal contamination [12], benthic environment, coral reefs and seagrass communities ([13] and references therein). Sediment budgets are thus of interest for many interdisciplinary studies [14].

Seasonally, the sediment discharge, transport and deposition in the Asian deltas are strongly affected by the variations in precipitation, wind and waves induced by the monsoons [15,16]. The monsoons highly influence coastal geomorphology of these deltas, which experience rapid changing shorelines. Understanding the influences of monsoons is essential for predicting changes in sedimentary environment and coastal geomorphology [16].

The Mekong River delta (MRD) is amongst the third largest delta plain in the world, with 16–20 million inhabitants in its Vietnamese part (density of ~460 people km^{-2}, [17]). It covers an area of 62,500 km^2 between Phnom Penh in the Cambodian lowlands and the southern Vietnamese coast [18]. The MRD lies entirely within three meters above sea level.

Recent studies focused on sediment variability in the Mekong River [19,20], on sedimentation in the Lower Mekong River [21,22], within the distributaries [23], and along the MRD coast at depths > 5 m [24]. The concentration of surface suspended matter in the Mekong River plume was shown to decrease by ~5% per year during the period 2002–2012 from the analysis of MERIS satellite data [25]. However, little is known on the fate of sediments reaching the sea that may be stored in estuarine or coastal systems, or dispersed over the continental shelf and slope ([26] and references therein).

With a tidal range up to 3 m in spring tide, sediment dynamics within the Mekong distributaries is strongly affected by tidal oscillations. From *in situ* measurements, Wolanski *et al.* [27,28] sketched a conceptual model of the present-day sediment dynamics in the Bassac estuary (the southern main distributary of the Mekong River), completed by Hein *et al.* [29]: most of the sediment brought by the River is deposited in the shallow coastal waters of depth <20 m (corresponding to the subaqueous delta) in the flood season, while a net flux of particles occurs upstream in the estuary during the low flow season. Alongshore, numerical simulations by Hein *et al.* [29] have shown that deposition dominates over the annual cycle directly off the mouths of the Mekong branches, where erosion prevails throughout the year in shallow waters to the North and to the South (along the Cape), while erosion and deposition alternate further offshore with the seasonal cycle. Therefore, there are specific needs to quantitatively assess the dispersal patterns of fluvial sediments close to the mouths and along the coasts for management purposes.

This paper aims at complementing the previous studies in the coastal area along the Mekong River delta. The variability of sediment transport along the delta is examined during the low flow and flood seasons from numerical modeling, under 50 scenarios based on the wave climate (25 scenarios per season). The resulting numerical simulations are discussed in view of recent changes in sediment deposition and erosion along the delta coastline.

2. Materials and Methods

2.1. The Mekong River Delta (MRD)

2.1.1. The Mekong River

The Mekong River originates from the Tibetan plateau at elevations mostly >5000 m, then flows through six countries along a 4880-km course down to the South China Sea (East Sea of Vietnam), making it the 12th longest river in the world [22]. The Mekong River, which drains a basin area of ~795 $\times$ 10^3 km^2 [19], was ranked as the 8th in terms of water discharge with an average flow of

15,000 m$^3 \cdot$s^{-1} [30] and as the 10th in terms of sediment load with an average value estimated to be 160 $\times$ 10^6 t yr^{-1} [31], corrected to 144 $\pm$ 36 $\times$ 10^6 t yr^{-1} over the last 3 k yr by Ta *et al.* [32].

The Mekong River water and sediment discharges are distributed amongst a network of two main distributaries which divide near Phnom Penh: the Mekong—called the Tiền River after entering Vietnam—to the North, and the Bassac—called the Hậu River in Vietnam—to the South. The Mekong (Tien) distributary divides itself into three main estuaries with Tieu and Dai mouths for the further North, Ham Luong for the central branch, and Co Chien and Cung Hau mouths for the further South. The Bassac (Hau) River divides mainly into the Tran De mouth and the Dinh An mouth (see Figure 1).

The Dong Nai–Saigon River system flows to the South China Sea just to the North of the MRD. The Saigon River is called Soai Rap River in its lower basin. The Vamco River is a tributary that joins the Soai Rap River just upstream of its mouth (Figure 1a).

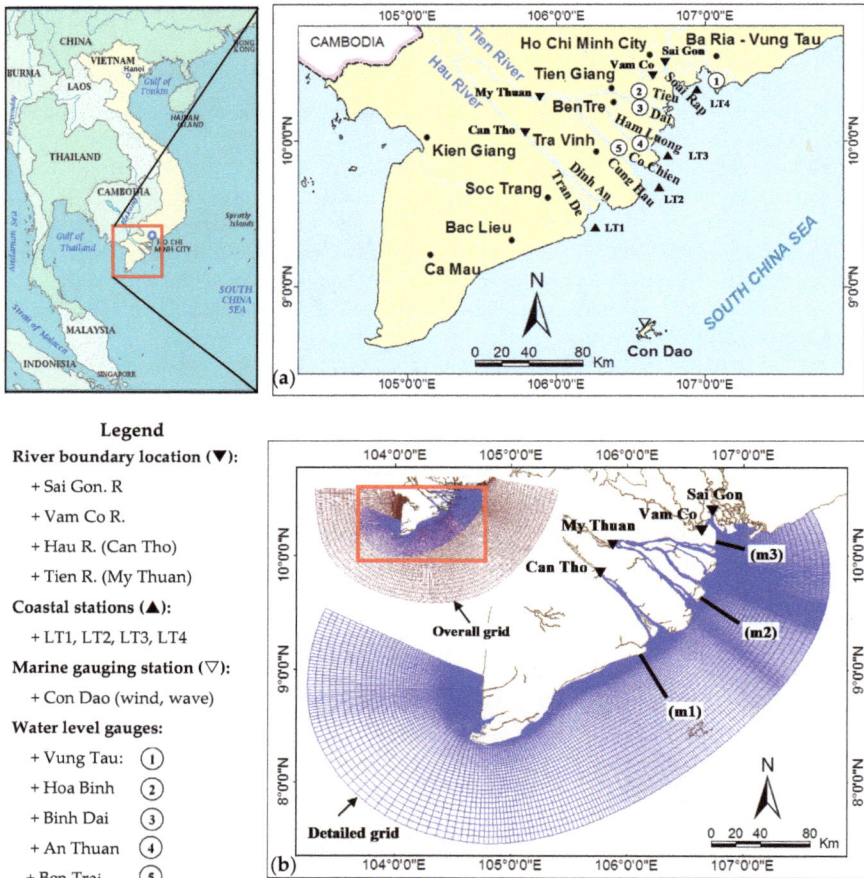

Figure 1. The Mekong River Delta and its region of freshwater influence. (**a**) General map with locations of measurements (coastal stations LT1, LT2, LT3 and LT4; marine gauging station of Con Dao; water level gauges); (**b**) The parent grid outside and the child grid over the study area, with locations of the river boundaries of the model, and cross sections (m1, m2 and m3) used in the calculations.

During the last 4500 years, the MRD prograded more than 250 km to the southeast [18]. Approximately 3000 years ago, the prodelta was located near Ben Tre and Tra Vinh cities (see [32], their Figure 4). During this recent progradation (~3 k yr BP), the southern coast of the MRD became more sensitive to waves from the East and North-East during the dry season, and the MRD turned from a "tide-dominated delta" into a "tide-and-wave dominated" delta [32]. The continental shelf is very shallow (most of the plume flows over depths <10–15 m), the river mouths being located in a sedimental plain. Despite the low Coriolis number, the large freshwater discharge still creates a baroclinic coastal current flowing in the direction of the Kelvin wave [33].

2.1.2. Climate and Rainfall

The Mekong basin ranges from cool temperate to tropical climates. The Lower Mekong River basin is subject to a tropical climate that is characterized by a summer monsoon from the Southwest (from May to October) and a winter monsoon from the Northeast (from November to April). The wet season (from June to October) alternates with a dry season (November–May) and accounts for 85%–90% of the total yearly rainfall. In South Vietnam, the annual rainfall ranges between 1600 and 2000 mm [34,35]. Rainfall is higher (2000–2400 mm per year) in the Western region than in the Eastern (1600–1800 mm) and in the central delta (1200–1600 mm) [36]. According to the Vietnamese National Centre for Hydro-Meteorological Forecasts [37], 18 typhoons impacted on the MRD during the 52-year period of 1961–2012. Amongst them, 13 typhoons occurred recently, during the last 20 years (1992–2012).

2.1.3. Hydrological Regimes and Sediment Transport

The total water discharge of the Mekong is ~500 km^3 yr^{-1}, of which 85% flows in flood season (from September to November) and 15% in low flow season (from December to August) [21,35]. In the delta, the flood season occurs later than the local rainy season, since the water flux mainly comes from the upper and central Mekong basin and is partially regulated by the dynamics of the Tonle Sap in Cambodia. Discharges at the Can Tho and My Thuan stations are almost the same, and vary between 25,000 m$^3 \cdot$ s^{-1} in September-October and a minimum in April (typically 2000 m$^3 \cdot$ s^{-1}) [38]. The total suspended sediment discharge entering the delta is about 145 × 10^6 t yr^{-1}, and the average sediment concentration in the river is about 60 mg$\cdot$ L^{-1}; maximum values can reach 500 mg$\cdot$ L^{-1} in the wet season [39].

The suspended sediments in the distributaries of the Mekong are mostly composed of fine silt with about 15% clay [27].

The coastal current along the MRD shifts with the monsoon forcing; it is oriented Southwestward during the Northeast (or winter) monsoon and Northeastward during the Southwest (or summer) monsoon [40].

2.1.4. Tides

The MRD is affected by tides of a mixed diurnal and semi-diurnal character. Semi-diurnal lunar tide M2 and solar tide S2 amplitudes are up to 0.9 m and 0.5, respectively, while diurnal solar tide K1 and lunar tide O1 amplitudes are up to 0.7 m and 0.5 m, respectively, along the MRD [41]. The resulting amplitude is gradually decreasing from 3.8 m in the Northeast (NE) area to approximately 2 m in the Southwest (SW) along the Ca Mau peninsula [26].

Tidal mechanisms are key processes acting on water distribution in deltas and on sediment transport in estuaries (e.g., [42–44]). In the middle and lower estuaries, deposition is mainly driven by the dynamics of the turbidity maximum zone, whose presence and dynamics are governed by the coupling between river discharge and tidal propagation (e.g., tidal pumping and/or density gradients; [27,28,45,46]). In this study, tidal propagation within the estuaries is included in the numerical model and the tide is taken into account through its boundary conditions in the river mouths.

2.2. Data

Data used to set up, calibrate and validate the model are the following.

Bathymetry and coastline in the MRD coastal area were digitized from topography maps in VN2000 coordinates (national coordinate system of Vietnam corresponding to UTM projection with a WGS84 reference ellipsoid and specified local parameters) with scale 1:50,000 in the coastal zone and 1:25,000 in the estuary. These maps were published by the Department of Survey and Mapping in 2004, now belonging to the Ministry of Natural Resources and Environment of Vietnam. Bathymetry offshore was extracted from GEBCO-1/8 with 30 arc-second interval grid [47].

River water discharge and suspended sediment concentration (SSC) measured in the period 2007–2014 at My Thuan (Tien River) and Can Tho (Hau River) were used to set-up river boundary condition in these rivers. In the absence of gauging station in the Vamco River and Soai Rap River (to the North of the MRD), the averaged river discharges and SSCs in low flow and flood seasons provided by a previous project [48] were used to set up the river boundary conditions in these rivers.

Wind and wave data measured at Con Dao Island and Vung Tau were analyzed and used in the simulations. Data measured with interval of 6 hours in 2012, 2013 and 2014 were considered in this study for calibration/validation. Wave and wind average values over 20 years (1992–2013) from the wave climate [49] were considered along the MRD coastal zone to setup different scenarios of simulations.

Sea level elevations measured at Vung Tau, Hoa Binh, Binh Dai, An Thuan and Ben Trai stations (Figure 1a) were used for model calibration and validation. Moreover, measured sea level near the coast was analyzed to determine the harmonic constants of 8 tidal constituents (M2, S2, K2, N2, O1, K1, P1, Q1) to be imposed at sea boundaries in the refined grid. The tidal harmonics constants imposed offshore were extracted from FES2014 [50,51]. The World Ocean Atlas [52] with 0.25 degree-grid resolution was used for open sea boundary transport condition of the parent model.

Current velocity, grain size of bottom sediment and SSC were measured in the framework of the project "Interaction between hydrodynamics of the Bien Dong (East Sea of Vietnam) and Mekong River water" at 4 *in situ* stations (LT1, LT2, LT3 and LT4, see Figure 1a) in September 2013 (flood season) and April 2014 (low flow season). These data were used for model calibration and validation.

2.3. Modelling Strategy

2.3.1. The Delft3D Model

In this study, hydrodynamics (resulting from tides, currents and waves) and suspended sediment transport were simulated using the Delft3D-Flow module and the Delft3D-Wave module (based on the SWAN model). The Delft3D-Flow model developed by Deltares (Delft, The Netherlands) is a 3D modeling suite to investigate hydrodynamics, sediment transport and morphology, and water quality for fluvial, estuarine and coastal environment [53]. Delft3D solves the Reynolds-averaged Navier-Stokes equations, including the k-ε turbulence closure model, and applies a horizontal curvilinear grid with σ-layers for vertical grid resolution.

SWAN is used as wave model [54–56]. In the action balance equation (see e.g., [57]), the JONSWAP expression [58] is chosen to express the bottom friction dissipation, with the parameter $C_{jon} = 0.067 \ \mathrm{m^2 \cdot s^{-3}}$ proposed by Bouws and Komen [59] for fully developed wave conditions in shallow water. The model from Battjes and Janssen [60] is used to model the energy dissipation in random waves due to depth-induced breaking waves.

2.3.2. Model Setup

The NESTING method (Delft3D-NESTHD) was used to create sea boundaries condition for a refined (child) grid within a coarse (parent or overall) model. The parent model grid is orthogonal, curvilinear with 210 × 156 grid cells. Horizontal grid size changes from 166 to 22,666 m (Figure 1b). Along the vertical, 4 layers in σ-coordinate are considered, each of them accounting for 25% of the water depth. Open sea boundary conditions of the overall model are provided by FES2004 and the World Ocean Atlas 2013.

The grid of the detailed model is also orthogonal curvilinear. Curvilinear grids are applied in the models to provide a high grid resolution in the area of interest and a low resolution elsewhere, thus saving computational effort. The model frame includes all the coastal zone of Ba Ria-Vung Tau to Ca Mau cape (Figure 1a) along circa 485 km in the long-shore direction and 100 km in the cross-shore direction. The horizontal grid encompasses 424×296 points with grid size between 44 m and 11,490 m. Along the vertical, 4 layers in σ-coordinate are considered, similar to the parent model. The hydrodynamics model takes into account the influences by water temperature, salinity, sediment transport and wave actions.

The model was setup and run for different periods of time, with time steps of 0.2 min (12 s). Its calibration and validation were performed during 4 periods: March–May 2012 (low flow season), August–October 2012 (flood season), September 2013 (flood season) and April 2014 (low flow season).

There are four open river boundaries: Soai Rap, Vamco, My Thuan and Can Tho (Figure 1b). The Soai Rap and Vamco boundaries belong to the Dong Nai–Saigon River catchment. River water discharge measured every hour in Can Tho and My Thuan stations were imposed as river boundary conditions. Others river boundaries were set to 546.9 $m^3 \cdot s^{-1}$ (Soai Rap R.) and 52.5 $m^3 \cdot s^{-1}$ (Vamco R.) in low flow season; 1310 $m^3 \cdot s^{-1}$ in the Soai Rap River and 177.8 $m^3 \cdot s^{-1}$ in the Vamco River in flood season. The averaged SSC values during flood tide and ebb tide at Can Tho and My Thuan stations were imposed as boundary condition for these two distributaries. SSC in Vamco and Soai Rap rivers were almost the same and considered in this study to be worth 55 $mg \cdot L^{-1}$ in low flow season and 70 $mg \cdot L^{-1}$ in flood season. At each river boundary, the same SSC values were imposed from the surface to the bottom.

The averaged water temperature and salinity values in the low flow and flood seasons were imposed at the river boundaries (T = 27.5 °C in low flow season and 27.2 °C in flood season; S = 0). Wind velocity and direction, which were measured every 6 h in Con Dao in the period 2012–2014, were imposed on the model. The wave module was setup with online coupling with hydrodynamics and sediment transport. Sea boundary conditions of the wave model were extracted from the wave climate [49].

The bottom roughness was specified by a spatial distribution of Manning (n) coefficients with values in the range 0.018–0.023 $m^{-1/3} \cdot s$ [61,62]. The background horizontal eddy viscosity and horizontal eddy diffusivity were considered to be, after calibration, equal to 8 $m \cdot s^{-2}$. The Horizontal Large Eddy Simulation (HLES) sub-grid, which is integrated in Delf3D-Flow, is based on theoretical considerations presented by Uittenbogaard [63] and Van Vossen [64]. In this study, the HLES sub-grid was activated to add calculated results to background values. Two sediment fractions were simulated in the model: one non-cohesive and one cohesive.

Measured grain sizes of non-cohesive particles in the MRD coastal area range from 29 to 252 μm (average 113 μm) in flood season (September 2013) and from 15 to 262 μm (averaged 103 μm) in low flow season (April 2014). Therefore, median sand-sized particles of 113 μm and 103 μm were considered in our simulations in flood and low flow season, respectively. A specific density of 2650 $kg \cdot m^{-3}$ and dry bed density of 1600 $kg \cdot m^{-3}$ were considered; all other sand transport calibration parameters were kept at the default values proposed by Delft3D. Sand fraction transport was modeled with the van Rijn TR2004 formulation [65], which has been shown to successfully represent the movement of non-cohesive sediment ranging in size from 60 μm to 600 μm.

Previous observational studies in the Mekong estuary indicated that most of Mekong sediments are flocculated fine particles. In low flow season, measured flocsize was 30–40 μm with about 20%–40% of clay content in volume [28]. In the flood season, the flocsize was 50–200 μm with a 20%–30% clay content in volume [27]. For the cohesive sediments, if the bed shear stress is larger than a critical value, erosion is modeled following the Partheniades' formulation [66], whereas if the bed shear stress is less than a critical value for deposition, Krone's formula [67] is used to quantify the deposition flux. The parameters required to simulate the cohesive sediment transport include critical bed shear stresses for erosion τ_{cre} and deposition τ_{crd}, the erosion parameter M and the particle settling velocity w_s [68]. After calibration, τ_{crd} was set to 1000 N·m^{-2}, which effectively implied that deposition was a function of concentration and fall velocity [69]. τ_{cre} was set to 0.2 N·m^{-2}, and M was set to 2 × 10^{-5} kg·m^{-2}·s^{-1}.

In salt water, cohesive sediments flocculate, the degree of flocculation depending on salinity. Flocs are much larger than the individual sediment particles and settle at a faster rate. In order to model this salinity dependency, Delft3D-Flow considers two settling velocities and a maximum salinity. The velocity $w_{s,f}$ is the settling velocity of the sediment fraction in fresh water (salinity S = 0), while the velocity $w_{s,max}$ is the settling velocity of the cohesive fraction in water having a salinity equal to salmax (S = 30 over our area). For the Mekong River plume and based on Stoke's law, Hung *et al.* [70] reported that grain size in the range of 29.4–40 μm has settling velocities in the range 0.9–1.7 mm·s^{-1}. Manh *et al.* [23], considering an extended range of grain sizes from 2.5 to 80 μm, evaluated a calibration range of settling velocities between 0.01 and 7 mm·s^{-1}. Portela *et al.* [71] reported settling velocities increasing by a factor of 6.5 between freshwater conditions (S = 0) and marine conditions (S = 30). In this study, the settling velocity of the cohesive sediments was set to 0.05 mm·s^{-1} in fresh water and to 0.325 mm·s^{-1} at S = 30.

In our simulations, we chose to set the input of sand concentration into suspension to 0 at the river boundaries (inside the delta, where the slope is almost flat) and specified the values of SSC for the cohesive suspended sediments. Simulations in the lower estuary and coastal areas are not sensitive to the sand concentrations at the upper river boundary since sand particles settle very fast; the sand profile is in equilibrium over very short distances, depending on the local capacity of transport and the available particles at the bed [72,73]. Sand particles can be eroded within the model area, in the estuary and along the coast as well, under the combined action of waves and currents.

The bottom-boundary layer calculation accounts for the interaction of wave and current over a moveable bed [65,74–76].

2.3.3. Calibration and Validation Process

Model calibration and validation were conducted simulating low flow and flood seasons of 2012, flood of 2013 and low flow season 2014. The discrepancy between results and measurements was quantified, for each simulation, using the Nash-Sutcliffe efficiency number E [77], calculated as follows:

$$E = 1 - \frac{\sum(obs - calc)^2}{\sum(obs - mean)^2} \tag{1}$$

in which the sum of the absolute squared differences between the predicted and observed values is normalized by the variance of the observed values during the period under investigation. E varies from 1.0 (perfect fit) to $-\infty$, a negative value indicating that the mean value of the observed time series would have been a better predictor than the model [78].

2.3.4. Scenario Simulation

In order to describe characteristics of sediment transport in the MRD coastal area, scenarios were set up with different conditions of wave, wind and river discharge. Initial conditions of the scenarios result from a previous simulation (either August for flood season or March for low flow season). At the river boundaries, temperature was fixed at 27.5 °C in the low flow season and at 27.2 °C in the flood season, and salinity was fixed to 0 (there is no income of seawater at the upstream boundaries). The average temperatures do not differ considerably since the low flow season includes both hot months (April to August) and cool months (December, January). The resulting average value is close to the water temperature in the flood season (September, October and November). Temperature and salinity at the open sea boundaries vary from one scenario to another following the climatology.

The average values of river discharge and SSC measured at Can Tho and My Thuan stations during the period 2007–2014 were imposed as boundary conditions. Two groups of scenarios were then considered, referred to as md (low flow season, mostly dry) and mf (flood season):

- during the low flow season (from December to August), the average water river discharges of 3054 $m^3 \cdot s^{-1}$ at My Thuan in the Tien River, 3739 $m^3 \cdot s^{-1}$ at Can Tho in the Hau River, 52.5 $m^3 \cdot s^{-1}$ in the Vamco River and 546.9 $m^3 \cdot s^{-1}$ in the Soai Rap River were imposed, with SSC = 50 $mg \cdot L^{-1}$ at Can Tho, 53.6 $mg \cdot L^{-1}$ at My Thuan, 55 $mg \cdot L^{-1}$ in the Vamco and Soai Rap rivers (tab:water-08-00255-t001);
- during the flood season (from September to November), the average water river discharges of 12,530 $m^3 \cdot s^{-1}$ at My Thuan, 13,130 $m^3 \cdot s^{-1}$ at Can Tho, 177.8 $m^3 \cdot s^{-1}$ in the Vamco River and 1310 $m^3 \cdot s^{-1}$ in the Soai Rap River were considered, with SSC = 67.8 $mg \cdot L^{-1}$ at Can Tho, 86.5 $mg \cdot L^{-1}$ at My Thuan and 70 $mg \cdot L^{-1}$ at Vamco and Soai Rap (tab:water-08-00255-t002).

Table 1. Scenario simulations in the low flow season (December to August).

Scenario	Wave Direction	Duration (Days)	Wave H_s (m)	T_p (s)	Wind Velocity (m·s⁻¹)
md0		29.87			
md1		0.27	0.5	6.5	4.5
md2	NE	5.48	2	8.5	7.5
md3		1.10	4	10.5	10.5
md4		1.64	0.5	6.5	4.5
md5		23.29	2	8.5	8
md6	E	11.78	4	10.5	12.5
md7		4.38	6	11.5	14.5
md8		0.55	8	12.5	16.5
md9		1.64	0.5	6.5	4.5
md10		18.36	2	8.5	7.5
md11	SE	9.59	4	10.5	10.5
md12		3.29	6	11.5	12.5
md13		0.27	8	12.5	14.5
md14		2.47	0.5	6.5	4.5
md15		25.21	2	8.5	6.5
md16	S	14.80	4	10.5	9.5
md17		7.67	6	11.5	12.5
md18		1.10	8	12.5	14.5
md19		3.29	0.5	6.5	4.5
md20		45.76	2	8.5	7.5
md21	SW	31.51	4	10.5	10.5
md22		21.65	6	11.5	12.5
md23		6.30	8	12.5	14.5
md24		2.74	10.5	13.5	16.5

Table 2. Scenario simulations in the flood season (September to November).

Scenario	Wave Direction	Duration (Days)	Wave H_s (m)	T_p (s)	Wind Velocity (m·s^{-1})
mf0		13.01			
mf1		0.09	0.5	6.5	4.5
mf2	NE	2.82	2	9	7.5
mf3		0.55	4	10.5	9.5
mf4		0.09	6	11.5	12.5
mf5		0.46	0.5	6.5	4.5
mf6		7.74	2	9	7.5
mf7	E	4.64	4	10.5	10.5
mf8		2.00	6	11.5	12.5
mf9		0.09	8	12.5	14.5
mf10		0.09	0.5	6.5	4.5
mf11	SE	6.19	2	9	7.5
mf12		4.55	4	10.5	11
mf13		1.00	6	11.5	12.5

Table 2. Scenario simulations in the flood season (September to November).

Scenario	Wave Direction	Duration (Days)	Wave H_s (m)	T_p (s)	Wind Velocity (m·s^{-1})
mf14		0.18	0.5	6.5	4.5
mf15		7.46	2	9	7.5
mf16	S	5.92	4	10.5	11.5
mf17		2.09	6	11.5	13
mf18		0.46	8	12.5	15
mf19		0.27	0.5	6.5	4.5
mf20		12.56	2	9	7.5
mf21	SW	11.01	4	10.5	11.5
mf22		6.28	6	11.5	13
mf23		1.18	8	12.5	15
mf24		0.27	10.5	13.5	17

The variation in SSC values at My Thuan and Can Tho between low flow and flood seasons is not large since these values are constrained by erosion, transport, deposition all along the Mekong River, and by storage and release by the Tonle Sap. However, the seasonal variations of suspended sediment discharge are higher than the seasonal variations of the flow.

The wave and wind conditions were derived from the wave climate obtained over the 22-year period of 1992–2013 by the BMT ARGOSS [49]. For each class of wave direction (NE, E, SE, S, SW) and significant wave height (0.5–1 m, 1–3 m, 3–5 m, 5–7 m, above 7 m), the average durations during the 274 days of the low flow season (from December to August) and the 91 days of the flood season (from September to November) were defined. Adding periods of calm weather (*i.e.*, with significant wave height <0.5 m), we obtained 25 typical scenarios for the low flow season (tab:water-08-00255-t001) and 25 typical scenarios for the flood season (tab:water-08-00255-t002). For each scenario, significant wave heights H_s representative of the above described classes were defined as: 0.5 m, 2 m, 4 m, 6 m, 8 m, respectively. Wave peak periods T_p were determined from statistics established over 20 years of measurements, available online at the BMT group website. In each scenario, we assumed that wind was coming from the same direction as the waves, which is mostly the case since the gauging station is located offshore.

Each of the above-defined scenarios was run for a 14.75-day period corresponding to one full neap-spring tide cycle [79]. Each simulation considered real tidal boundary conditions from one typical cycle, which was chosen as the period 11–25 September, 2012.

3. Results

3.1. Model Validation from Field Measurements

Water elevation at hydrometeorological stations of Vung Tau, Binh Dai, An Thuan and Ben Trai (e.g., Figure 2a,b) was used to calibrate (e.g., for the choice of the Manning coefficient) and then validate the model. After calibration, comparisons show good agreement between model results and measurements with E coefficients between 0.68 and 0.85. Current velocity measured in LT1 (Figure 2c), LT2 (Figure 2d), LT3 and LT4 in September 2013 and April 2014 were analyzed to provide horizontal velocity components U and V, which were compared with model simulations. The comparison showed an acceptable agreement between model and measurements with E coefficients between 0.65 and 0.89. Comparisons between measured SSC in LT1 (Figure 2e), LT2 (Figure 2f), LT3, LT4 and modeled SSC also resulted in acceptable E coefficients in the range 0.66–0.78.

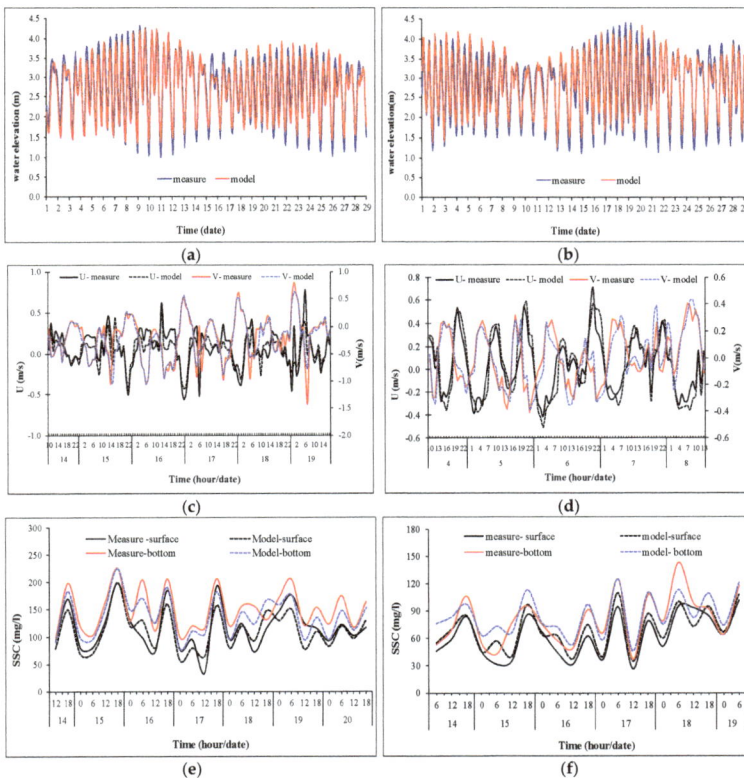

Figure 2. Comparison between simulations and measurements: (**a**) water elevation at Bền Trại (Cồ Chiên mouth) in April 2012; (**b**) water elevation at Bền Trại (Cồ Chiên mouth) in September 2012; (**c**) current components in the surface layer at LT1, flood season (14–19 September 2013); (**d**) current components in the middle of water column at LT2, low flow season (4–8 April 2014); (**e**) SSC at LT1, flood season (14–20 September 2013); (**f**) SSC at LT2, flood season (14–19 September 2013).

3.2. Spatial Distribution of SSC with or without Waves

Figure 3 (in low flow season) and Figure 4 (in flood season) show the distribution of SSC below the surface in the absence of wind and waves (Figures 3a and 4a), and with waves of significant height of 2 m, coming from 5 different directions from NE to SW. These distributions are provided in Figures 3 and 4 one hour before low tide, which correspond to the maximum extent of the plume to the ocean.

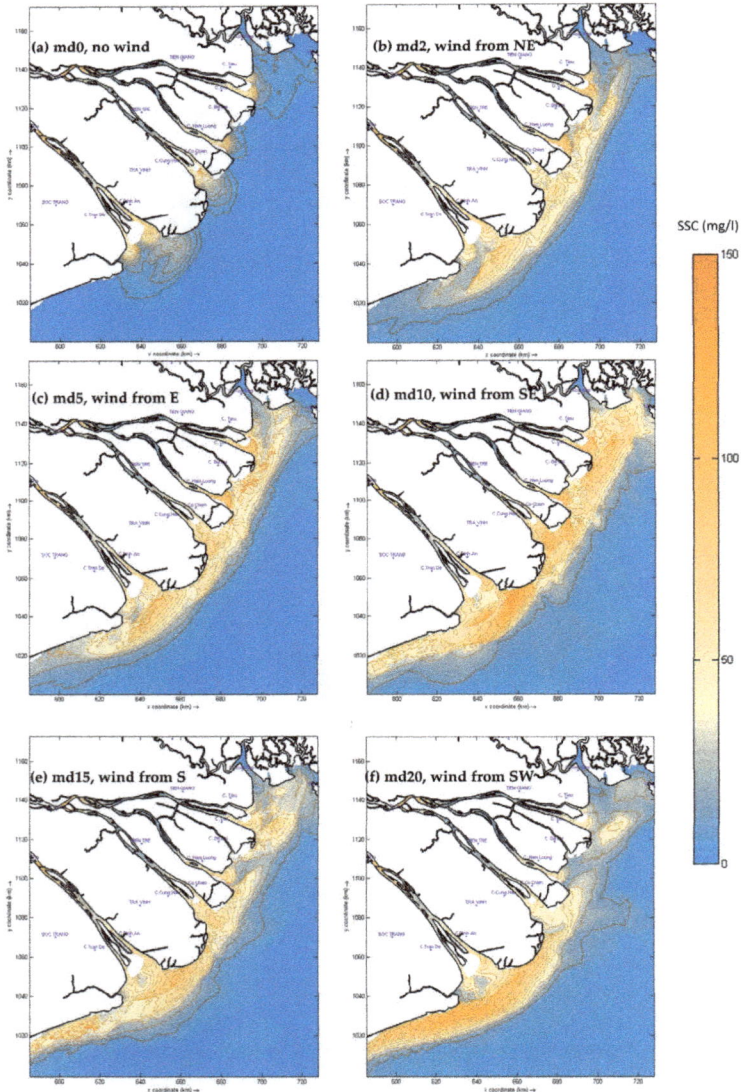

Figure 3. Distribution of suspended sediment concentration in the surface layer in the low flow season in the MRD coastal area (1 h before low tide), with no wave or wave heights = 2 m from different directions.

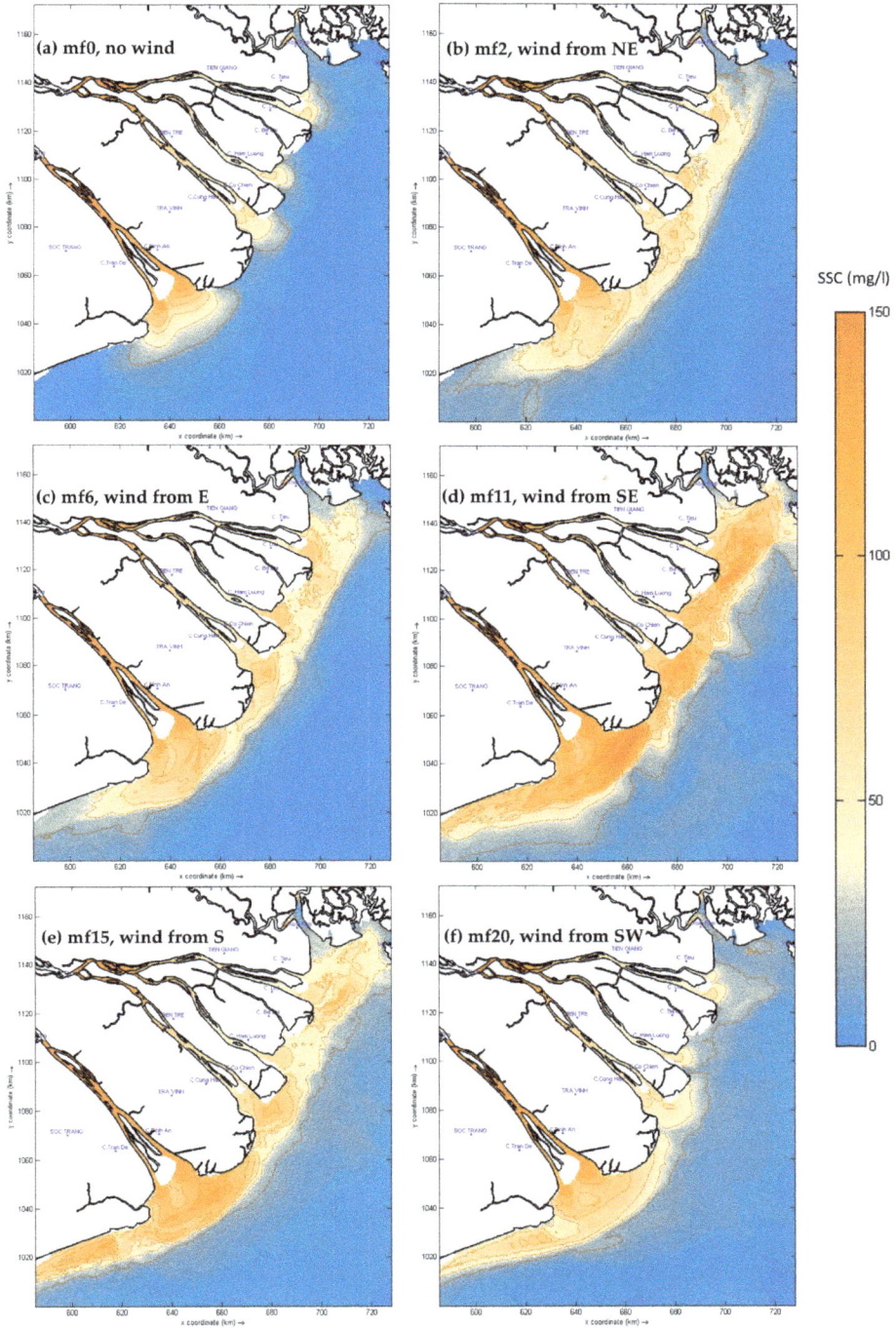

Figure 4. Distribution of suspended sediment concentration in the surface layer in the flood season in the MRD coastal area (1 h before low tide), with no wave or wave heights = 2 m from different directions.

3.2.1. Without Waves

It appears from the wave climate [49] that calm weather conditions (without wave) in the MRD coastal area occurs about 25.2% of the year (14.3% during the flood season and 10.9% during the low flow season). Therefore, annually, there are about 43 days without waves (13 days in flood season and 30 days in the low flow season; see Tables 1 and 2).

In this case with no waves, the distribution of suspended sediment concentration derives from the interaction between sediments originating from the river and tidal oscillation. In the low flow season, sediments from the river settle in the estuary or very close to the mouths (not further than 10–15 km seawards from the estuaries) with SSC in the range 40–60 mg·L^{-1} (Figure 3a). There are relevant differences amongst SSC values along the delta: SSC is higher in Southwestern estuaries (Tran De, Dinh An mouths) than in Northeastern estuaries (Tieu, Dai, Ham Luong mouths). In the flood season, SSC varies between 40 and 80 mg·L^{-1}, with higher values in the Southwestern estuaries and river mouth area as well (Figure 4a).

3.2.2. With Waves

With increasing wave height, the combination of waves and tides strengthens resuspension over increasing depths along the delta. As a result, SSC values increase locally and high SSCs also extend further offshore. However, their distribution strongly depends on wave height and direction.

Low flow season. In the low flow season, NE waves extend the plume up to around 15 km from the coast with SSC values of 30–50 mg·L^{-1} inside and <10 mg·L^{-1} further offshore with the md2 scenario (Figure 3b). Higher SSC values due to erosion are observed with higher waves (md3 scenario), but the plume pattern remains the same (and its time evolution during the tidal cycle—not shown in this paper—is very similar as well). In the case of waves coming from the East, the turbid plume is slightly less extended offshore than with NE waves but with sensitively higher SSC values (scenario md5, SSC about 40–60 mg·L^{-1}, Figure 3c). SSC values and the plume extent increase with increasing wave height (scenarios md6-md8). Waves from SE have the highest impact on suspended sediment transport because they travel perpendicularly toward the coast. With wave height in the 1–3 m range (scenario md10, Figure 3d), higher SSC values (40–70 mg·L^{-1}) covered the all coastal zones up to 15–20 km from the coast. Outside of 20 km from the coast, SSC decreases to below 20 mg·L^{-1}. Waves from the South have a higher impact on Southwestern estuaries (Dinh An, Tran De, Cung Hau) which experienced higher SSC values (30–50 mg·L^{-1}) than on the Northeastern estuaries (with SSC values in the range 20–40 mg·L^{-1}) (scenario md15, Figure 3e). This trend remains the same with higher waves. Waves from the SW direction cause strong erosion in the southern part of the delta, near Tran De and Dinh An mouths. As a result, SSC is higher (about 60–120 mg·L^{-1}) in the Southwestern part than in the Northeastern coastal area with values almost below 30 mg·L^{-1} (scenario md20, Figure 3f).

Flood season. During the flood season, mean water river discharge increased about 4 times compared to the low flow season, and SSC in the river increased by 30%–60%. The general patterns of the turbid plume are similar to the ones observed in the low flow season, with almost the same shape and horizontal extension, but with higher SSC values. With waves from the NE, the highest SSC values reach 70–100 mg·L^{-1} in the Southwestern estuaries (scenario mf2, Figure 4b). With waves from the East, the sediment distribution is quite similar but exhibits slightly higher SSC values around the Northeastern mouths (scenario mf6, Figure 4c). Waves from SE have the higher impact on resuspension and SSC values range from 80 and 150 mg·L^{-1} all along the delta (scenario mf11, Figure 4d). Waves from the South and the SW induce higher SSC values in the Southwestern estuaries up to 80–120 mg·L^{-1} (scenario mf15, Figure 4e). High SSC values (80–100 mg·L^{-1}) are also observed around the Southwestern delta when waves originate from the SW, with lower values in the Northern coastal zone (scenario mf20, Figure 4f). Surprisingly, the impact of waves from the SW seems strongly reduced by fresh water flux in the flood season, through the interaction of waves and fresh water (see Figures 3f and 4f). We observed one noticeable difference between the low flow and flood seasons: except for waves from the SE, the highest values of SSC were observed in the Southern estuaries in the case of SW waves in the low flow season (Figure 3f) and in the case of Southern waves in the flood season (Figure 4e).

Globally, the river plume is very narrow and its extent is reduced throughout the year (as also shown by Loisel *et al.* [25]) for two reasons: (1) most of the particles carried by the river settle in the lower estuary or around the river mouths; (2) the plume is constrained along a baroclinic coastal current as demonstrated by Hordoir *et al.* [33]. Differences between low flow and flood seasons are small because the turbidity patterns are driven by wave action, more than by the river flow.

3.3. Temporal Variation of SSC

For a given river discharge, temporal variation of SSC in the MRD coastal area depends on tidal oscillation and wave conditions. Figure 5 shows the water elevation and SSC values at two stations during 11 days over 14.75 of the full spring-neap tide cycle (for readability purpose). The highest SSC values are observed at low tide or at the beginning of the flood. The lowest SSC values are seen 2–3 h after the high tide, during the ebb period (Figure 5). SSC values are also higher during the spring tide (days 18–22) than the neap tide (day 23 and after), but the difference is quite small: less than 10 mg· L^{-1} at the Northeastern coastal station (Figure 5a), and less than 20 mg· L^{-1} at the Southwestern coastal station (Figure 5b).

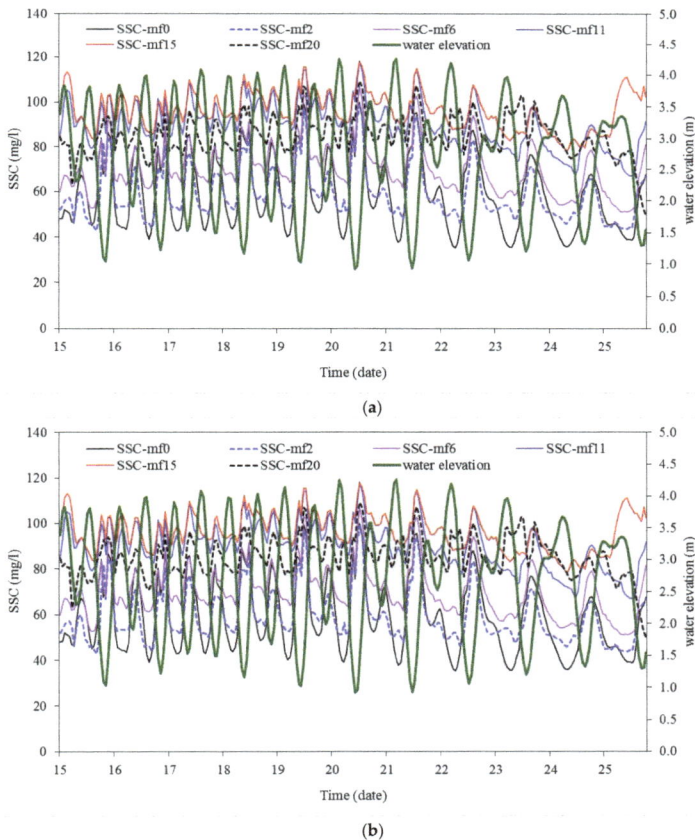

Figure 5. Temporal variation of SSC under some scenario simulations during 11 days of the 14.75 neap-spring cycle, in the flood season: (**a**) at station LT3 to the NE; (**b**) at station LT1 to the SW.

The temporal variation in SSC is similar with or without waves, but SSC values are much higher with waves. In the Northeastern coastal zone, waves from E and SE directions (scenarios mf6 and mf11) induce SSC values 3–4 times higher than under calm conditions, and higher than any other wave direction (Figure 5a). In the Southwestern coastal zone, the highest SSC values occur when waves originate from the S and SW (scenarios mf15 and mf20), 20%–50% higher than when waves come from other directions (Figure 5b).

3.4. Alongshore Sediment Transport

3.4.1. Without Waves

Under calm conditions, a net sediment transport occurs from NE to SW in the low flow season with fluxes across sections m3, m2 and m1 of 0.15, 0.65 and 2.32 × 10^3 tons, respectively, during a spring-neap tide cycle (tab:water-08-00255-t003). In the flood season, with a much higher river discharge, the sediment flux to the Southwest across the Southern transect m1 more than doubled (5.04 × 10^3 tons) as compared to the low flow season, the sediment flux across section m2 decreased by about 40%, and a flux of 0.71 × 10^3 tons was generated towards the Northeastern direction across section m3 directly North to the Northern mouths (tab:water-08-00255-t003).

Table 3. Total sediment flux (10^3 tons) transported across the sections under the 50 scenarios. (Positive values represent sediment transport from the SW to NE, negative values from the NE to SW).

	Low Flow Season					Flood Season			
Scenario	Wave Direction	Cross Section			Scenario	Wave Direction	Cross Section		
		m1	m2	m3			m1	m2	m3
md0		−2.32	−0.65	−0.15	mf0		−5.04	−0.37	0.71
md1		−0.05	−0.02	−0.01	mf1		−0.07	−0.01	0.00
md2	NE	−7.02	−4.44	−1.52	mf2	NE	−7.28	−3.04	−0.89
md3		−6.38	−4.43	−1.23	mf3		−3.65	−2.19	−0.49
md4		−0.30	−0.10	−0.06	mf4		−1.50	−1.00	−0.21
md5		−39.60	−15.97	−5.49	mf5		−0.31	−0.04	0.01
md6	E	−117.58	−51.71	−5.98	mf6		−21.38	−6.11	−1.54
md7		−81.66	−37.03	−3.30	mf7	E	−38.83	−15.98	−1.35
md8		−16.91	−7.61	−0.80	mf8		−29.81	−13.39	0.26
md9		−0.39	−0.07	−0.07	mf9		−0.02	−1.08	0.02
md10		−24.16	2.63	11.12	mf10		−0.08	0.00	0.00
md11	SE	−53.18	17.96	37.14	mf11		−19.14	1.47	6.24
md12		−36.69	15.49	26.06	mf12	SE	−41.88	13.45	27.64
md13		−5.53	2.12	3.23	mf13		−15.00	5.97	10.29
md14		−0.23	0.02	0.01	mf14		−0.07	0.01	0.02
md15		2.02	9.39	3.83	mf15		1.89	0.29	3.79
md16	S	30.21	47.87	32.25	mf16	S	38.13	6.13	41.94
md17		62.96	80.14	67.92	mf17		23.29	22.69	25.24
md18		38.93	44.46	36.71	mf18		10.61	12.48	9.27
md19		−0.13	0.02	0.05	mf19		−0.06	0.01	0.03
md20		67.09	23.05	6.33	mf20		10.42	6.02	2.51
md21	SW	229.37	125.38	51.66	mf21	SW	85.28	55.33	19.94
md22		361.55	201.51	95.72	mf22		95.57	58.32	21.96
md23		234.97	128.43	63.25	mf23		37.57	20.87	8.62
md24		187.43	104.88	54.69	mf24		15.46	8.40	3.65

3.4.2. Sensitivity to Wave Height and Direction

Low flow season. Waves from the NE cause a net alongshore sediment transport to the SW (tab:water-08-00255-t003). This flux increases over each cross section from the Northern estuaries to the Southern ones, with net transports of 2.76×10^3 tons across m3, 8.89 across m2 and 13.45 across m1 (tab:water-08-00255-t004). They are mainly due to waves higher than 2 m (scenarios md2 and md3, tab:water-08-00255-t003). Waves from the East globally induce a higher sediment flux Southwestward, which also increases from across section m3 (15.6×10^3 tons), to m2 (112×10^3 tons) and m1 (256×10^3 tons). These fluxes are higher than with NE waves by a factor of 20 in m1, 12 in m2 and 5 in m3 (tab:water-08-00255-t004). In the low flow season, with waves from the SE (*i.e.*, perpendicular to the coastline), the alongshore sediment transport is globally oriented to the NE direction in the Northern and central area (m3 and m2), and to the SW in the Southern area (1). The global flux to the SW across m1 is much lower than when waves come from the East (120×10^3 tons instead of 256×10^3 tons, see tab:water-08-00255-t004). Waves coming from the S or SW globally induce alongshore sediment fluxes in a Northwestward direction. When waves come from the South, the flux is higher in the central zone (m2, 182×10^3 tons), while it is higher in the South (m1, 1080×10^3 tons) in the case of SW waves. The highest values of sediment flux occur when waves come from SW, with decreasing fluxes from the Southwest to the Northeastern area (1080×10^3 tons across m1, 583 across m2 and 272 across m3). A total of 89% of the total flux of sediment across m1 section is due to SW waves. This percentage drops to 73% and 55% across sections m2 and m3, respectively.

Table 4. Total sediment flux (10^3 tons) transported across sections per year. (Positive values represent sediment transport from the SW to NE, negative values from the NE to SW).

Wave Direction	Low Flow Season			Flood Season			Total Year		
	Cross Section			Cross Section			Cross Section		
	m1	m2	m3	m1	m2	m3	m1	m2	m3
calm	−2.32	−0.65	−0.15	−5.04	−0.37	0.71	−7.356	−1.013	0.555
NE	−13.45	−8.89	−2.76	−12.51	−6.24	−1.59	−25.95	−15.12	−4.35
E	−256.06	−112.43	−15.64	−90.35	−36.59	−2.59	−346.41	−149.02	−18.24
SE	−119.95	38.13	77.48	−76.11	20.90	44.18	−196.06	59.03	121.66
S	133.88	181.88	140.72	73.84	41.61	80.26	207.73	223.49	220.98
SW	1080.28	583.25	271.69	244.25	148.95	56.70	1324.53	732.20	328.39
NEward	1214.52	803.34	489.96	318.22	211.45	182.13	1532.26	1014.72	671.59
SWward	−392.13	−122.03	−18.62	−184.14	−43.20	−4.47	−575.78	−165.15	−22.58
Net transport	822.39	681.31	471.34	134.08	168.25	177.66	956.48	849.57	649.01

Flood season. Although the river discharge is much higher than in the low flow season, the alongshore sediment transport qualitatively occurs in the same way in both seasons. In the case of waves from the NE, the net alongshore sediment transport is still oriented to the Southwest, with increasing values from the North to the South: 1.59×10^3 tons across m3, 6.24×10^3 tons across m2, and 12.51×10^3 tons across m1 (tab:water-08-00255-t004). When waves come from the East, sediment is mainly transported alongshore to the SW, with increasing values from m3 to m1. However, we observe that the net transport is oriented Northeastward at m3 (in the North) in case of wave height less than 2.0 m or higher than 4.0 m (tab:water-08-00255-t003). Similar to the low flow season, waves from the SE induce alongshore sediment transport Northeastward in the Northern (m3) and central delta (m2), and Southwestward in the South (m1) in the flood season. The highest flux occurs across section m1 from the NE to SW with about 76×10^3 tons (tab:water-08-00255-t004). With waves from the South, the net alongshore sediment transport is significantly higher in the Northeast (80×10^3 tons across m3, see tab:water-08-00255-t004) and Southwest (74×10^3 tons across m1) than in the middle delta (42×10^3

tons across m2). Sediment transport is oriented to the NE, except across section m1 with waves less than 0.5 m (tab:water-08-00255-t003).   The same behavior is observed in the case of waves from the SW, with decreasing flux from the South to the North: ~244 × 10^3 tons (m1), 146 × 10^3 tons (m2) and 57 × 10^3 tons (m3). In the Southwestern estuaries of Dinh An-Tran De, the net sediment transport is oriented Southwestward in the case of small waves, and Northeastward with waves higher than 0.5 m (tab:water-08-00255-t003). In that case, the net alongshore sediment flux is thus a consequence of resuspension of sediment along the Southern delta, along the Ca Mau peninsula.

Globally, the net alongshore sediment transport is oriented to the Northeast of the MRD coastal area in both seasons (tab:water-08-00255-t004). In the MRD coastal area, the flood season lasts about 3 months. Even if the instantaneous sediment discharge is much higher, the net alongshore sediment transport is much higher in the low flow season than in the flood season: 86% of the net annual alongshore transport across m1 occurs in the low flow season (822 × 10^3 tons *vs.* 134, see   tab:water-08-00255-t004), 80% across m2 and 73% across m3. Much of the sediment brought by the river settle and is deposited in the estuaries and close to the river mouths.

In the Southwestern area (across section m1), the annual alongshore sediment transport to the Northeast (1532 × 10^3 tons,   tab:water-08-00255-t004) is due to the waves coming from the S and SW and thus to resuspension along the Ca Mau peninsula. Waves from the NE, E and SE, as well as calm conditions, induce a sediment transport of about 575 × 10^3 tons to the Southwest.

In the middle coastal part (across section m2 between the Tien River mouth and Hau River mouth), waves from the SE, S and SW generate an annual sediment transport to the Northeast of 1014 × 10^3 tons, while calm conditions and waves from the NE and E generate alongshore transport of 165 × 10^3 tons.

In the Northern coastal part (across section m3), the total flux to the Northeast of 672 × 10^3 tons is generated in the case of waves from the SE, S and SW, while the sediment transport to the Southwest (in case of waves from NE and E) is restricted to about 23 × 10^3 tons.

As the net alongshore sediment transport decreases from the Southwest (m1) to the Northeast (m3), it shows that the source of global alongshore sediment transport along the MRD is double, generated both by the sediments settling from the river plume, and by erosion, which seems to occur mainly along the Ca Mau peninsula to the Southwest, while a net deposition occurs to the North of section m1.

4. Discussion and Conclusions

In this study, the cumulative alongshore transport of suspended sediments across three sections was estimated to be less than 1.5 × 10^6 tons per year in a given direction. A comment is necessary since the average sediment load of the Mekong River was estimated to be around 145 × 10^6 t yr^{-1} over the last 3 k yr [32,39]. In fact, using a detailed numerical model of sediment transport, erosion and deposition within the Mekong delta, Manh *et al.* [23] reported that total sediment load to the coast was about 50%–60% of the sediment flux at the gauging station of Kratie in Cambodia during a year of low flow (like 2010) or normal condition (like 2009), and that this percentage was about 48% during a year with extreme flooding (like in 2011). In a normal flood year, they estimated that the sediment load to the coast was about 42 × 10^6 tons per year [23]. A complementary study by Xue *et al.* [24] reported that 86.9% of the sediment coming from the Mekong River system was trapped in the MRD estuaries and nearshore area. As a result, in normal years, around 5.5 × 10^6 tons of sediments are transported alongshore and seaward. As most of these sediments seem to settle and deposit very near to the mouths, the estimations obtained for the cumulated alongshore transports in this study are thus consistent with previous studies.

The present study underlined the role of waves in the sediment redistribution along the delta. Under the present configuration and shoreline, the main alongshore flux is not due to the fate of the river plume but to the action of waves coming from the SW and S, which induce resuspension along the Southern delta and sediment transport to the Northeast. Under small wave conditions, the highest SSC values are seen in the estuaries or very close to the mouths. With increasing wave height, SSC values increase and zones with high SSC values extend a little bit offshore. However, the plume never extends further than 20 km from the coast, which is the limit of the subaqueous delta. Our results are consistent with those from Hordoir *et al.* [33], who showed that the plume of the Mekong River is dominated by geostrophy from a dynamical perspective, with freshwater advected by a coastally trapped baroclinic current. This behavior was confirmed by a Rossby number less than 1 throughout the year, due to the large freshwater discharge.

Waves from the SW direction (and to a lesser extent waves from the South) are mainly responsible of the net alongshore transport to the Northeast of the delta. Waves from the SW generate about 86% of the Northeastward flux in the Southern delta (1324×10^3 tons over 1532 across section m1), 72% along the central delta (m2) and 49% in the Northern part (across m3).

A companion paper [80] provided the maps of deposition and erosion (in mm) in the study area, per season, and per year. It showed that: (1) deposition occurs in the lower estuary at a rate reaching 40–50 mm yr^{-1}; (2) deposition is decreasing fast with decreasing water depth in front of the mouths up to 5 m depth; (3) patches of erosion are found along the MRD in water between 5 and 10 m depth; (4) further offshore at 10–15 m depths, deposition occurs at rates <10 mm yr^{-1}. These values are consistent with the accumulation rates determined by Ta *et al.* [32] in front of the MRD for the most recent period. Erosion zones and rates differ slightly from these given by Xue *et al.* [25] over water depth between 5 and 10 m since their study did not take into account the action of the highest waves.

The net sediment transport to the Northeast observed in the Southern part of the delta (along Ca Mau cape) is the main original result of our calculations. This trend is consistent with simulations of the Mekong shelf circulation and sediment transport and dispersal performed for 2005 by Xue *et al.* [22], who reported that the Southwestern part of the MRD (between Tran De mouth and Ca Mau Cape) receives very few sediments from the Mekong River. Deposition in that area should almost result from a slow but persistent deposition process occurring over a much longer (millennial) time scale. Regardless, the sediment deficit along the Southern coastal area of the delta, which results mainly from the action of waves coming from the S and SW in low flow season, is consistent with the geological studies and with the severe erosion along the Ca Mau peninsula, estimated to be an average of 1.1 km^2 yr^{-1} since 1885 [81].

Remote sensing data can be used in complement to numerical simulations to study the river plume dynamics and sediment transport [82–86]. Satellite data are available under clear sky conditions only; such data are numerous during the dry season but less abundant during the wet and mostly cloudy season, which makes it difficult to estimate sediment budgets from remote sensing alone. Continuous and costly measurements or modeling (after validation from *in situ* data) are the two ways to estimate annual fluxes such as given in this study (tab:water-08-00255-t004). However, such estimates could greatly benefit from the available remote sensing data (e.g., [25]) to offer a synoptic view to complement local measurements and further constraint the model. Such a coupling between modeling, satellite data analysis and field measurements may be envisaged in a future step over the Mekong coastal zone.

Acknowledgments: This work was financed by the science and technological cooperation program between the Vietnam Academy of Sciences and Technology (VAST) and the French Institut de Recherche pour le Développement (IRD) within the space science and technological program, through the VT/CB-01/14-15 project: "Study on method for analyzing, assessing and monitoring coastal water quality by using the high and average resolution and multi-temporal remote sensing data; testing the VNRedsat–1", by the project "Interaction between hydrodynamic processes of the Bien Dong (East Sea of Vietnam) and water Mekong River", by the project VAST.ĐLT.06/15-16, and by the project VIETNAMINS from the University of Science and Technology of Hanoi.

Author Contributions: V.D.V. and N.V.T. conceived the study; V.D.V. and N.N.T. organized the field trip and designed the numerical experiments; V.D.V. performed the simulations; V.D.V., N.V.T. and S.O. analyzed the data and simulations; V.D.V. and S.O. wrote the paper.

Conflicts of Interest: The authors declare no conflict of interest.

References

1. Liu, J.P.; Xue, Z.; Ross, K.; Wang, H.J.; Yang, Z.S.; Li, A.C.; Gao, S. Fate of sediments delivered to the sea by Asian large rivers: Long-distance transport and formation of remote alongshore clinothems. *Sediment. Rec.* **2009**, *7*, 4–9.

2. Saito, Y.; Chaimanee, N.; Jarupongsakul, T.; Syvitski, J.P.M. Shrinking megadeltas in Asia: Sea-level rise and sediment reduction impacts from case study of the Chao Phraya delta. *Newsl. IGBP/IHDP Land Ocean Interact. Coast. Zone* **2007**, *2*, 3–9.

3. Vörösmarty, C.J.; Meybeck, M.; Fekete, B.; Sharma, K.; Green, P.; Syvitski, J.P.M. Anthropogenic sediment retention: Major global impact from registered river impoundments. *Glob. Planet. Chang.* **2003**, *39*, 169–190. [CrossRef]

4. Syvitski, J.P.M.; Saito, Y. Morphodynamics of Deltas under the Influence of Humans. *Glob. Planet. Chang.* **2007**, *57*, 261–282. [CrossRef]

5. Wang, H.J.; Yang, Z.S.; Saito, Y.; Liu, J.P.; Sun, X. Interannual and seasonal variation of the Huanghe (Yellow River) water discharge over the past 50 years: Connections to impacts from ENSO events and dams. *Glob. Planet. Chang.* **2006**, *50*, 212–225. [CrossRef]

6. Wang, H.J.; Yang, Z.S.; Wang, Y.; Saito, Y.; Liu, J.P. Reconstruction of sediment flux from the Changjiang (Yangtze River) to the sea since the 1860s. *J. Hydrol.* **2008**, *349*, 318–332. [CrossRef]

7. Vinh, V.D.; Ouillon, S.; Tanh, T.D.; Chu, L.V. Impact of the Hoa Binh dam (Vietnam) on water and sediment budgets in the Red River basin and delta. *Hydrol. Earth Syst. Sci.* **2014**, *18*, 3987–4005. [CrossRef]

8. Schlünz, B.; Schneider, R.R. Transport of terrestrial organic carbon to the oceans by rivers: Re-estimating flux- and burial rates. *Int. J. Earth Sci.* **2000**, *88*, 599–606. [CrossRef]

9. Viers, J.; Dupré, B.; Gaillardet, J. Chemical composition of suspended sediments in World Rivers: New insights from a new database. *Sci. Total Environ.* **2009**, *407*, 853–868. [CrossRef] [PubMed]

10. Rochelle-Newall, E.J.; Chu, V.T.; Pringault, O.; Amouroux, D.; Arfi, R.; Bettarel, Y.; Bouvier, T.; Bouvier, C.; Got, P.; Nguyen, T.M.; *et al.* Phytoplankton diversity and productivity in a highly turbid, tropical coastal system (Bach Dang Estuary, Vietnam). *Mar. Pollut. Bull.* **2011**, *62*, 2317–2329. [CrossRef] [PubMed]

11. Mari, X.; Torréton, J.P.; Chu, V.T.; Lefebvre, J.P.; Ouillon, S. Seasonal aggregation dynamics along a salinity gradient in the Bach Dang estuary, North Vietnam. *Estuar. Coast. Shelf Sci.* **2012**, *96*, 151–158. [CrossRef]

12. Navarro, P.; Amouroux, D.; Duong, T.N.; Rochelle-Newall, E.; Ouillon, S.; Arfi, R.; Chu, V.T.; Mari, X.; Torréton, J.P. Butyltin and mercury compounds fate and tidal transport in waters of the tropical Bach Dang estuary (Haiphong, Vietnam). *Mar. Pollut. Bull.* **2012**, *64*, 1789–1798. [CrossRef] [PubMed]

13. Syvitski, J.P.M.; Vörösmarty, C.J.; Kettner, A.J.; Green, P. Impact of Humans on the Flux of Terrestrial Sediment to the Global Coastal Ocean. *Science* **2005**, *308*, 376. [CrossRef] [PubMed]

14. Ouillon, S. Erosion and sediment transport: Width and stakes. *Houille Blanche* **1998**, *53*, 52–58. [CrossRef]

15. Clift, P.D.; Plumb, R.A. *The Asian Monsoon: Causes, History and Effects*; Cambridge University Press: Cambridge, UK, 2008; p. 288. [CrossRef]

16. Tamura, T.; Horaguchi, K.; Saito, Y.; Nguyen, V.L.; Tateishi, M.; Ta, T.K.O.; Nanayama, F.; Watanabe, K. Monsoon-influenced variations in morphology and sediment of a mesotidal beach on the Mekong River Delta coast. *Geomorphology* **2010**, *116*, 11–23. [CrossRef]

17. Eastham, J.; Mpelasoka, F.; Mainuddin, M.; Ticehurst, C.; Dyce, P.; Hodgson, G.; Ali, R.; Kirby, M. *Mekong River Basin Water Resources Assessment: Impacts of Climate Change*; CSIRO National Research Flagship, Water for a Healthy Country: Canberra, Australia, 2008.

18. Nguyen, V.L.; Ta, T.K.O.; Tateishi, M. Late Holocene depositional environments and coastal evolution of the Mekong River Delta, Southern Vietnam. *J. Asian Earth Sci.* **2000**, *18*, 427–439. [CrossRef]

19. Walling, D.E. The changing sediment load of the Mekong River. *AMBIO* **2008**, *37*, 150–157. [CrossRef]

20. Xue, Z.; Liu, J.P.; Ge, Q. Changes in hydrology and sediment delivery of the Mekong River in the last 50 years: Connection to damming, monsoon, and ENSO. *Earth Surf. Process. Landf.* **2011**, *36*, 296–308. [CrossRef]

21. Le, T.V.H.; Nguyen, H.N.; Wolanski, E.; Tran, T.C.; Haruyama, S. The combined impact on the flooding in Vietnam's Mekong River Delta of local man-made structures, sea level rise, and dams upstream in the river catchment. *Estuar. Coast. Shelf Sci.* **2007**, *71*, 110–116. [CrossRef]

22. Wang, J.J.; Lu, X.X.; Kummu, M. Sediment load estimates and variations in the lower Mekong River. *River Res. Appl.* **2011**, *27*, 33–46. [CrossRef]

23. Manh, N.V.; Dung, N.V.; Hung, N.N.; Merz, B.; Apel, H. Large-scale suspended sediment transport and sediment deposition in the Mekong delta. *Hydrol. Earth Syst. Sci.* **2014**, *18*, 3033–3053. [CrossRef]

24. Xue, Z.; He, R.; Liu, J.P.; Warner, J.C. Modeling transport and deposition of the Mekong River sediment. *Cont. Shelf Res.* **2012**, *37*, 66–78. [CrossRef]

25. Loisel, H.; Mangin, A.; Vantrepotte, V.; Dessailly, D.; Dinh, N.D.; Garnesson, P.; Ouillon, S.; Lefebvre, J.P.; Mériaux, X.; Phan, M.T. Analysis of the suspended particulate matter concentration variability of the coastal waters under the Mekong's influence: A remote sensing approach. *Remote Sens. Environ.* **2014**, *150*, 218–230. [CrossRef]

26. Szczuciński, W.; Jagodziński, R.; Hanebuth, T.J.J.; Stattegger, K.; Wetzel, A.; Mitręga, M.; Unverricht, D.; Phach, P.V. Modern sedimentation and sediment dispersal pattern on the continental shelf off the Mekong River delta, South China Sea. *Glob. Planet. Chang.* **2013**, *110*, 195–213. [CrossRef]

27. Wolanski, E.; Ngoc Huan, N.; Trong Dao, L.; Huu Nhan, N.; Ngoc Thuy, N. Fine sediment dynamics in the Mekong River Estuary, Vietnam. *Estuar. Coast. Shelf Sci.* **1996**, *43*, 565–582. [CrossRef]

28. Wolanski, E.; Nhan, N.H.; Spagnol, S. Sediment dynamics during low flow conditions in the Mekong River Estuary, Vietnam. *J. Coast. Res.* **1998**, *14*, 472–482.

29. Hein, H.; Hein, B.; Pohlmann, T. Recent dynamics in the region of Mekong water influence. *Glob. Planet. Chang.* **2013**, *110*, 183–194. [CrossRef]

30. Mekong River Commission (MRC). *State of the Basin Report: 2003*; MRC: Phnom Penh, Cambodia, 2003.

31. Milliman, J.D.; Meade, R.H. World-wide delivery of river sediment to the oceans. *J. Hydrol.* **1983**, *91*, 1–21. [CrossRef]

32. Ta, K.T.O.; Nguyen, V.L.; Tateishio, M.; Kobayashi, I.; Tanabe, S.; Saito, Y. Holocene delta evolution and sediment discharge of the Mekong River, southern Vietnam. *Quat. Sci. Rev.* **2002**, *21*, 1807–1819. [CrossRef]

33. Hordoir, R.; Nguyen, K.D.; Polcher, J. Simulating tropical river plumes, a set of parametrizations based on macroscale data: A test case in the Mekong Delta region. *J. Geophys. Res.* **2006**, *111*, C09036. [CrossRef]

34. Snidvongs, A.; Teng, S.K. *Global International Waters Assessment, Mekong River GIWA Regional Assessment 55*; University of Kalmar: Kalmar, Sweden, 2006.

35. Unverricht, D.; Szczuciński, W.; Stattegger, K.; Jagodziński, R.; Le, X.T.; Kwong, L.L.W. Modern sedimentation and morphology of the subaqueous Mekong delta, Southern Vietnam. *Glob. Planet. Chang.* **2013**, *110*, 223–235. [CrossRef]

36. Deltares. *Mekong Delta Water Resources Assessment Studies*; Report of the Vietnam-Netherlands Mekong Delta Masterplan Project; Deltares: Delft, The Netherlands.

37. Vietnamese National Centre for Hydro-Meteorological Forecasts. Statistics on Typhoon Occurrence in Vietnam. 2016. Available online: http://ttnh.vnea.org (accessed on 20 April 2016).

38. Tri, V.K. Hydrology and hydraulic infrastructure systems in the Mekong delta, Vietnam. In *The Mekong Delta System*; Renaud, F.G., Kuenzer, C., Eds.; Springer: Dordrecht, Germany, 2012; pp. 49–82. [CrossRef]

39. Lu, X.X.; Siew, R.Y. Water discharge and sediment flux changes over the past decades in the Lower Mekong River: Possible impacts of the Chinese dams. *Hydrol. Earth Syst. Sci.* **2006**, *10*, 181–195. [CrossRef]

40. Kubicki, A. Large and very large subaqueous delta dunes on the continental shelf off southern Vietnam, South China Sea. *Geo-Mar. Lett.* **2008**, *28*, 229–238. [CrossRef]

41. Nguyen, C.T. Processes and Factors Controlling and Affecting the Retreat of Mangrove Shorelines in South Vietnam. Ph.D. Thesis, Kiel University, Kiel, Germany, 2012.

42. Allen, G.P.; Salomon, J.C.; Bassoulet, P.; du Penhoat, Y.; de Grandpre, C. Effects of tides on mixing and suspended sediment transport in macrotidal estuaries. *Sediment. Geol.* **1980**, *26*, 69–90. [CrossRef]

43. Dyer, K.R. *Coastal and Estuarine Sediment Dynamics*; Wiley: Chichester, UK, 1986; p. 342. [CrossRef]

44. Dronkers, J. Tide-induced residual transport of fine sediment. In *Physics of Shallow Estuaries and Bays*; van de Kreeke, J., Ed.; Lecture Notes Coastal Estuarine Studies, Volome 16; Springer: Berlin, Germany, 1986; pp. 228–244. [CrossRef]

45. Sottolichio, A.; Le Hir, P.; Castaing, P. Modeling mechanisms for the turbidity maximum stability in the Gironde estuary, France. *Proc. Mar. Sci.* **2001**, *3*, 373–386. [CrossRef]

46. Lefebvre, J.P.; Ouillon, S.; Vinh, V.D.; Arfi, R.; Panche, J.Y.; Mari, X.; Van Thuoc, C.; Torréton, J.P. Seasonal variability of cohesive sediment aggregation in the Bach Dang-Cam Estuary, Haiphong (Vietnam). *Geo-Mar. Lett.* **2012**, *32*, 103–121. [CrossRef]

47. Weatherall, P.; Marks, K.M.; Jakobsson, M.; Schmitt, T.; Tani, S.; Arndt, J.E.; Rovere, M.; Chayes, D.; Ferrini, V.; Wigley, R. A new digital bathymetric model of the world's oceans. *Earth Space Sci.* **2015**, *2*, 331–345. [CrossRef]

48. Lanh, D.T. *Integrated Water Resources Management and Sustainable Use for the Dong Nai River System*; Technical Report of the Project KC.08.18/06-10; The Southern Institute of Water Resources Research: Ho Chi Minh City, Vietnam, 2010. (In Vietnamese)

49. Groenewoud, P. *Overview of the Service and Validation of the Database*; Reference: RP_A870; BMT ARGOSS: Marknesse, The Netherlands, 2011.

50. Lefevre, F.; Lyard, F.; Le Provost, C.; Schrama, E.J.O. FES99: A global tide finite element solution assimilating tide gauge and altimetric information. *J. Atmos. Ocean. Technol.* **2002**, *19*, 1345–1356. [CrossRef]

51. Lyard, F.; Lefevre, F.; Letellier, T.; Francis, O. Modelling the global ocean tides: Modern insights from FES2004. *Ocean Dyn.* **2006**, *56*, 394–415. [CrossRef]

52. World Ocean Atlas 2013 Version 2 (WOA13 V2). Available online: https://www.nodc.noaa.gov/OC5/woa13/ (accessed on 20 April 2016).

53. Deltares Systems. *Delft3D-FLOW User Manual: Simulation of Multi-Dimensional Hydrodynamic Flows and Transport Phenomena, including Sediments*; Technical Report; Deltares: Delft, The Netherlands, 2014.

54. Booij, N.; Ris, R.C.; Holthuijsen, L.H. A third-generation wave model for coastal regions: 1. Model description and validation. *J. Geophys. Res.* **1999**, *104*, 7649–7666. [CrossRef]

55. Ris, R.C.; Holthuijsen, L.H.; Booij, N. A third-generation wave model for coastal regions: 2. Verification. *J. Geophys. Res.* **1999**, *104*, 7667–7681. [CrossRef]

56. Deltares Systems. *Delft3D-WAVE User Manual: Simulation of Short-Crested Waves with SWAN*; Technical Report; Deltares: Delft, The Netherlands, 2014.

57. Jouon, A.; Lefebvre, J.P.; Douillet, P.; Ouillon, S.; Schmied, L. Wind wave measurements and modelling in a fetch-limited semi-enclosed lagoon. *Coast. Eng.* **2009**, *56*, 599–608. [CrossRef]

58. Hasselmann, K.; Barnett, T.P.; Bouws, E.; Carlson, H.; Cartwright, D.E.; Enke, K.; Ewing, J.; Gienapp, H.; Hasselmann, D.E.; Kruseman, P.; *et al.* *Measurements of Wind Wave Growth and Swell Decay during the Joint North Sea Wave Project (JONSWAP). Deutsche Hydrographische Zeitschrift 8 (12)*; Deutsches Hydrographisches Institut: Hamburg, Germany, 1973.

59. Bouws, E.; Komen, G. On the balance between growth and dissipation in an extreme, depth-limited wind-sea in the southern North Sea. *J. Phys. Oceanogr.* **1983**, *13*, 1653–1658. [CrossRef]

60. Battjes, J.; Janssen, J. Energy loss and set-up due to breaking of random waves. In Proceedings of the 16th International Conference Coastal Engineering, ASCE, Hamburg, Germany, 26 August–6 September 1978; pp. 569–587.

61. Arcement, G.J., Jr.; Schneider, V.R. Guide for Selecting Manning's Roughness Coefficients for Natural Channels and Flood Plains. U.S. Geological Survey Water Supply Paper 2339; 1989. Available online: http://www.fhwa.dot.gov/bridge/wsp2339.pdf (accessed on 20 April 2016).

62. Simons, D.B.; Senturk, F. *Sediment Transport Technology—Water and Sediment Dynamics*; Water Resources Publications: Littleton, CO, USA, 1992.

63. Uittenbogaard, R.E. *Model for Eddy Diffusivity and Viscosity Related to Sub-Grid Velocity and Bed Topography*; Technical Report; WL Delft Hydraulics: Delft, The Netherlands, 1998.

64. Van Vossen, B. *Horizontal Large Eddy Simulations; Evaluation of Computations with DELFT3D-FLOW*; Report MEAH-197; Delft University of Technology: Delft, The Netherlands, 2000.

65. Van Rijn, L.C. Unified view of sediment transport by currents and waves, Part II: Suspended transport. *J. Hydraul. Eng.* **2007**, *133*, 668–689. [CrossRef]

66. Partheniades, E. Erosion and deposition of cohesive soils. *J. Hydraul. Div.* **1965**, *91*, 105–139.

67. Krone, R.B. *Flume Studies of the Transport of Sediment in Estuarial Shoaling Processes*; Hydraulic Engineering Laboratory and Sanitary Engineering Research Laboratory, University of California: Berkeley, CA, USA, 1962.

68. Douillet, P.; Ouillon, S.; Cordier, E. A numerical model for fine suspended sediment transport in the south-west lagoon of New-Caledonia. *Coral Reefs* **2001**, *20*, 361–372. [CrossRef]

69. Winterwerp, J.C.; van Kesteren, W.G.M. *Introduction to the Physics of Cohesive Sediment in the Marine Environment*; Elsevier: Amsterdam, The Netherlands, 2004; pp. 1–466.

70. Hung, N.N.; Delgado, J.M.; Güntner, A.; Merz, B.; Bárdossy, A.; Apel, H. Sedimentation in the floodplains of the Mekong Delta, Vietnam Part II: Deposition and erosion. *Hydrol. Process.* **2004**, *28*, 3145–3160. [CrossRef]

71. Portela, L.I.; Ramos, S.; Rexeira, A.T. Effect of salinity on the settling velocity of fine sediments of a harbour basin. *J. Coast. Res.* **2013**, *2*, 1188–1193. [CrossRef]

72. Van Rijn, L.C. Mathematical modeling of suspended sediment in non-uniform flows. *J. Hydraul. Eng.* **1986**, *112*, 433–455. [CrossRef]

73. Ouillon, S.; Le Guennec, B. Modelling non-cohesive suspended sediment transport in 2D vertical free surface flows. *J. Hydraul. Res.* **1996**, *34*, 219–236. [CrossRef]

74. Van Rijn, L. *Principles of Sediment Transport in Rivers, Estuaries and Coastal Seas*; Aqua Publications: Amsterdam, The Netherlands, 1993.

75. Madsen, O.S. Spectral wave–current bottom boundary layer flows. In Proceedings of the 24th International Conference on Coastal Engineering Research Council, Kobe, Japan, 23–28 October 1994; pp. 384–398. [CrossRef]

76. Walstra, D.J.R.; Roelvink, J.A. 3D Calculation of Wave Driven Cross-shore Currents. In Proceedings of the 27th International Conference on Coastal Engineering, Sydney, Australia, 16–21 July 2000; pp. 1050–1063. [CrossRef]

77. Nash, J.E.; Sutcliffe, J.V. River flow forecasting through conceptual models, Part I—A discussion of principles. *J. Hydrol.* **1970**, *10*, 282–290. [CrossRef]

78. Krause, P.; Boyle, D.P.; Bäse, F. Comparison of different efficiency criteria for hydrological model assessment. *Adv. Geosci.* **2005**, *5*, 89–97. [CrossRef]

79. Kvale, E.P.; Archer, A.W.; Johnson, H.R. Daily, monthly, and yearly tidal cycles within laminated siltstones of the Mansfield Formation (Pennsylvanian) of Indiana. *Geology* **1989**, *17*, 365–368. [CrossRef]

80. Vinh, V.D.; Lan, T.D.; Tu, T.A.; Anh, N.K.; Tien, N.N. Influence of dynamic processes on morphological change in the coastal area of Mekong river mouth. *J. Mar. Sci. Technol.* **2016**, *16*, 32–45. [CrossRef]

81. Saito, Y. Deltas in Southeast and East Asia: Their evolution and current problems. In *APN/SURVAS/LOICZ Joint Conference on Coastal Impact of Climate Change and Adaption in the Asia-Pacific Region*; Mimura, N., Yokoki, H., Eds.; APN: Kobe, Japan, 2000; pp. 185–191.

82. Durand, N.; Fiandrino, A.; Fraunie, P.; Ouillon, S.; Forget, P.; Naudin, J.J. Suspended matter dispersion in the Ebro ROFI: An integrated approach. *Cont. Shelf Res.* **2002**, *22*, 267–284. [CrossRef]

83. Ouillon, S.; Douillet, P.; Andréfouët, S. Coupling satellite data with *in situ* measurements and numerical modeling to study fine suspended sediment transport: A study for the lagoon of New Caledonia. *Coral Reefs* **2004**, *23*, 109–122. [CrossRef]

84. Stroud, J.R.; Lesht, B.M.; Schwab, D.J.; Beletsky, D.; Stein, M.L. Assimilation of satellite images into a sediment transport of Lake Michigan. *Water Resour. Res.* **2009**, *45*, W02419. [CrossRef]

85. Carniello, L.; Silvestri, S.; Marani, M.; D'Alpaos, A.; Volpe, V.; Defina, A. Sediment dynamics in shallow tidal basins: *In situ* observations, satellite retrievals, and numerical modeling in the Venice Lagoon. *J. Geophys. Res. Earth Surf.* **2014**, *119*, 802–2015. [CrossRef]

86. Yang, X.; Mao, Z.; Huang, H.; Zhu, Q. Using GOCI retrieval data to initialize and validate a sediment transport model for monitoring diurnal variation of SSC in Hangzhou Bay, China. *Water* **2016**, *8*, 108. [CrossRef]

water

MDPI

Article

Spatiotemporal Variation of Turbidity Based on Landsat 8 OLI in Cam Ranh Bay and Thuy Trieu Lagoon, Vietnam

Nguyen Hao Quang [1,*], Jun Sasaki [2], Hiroto Higa [3] and Nguyen Huu Huan [4]

[1] Institute of Marine Science and Fishing Technology, Nha Trang University, 02 Nguyen Dinh Chieu Street, Nha Trang 650000, Vietnam

[2] Sasaki Laboratory, Department of Socio-Cultural Environmental Studies, Graduate School of Frontier Sciences, the University of Tokyo, 5-1-5 Kashiwanoha Kashiwa, Chiba 277-8563, Japan; jsasaki@k.u-tokyo.ac.jp

[3] Yokohama National University, 79-5 Tokiwadai, Hodogaya-ku, Yokohama 240-8501, Japan; higa-h@ynu.ac.jp

[4] Institute of Oceanography, Vietnam Academy of Science and Technology (VAST), 01 Cau Da, Nha Trang 650000, Vietnam; nghhuan@gmail.com

* Correspondence: ri.nguyenri@gmail.com; Tel.: +84-933-566-290

Received: 12 April 2017; Accepted: 15 June 2017; Published: 8 August 2017

Abstract: In recent years, seagrass beds in Cam Ranh Bay and Thuy Trieu Lagoon have declined from 800 to 550 hectares, resulting insignificantly reducing the number of fish catch. This phenomenon is due to the effect of the degradation of water environment. Turbidity is one of the most important water quality parameters directly related to underwater light penetration which affects the primary productivity. This study aims to investigate spatiotemporal variation of turbidity in the waters with major factors affecting its patterns using remote sensing data. An algorithm for turbidity retrieval was developed based on the correlation between in situ measurements and a red band of Landsat 8 OLI with $R^2 = 0.84$ ($p < 0.05$). Simulating WAves Nearshore (SWAN) model was used to compute bed shear stress, a major factor affecting turbidity in shallow waters. In addition, the relationships between turbidity and rainfall, and bed shear stress induced by wind were analyzed. It was found that: (1) In the dry season, turbidity was low at the middle of the bay while it was high in shallow waters nearby coastlines. Resuspension of bed sediment was a major factor controlling turbidity during time with no rainfall. (2) In the rainy season or for a short time after rainfall in the dry season, turbidity was high due to a large amount of runoff entering into the study area.

Keywords: remote sensing reflectance; turbidity; seagrass beds; bed shear stress; fresh water runoff; oceanic water intrusion

---

1. Introduction

Turbidity is an important water quality parameter and a surrogate for water clarity [1,2]. Turbidity can harm fish and other aquatic organisms by reducing feed supplies, degrading spawning grounds, and affecting gill function [3]. An increase or decrease in water clarity can negatively impact on biological components of the system that may be adapted to specific light-penetrating conditions [4,5]. In estuarine waters with high turbidity, the concentration of dissolved oxygen can dramatically decrease as a result of its imbalance between autotrophic and heterotrophic processes, which may lead to decline in marine organisms [6].

Traditionally, turbidity is estimated visually using a Secchi disk or measured directly with nephelometry [2,4]. In recent years, the technique of ocean color remote sensing has become a useful tool to map turbidity and suspended particulate matter concentration at surface in turbid coastal waters [7]. The strength of using remote sensing for water quality analysis is its ability to capture synoptic data of a whole study area to produce continuous surface data, often showing detailed spatial variability in water quality [8]. While a number of remote sensing studies have been devoted to retrieving total suspended solids (TSS), studies on retrieving turbidity have been limited [7]. Even though satellite remote sensing cannot detect near-bed concentration, it can be used to study spatiotemporal variation of surface turbidity, and satellite-derived turbidity maps are valuable tools to study the influence of turbidity in shallow waters influenced by resuspension of bottom sediment and river plume dynamics.

Cam Ranh Bay and Thuy Trieu Lagoon opening to Vietnam East Sea/South China Sea (Figure 1) were known as the second largest seagrass area in Vietnam with abundant biotic resources contributing to development of aquaculture and marine production. The area, however, has decreased from 800 to 550 ha due to degradation of water environment [9] and marine resources as well as the number of fish catch have significantly decreased [10]. The study site is surrounded by hillocks and rivers; therefore, the turbidity and total suspended solids significantly increase when heavy rains bring increased inputs of suspended substances. High concentration of turbidity negatively affects aquaculture (seaweed, lobster and grouper) farming in Cam Ranh Bay and Thuy Trieu Lagoon.

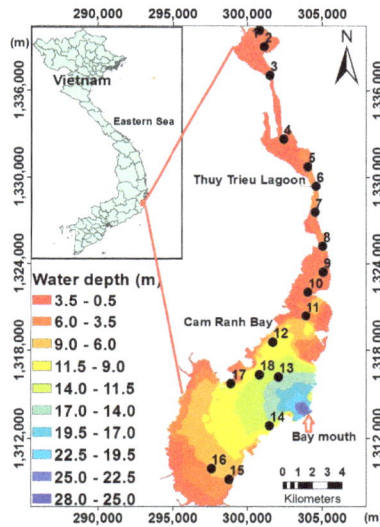

Figure 1. Location of the sampling stations (black points) and bathymetry of Cam Ranh Bay and Thuy Trieu Lagoon.

Estimation of turbidity distribution in complex environments like Cam Ranh Bay and Thuy Trieu Lagoon requires non-traditional approaches. Remote sensing techniques offer reliable advantages to observe and understand the variation in space and time and especially in large area with restricted access, e.g., military zones such as Cam Ranh Bay. Mapping turbidity as well as other water quality parameters has been routinely made using data from dedicated wide-swath ocean color instruments such as Orbview-2/SeaWiFS, Aqua/MODIS and ENVISAT/MERIS medium resolution images [11]. For small and narrow areas, however, such as the study site (the minimum width is just 250 m), these applications are inappropriate because their low spatial resolutions lead to numerous mixed pixels, resulting in an estimation with low precision. Compared with these medium resolution images, Landsat 8 Operational Land Imager (Landsat 8 OLI) has higher spatial resolution of 30 m. Landsat 8 has been widely used in coastal management and especially has been demonstrated to be a useful tool for monitoring of coastal sediment concentrations at high resolution and accuracy due to its higher signal-to-noise ratio (S/N) compared to previous Landsat imagers [11]. Besides, Landsat 8 OLI has been innovated with 12 bits of radiance resolution compared to 8 bits for the older ones, Landsat 5 TM and Landsat 7 ETM+.

Numerous studies on ocean color remote sensing for water quality parameter retrieval have been done, most of which follow three main methods: (i) using semi-analytic models based on the correlation between the inherent optical properties (IOPs) and the water quality parameters [7,11–15]; (ii) using empirical models between IOPs and these parameters [1,16,17]; and (iii) using empirical models between remote sensing reflectance and the same parameters [18–21]. The third approach has been applied in this study, which is based on the correlation between remote sensing reflectance values extracted from Landsat 8 OLI and field measurements.

Using Cam Ranh Bay and Thuy Trieu Lagoon as a typical case study for turbid deep waters, the aim of this study is to determine the variation as well as major factors affecting the spatiotemporal patterns of turbidity. To achieve this, the following specific objectives must be satisfied: (1) estimating turbidity based on Landsat 8 OLI images and in situ measurements; and (2) analyzing its spatial and temporal variability along with their physical mechanisms.

2. Materials and Methods

2.1. Study Site

Cam Ranh Bay is an enclosed water located in the south of Khanh Hoa Province with approximately 19 km in length, 7 km in width and the average depth of 15 m while Thuy Trieu Lagoon is a narrow enclosed water connected with the north of the bay at Long Ho Bridge with approximately 16 km in length, 250 m in width and 0.5 to 6 m in depth. This water system consisting of the bay and the lagoon has the total area of approximately 119 km^2,the total length of 35 km and the average depth of 10 m parallel to the coast of central Vietnam (Figure 1) and connected to the East China Sea through the bay entrance with 1.2 km in width and 30 m in depth [22]. These waters are surrounded by sandy hill sand sheltered from winds. The study site represents aquaculture and fishing areas for sustaining more than 150,000 local people.

There are no major rivers entering into the bay and lagoon, but many streams or small rivers discharging into this area (Figure 2). Most of these streams/rivers have dams upstream to control and distribute the water in both dry and rainy season. Dams open during droughts (dry season) or when the water volume raises significantly (rainy season). Therefore, turbidity contributed from rivers to the study area is strongly influenced by the operation of the dams.

Figure 2. Dams and rivers surrounding study area (Google Earth).

During the dry season, these rivers have almost no water flowing into the Bay and Lagoon. In contrast, throughout the rainy season with high rainfall, the waters of the study area are affected by fresh water in the upper regions, reservoirs and rivers [10].

2.2. Field Survey

The availability of concurrent remote sensing and in situ data is considered a key element in this study. Field surveys were performed in Cam Ranh Bay and Thuy Trieu Lagoon on 16–19 September during the rainy season in 2015 and on 12–13 February during the dry season in 2016. The latter was performed concomitantly with a Landsat 8 OLI image obtained at 10:00 a.m. on 14 February 2016 to minimize the difference between in situ measurements and remotely sensing data. Because this is an enclosed water system with the resident times of more than 10 and 4 days [10] in dry and rainy seasons, respectively, one day lag is acceptable to develop turbidity algorithm.

Surface water samples and surface sediment samples were collectedat 18 stations (Figure 1) using a water sampler, plastic cup, grab, and Petersen sediment samplers. Water quality parameters, including temperature, salinity, turbidity, and chlorophyll-a fluorescence, were measured using a multiple water quality sensor, AAQ 1186 (Alec Electronics), with the interval of 0.1 m in the vertical at the same 18 stations. Water and sediment samples were transferred to laboratory of Department of Marine Ecology, Institute of Oceanography, Nha Trang, Vietnam, immediately after the field measurements. Total suspended solids (TSS), chlorophyll-a, inorganic suspended solids (ISS) andorganic suspended solids (OSS) were analyzed using the water samples while grain size distribution was determined using 18 sediment samples.

2.3. Analysis of Water and Sediment Samples

Total suspended solids (TSS) was determined using a vacuum filtration technique. TSS of a water sample was calculated by pouring a suitable measured volume of water (typically 1.5 to 2 liters but less if the particulate density was high) through a pre-weighed filter of a 0.45 μm pore size placed in a millipore filter holder and attach a controlled vacuum source which does not exceed approximately 1/3 atm, and then weighing the filter again after drying (in an oven at 103–105 °C for at least one hour in aluminum dish) to remove all the water [23]. TSS is the retained material on a standard glass fiber filter after filtration of a well-mixed sample divided by the total volume of sampled water in mg/L.

Organic suspended solids (OSS) was determined using a similar method as TSS by taking the difference between the original TSS and TSS after burning in an oven at 500 °C. Inorganic suspended solids (ISS) was then calculated by subtracting OSS from TSS [23].

Chlorophyll-a (Chl-a) concentration was determined with an established in vitro laboratory analysis where a solvent was used to extract chlorophyll pigments from phytoplankton samples. The optical signals of the extracted pigments were then measured using a spectrophotometer [24,25]. In this study, absorbance of samples was measured at four wavelengths, 630, 647, 664 and 750 nm. Five or tenmilliliters ofacetone solution were used when concentration of algae was low or very high, respectively. Chl-a concentration was estimated by the formula of Jeffrey and Humphrey (1975):

$$\text{Chl-a (mg/m}^3) = (11.85E_{664} - 1.54E_{647} - 0.08E_{630}) \times V_{\text{acetone}}/V_{\text{water}} \tag{1}$$

where E_{664}, E_{647} and E_{630} are the absorptions of 664, 647 and 630 nm (after subtract absorption of E_{750}), respectively, V_{acetone} is the extracted volume of acetone (mL), and V_{water} is the volume of water sample [24].

Sediment grain size distribution was determined by sieve analysis (the mesh sizes in mm are 0.0001, 0.004, 0.016, 0.031, 0.083, 0.125, 0.25, 0.5, 1, 2, 4 and >4), commonly known as the gradation test. The top sieve has the largest screen openings and the screen opening sizes decrease with each sieve down to the bottom sieve which has the smallest opening size screen. During this process, the particles were compared with the apertures of every single sieve. The probability of a particle passing through the sieve mesh was determined by the ratio of the particle size to the sieve openings, the orientation of the particle and the number of encounters between the particle and the mesh openings [26].

2.4. Meteorological and Oceanographic Data Collections

Observed meteorological data, including wind and precipitation, are very limited in this study area and no complete datasets are available for the purpose of our analysis. Instead, regarding wind data, a meteorological reanalysis dataset provided by National Center for Atmospheric Research Final (NCEP-FNL) Operational Model Global Tropospheric Analyses was collected covering the period from 2013 to 2016 with the geospatial and temporal resolutions of $1° \times 1°$ and 6 h, respectively. Rainfall data were obtained from NOAA CPC Morphing Technique (CMORPH) Global Precipitation Analyses that is available on the website of National Center for Atmospheric Research (NCAR). Its geospatial resolution is $0.25° \times 0.25°$ from $0.125°$ E to $359.875°$ E and $59.875°$ N to $59.875°$ S (1440×480 grids) and the temporal resolution is 3 h.

Tidal levels were calculated using NAOTIDE model developed by National Astronomical Observatory, Japan, which is designed to predict ocean tidal heights at any given time and location using an ocean tide model by assimilating TOPEX/POSEIDON altimeter data. Since the study area is small and short in length, it was assumed that tidal levels are uniform in the bay and lagoon and a representative point at the bay mouth ($11.878142°$ latitude (lat.), $109.196341°$ longitude (long.)).

Besides, bathymetric data were provided by Institute of Oceanography with aspatial resolution of 200 m, measured in 2012.

2.5. Turbidity Retrieval Using Landsat 8 OLI

2.5.1. Image Acquisition

Remotely sensed data used for this study are Landsat 8 OLI at path 123 and row 52. Landsat 8 OLI images at L1T—Orthorectified products were downloaded freely from the website of United States Geological Survey (USGS) (http://earthexplorer.usgs.gov/).

Landsat 8 launched as the Landsat Data Continuity Mission on 11 February 2013, and is equipped with state-of-the-art imaging technology and other advancements to better collect data. Landsat 8 OLI has nine spectral bands ranging from 433 nm to 1390 nm at 30 m spatial resolution and one panchromatic channel at 15 m spatial resolution. This is a sun-synchronous orbiting satellite with an altitude of 705 km, an orbital inclination of $98.2°$, and a circle time of 98.8 min around the earth. Landsat 8 OLI has been innovated with 12 bit of radiance resolution compared to 8 bit forthe older ones, Landsat 5 TM and Landsat 7 ETM+. In addition, the OLI sensor employs push broom scanner technology that enables data acquisition with much better performance in terms of the signal-to-noise ratio (S/N) and higher radiometric resolution [15]. Specification of Landsat 8 OLI is summarized in Table 1.

Table 1. Specification of Landsat 8 OLI.

Landsat 8	
Band Description (30 m Native Resolution unless Otherwise Denoted)	**Wavelength (μm)**
Band 1—blue	0.43–0.45
Band 2—blue	0.45–0.51
Band 3—green	0.53–0.59
Band 4—red	0.64–0.67
Band 5—near infrared	0.85–0.88
Band 6—shortwave infrared	1.57–1.65
Band 7—shortwave infrared	2.11–2.29
Band 8—panchromatic (15 m)	0.50–0.68
Band 9—cirrus	1.36–1.38
Band 10—thermal infrared (100 m)	10.60–11.19
Band 11—thermal infrared (100 m)	11.50–12.51

Landsat 8 OLI images at L1T after system radiation correction and geometry correction [15] were collected with less cloud cover, no rain, high visibility and suitable conditions. Totally, 18 high quality images (Table 2) were used to estimate turbidity in Cam Ranh Bay and Thuy Trieu Lagoon. Each scene was downloaded as zipped folder containing TIFF images of the scene for each band and accompanying metadata file.

Table 2. Date acquisition and cloud cover (%) of Landsat 8 OLI images obtained.

No.	Cloud Cover (%)	Date Acquisition	No.	Cloud Cover (%)	Date Acquisition
1	5.00	14 February 2016	10	1.60	15 March 2015
2	5.50	9 October 2015	11	3.41	26 January 2015
3	2.50	23 September 2015	12	0.55	20 September 2014
4	0.49	7 September 2015	13	4.03	2 July 2014
5	4.52	22 August 2015	14	17.10	15 May 2014
6	15.63	6 August 2015	15	8.34	12 March 2014
7	8.23	21 July 2015	16	2.51	8 February 2014
8	6.03	3 June 2015	17	0.24	16 August 2013
9	3.66	16 April 2015	18	0.28	29 June 2013

2.5.2. Masking

Before atmospheric correction, only the water body should be retained and the rest needs to be masked. Masking of the Landsat imagery was performed using ArcGIS software Version 10.2.2. An area of interest was extracted as shapefiles using Google Earth and imported onto ArcGIS (.evf file) and overlaid on Landsat 8 OLI image. Using a masking tool embedded in ArcGIS, only the water area was retained. This process is required to obtain better contrast stretch of study area, to remove meaningless pixels and to eliminate the noises from other objects irrelevant to the main targets. Masking technique can also hide cloud cover that is the main source of noises and distortions of the satellite images.

2.5.3. Atmospheric Correction

There is a variety of atmospheric correction methods and associated models, such as DOS (Dark Object Subtraction), FLAASH (Fast Line-of-sight Atmospheric Analysis of Spectral Hypercubes), 6S (Second Simulation of the Satellite Signal in the Solar Spectrum), QUAC (Quick Atmospheric Correction), etc. Among them, FLAASH is one of the most widely used tools because of its higher accuracy and easier use compared to the others [14,20,27–30].

FLAASH model, which is a MODTRAN-based "atmospheric correction" software package, embedded in ENVI software is a strong tool for atmospheric correction [29]. FLAASH is a first-principles atmospheric correction tool that corrects wavelengths in the visible through near-infrared and shortwave infrared regions, up to 3 μm [31]. In this study, FLAASH method was applied to retrieve reflectance from the surface.

In the first step of atmospheric correction process, radiometric calibration, image data were converted from brightness values in digital number (D_{br}) into radiance at top of atmosphere, L_{TOA}. Gain and offset values from the image header files were used.

$$L_{TOA} = M_L D_{br} + A_L \tag{2}$$

where M_L is multiplicative factor (gain) and A_L is additive factor (offset) provided in the metadata file. Subsequently, radiance at the top of atmosphere was converted to the non-dimensional reflectance at the top of atmosphere, ρ_{TOA}, defined as the reflectance measured by a space-based sensor flying higher than the earth's atmosphere. The reflectance at the top of atmosphere was computed by normalizing L_{TOA} to the band averaged solar irradiance:

$$\rho_{TOA} = \frac{\pi L_{TOA} d^2}{\mathrm{ESUN}_\lambda \cos\theta_s} \tag{3}$$

where d is Earth–Sun distance in astronomical units; $ESUN_\lambda$, the direct solar radiation component, is mean solar exoatmospheric irradiances; and θ_s is solar zenith angle.

Converting reflectance at the top of the atmosphere (ρ_{TOA}) into surface reflectance is the next step in the atmospheric correction process. The key parameters used in the FLAASH module of ENVI were: *tropical* for atmospheric model, *tropospheric* for the aerosol model because the study area is far from urban or industrial locations belonging to tropical area, and 2-band (K-T) for the aerosol retrieval; initial visibility was chosen as 40 km, depending on the image quality, and *no water retrieval* was conducted because no specific band was configured with the Landsat instruments. After carrying out the atmospheric correction, radiance values were converted into surface reflectance (ρ_w) [30] which was further converted to remote sensing reflectance (R_{rs}) by dividing pi (π):

$$R_{rs} = \rho_w / \pi \tag{4}$$

Once the image is calibrated as R_{rs}, it is possible to apply directly the relationship between turbidity and R_{rs} computed with a training set of in situ point data to obtain a gridded image of turbidity [32]. In the current study, R_{rs} was used to develop turbidity model.

2.5.4. Regression Models

Using preprocessed satellite images, it is possible to identify which band's reflectance showing the most significant regression with turbidity. To avoid non-water features such as artificial structures (bridges) or eliminate distortion as well as disturbances, the pixel data within a 90-m distance (3 × 3 pixels) from structures were excluded from the analysis [33,34]. The average values of 3 × 3 pixels were extracted for statistical analyses by defining a region of interest (ROI) corresponding to that specific area. These ROIs were created as rectangles corresponding to the stations measured in the field.

Subsequently, simple linear regressions were exploited to develop a relation between in situ measurements and the space-based observations. Ritchie et al. developed algorithms for remote sensing retrieving water quality parameters, including suspended solids [35]. The general forms of these empirical equations were given by the following four types of expressions:

$$
\begin{aligned}
Y &= A + BX \\
Y &= AB^X \\
Y &= A + B \ln X \\
\ln Y &= A + BX
\end{aligned}
\tag{5}
$$

where X is the remote sensing measurement (i.e., radiance, reflectance, and energy) and Y is the water quality parameter of interest (i.e., suspended sediment, and chlorophyll). A and B are empirically derived factors and X could be reflectance, radiance, energy in a single band, or ratio of two bands. Most of the studies on remote sensing for retrieving water quality parameters have adopted this concept, which was also followed by the present study to develop an algorithm for turbidity retrieval based on the correlation between remote sensing reflectance and in situ turbidity.

2.6. Numerical Estimation of Bed Shear Stress

To discuss the effect of waves on resuspension of sediment and the resultant increase in turbidity, spatiotemporal bed shear stresses were estimated by combining numerical wave hindcasting using SWAN and the linear water wave theory.

2.7. Numerical Hindcasting of Wave Field Using SWAN

SWAN (Simulating WAves Nearshore) model developed by Delft University of Technology, the Netherlands, is a third-generation, discrete spectral wave model that describes the evolution of the two-dimensional wave energy spectrum in arbitrary conditions of wind, currents, and bathymetry [36,37]. This model has been widely used in communities of coastal engineering and physical oceanography. The governing equation is given by:

$$\frac{\partial N}{\partial t} + \frac{\partial C_x N}{\partial x} + \frac{\partial C_y N}{\partial y} + \frac{\partial C_\sigma N}{\partial \sigma} + \frac{\partial C_\theta N}{\partial \theta} = \frac{S}{\sigma} \tag{6}$$

where $N(\sigma, \theta; x, y, t)$ is the action density as a function of the intrinsic frequency σ, the wave direction θ, the horizontal coordinates of x and y, and the time t. The action density is defined as $N = E/\sigma$ using the energy density E. The first term on the left-hand side represents the local rate of change of action density in time, the second and third terms represent propagation of action in geographical (x, y) coordinates with the wave celerities of C_x and C_y. The fourth term denotes the shifting of the relative frequency due to variations in depths and currents and the fifth term represents depth and current-induced refraction. The source term $S(\sigma, \theta, ; x, y, t)$ on the right-hand side represents the generation, dissipation and non-linear wave-wave interactions [36].

Estimation of Bed Shear Stress

There are various models for predicting wave induced or combined wave and current induced bed shear stress [38]. In Cam Ranh Bay and Thuy Trieu Lagoon, the magnitudes of typical tide induced currents in dry and rainy seasons are 0.15 m/s and 0.25 m/s, respectively [39]. Considering these small current velocities and shallow water bodies, the bed shear stress due to wind waves can be much larger than that due to currents [2], and thus only the wave induced stress was included in this study. The maximum bed shear stress τ_{bm} is estimated by:

$$\tau_{bm} = \frac{1}{2} \rho f u_{bm}^2 \tag{7}$$

where ρ and f are the water density and the friction factor, respectively, and u_{bm} is the amplitude of the wave induced orbital velocity at the bottom determined by [40]:

$$u_{bm} = \frac{\pi H_s}{T_p \sinh(2\pi h/L)} \tag{8}$$

using the small amplitude water wave theory, where h is the water depth, H_s, T_p and L are the significant wave height, the peak wave period and the wavelength, respectively, obtained from the SWAN computation.

In the present study, owning to the lack of data on wave field and bottom roughness in the study site, f is assumed to have a constant value of 0.004 [41]. The description of adopted model parameters and computational conditions is presented in Table 3.

Table 3. Description of computational conditions for SWAN (Simulating WAves Nearshore) model.

Input Parameters	Input Values/Command
(a) Details of computational grid and input bottom grid	
Computational Grid	Regular and uniform
Origin (E, N), direction of positive x-axis of computational grid	292022.3 1306301 0. (the original grid cell on the lowest left corner/UTM)
Length of computational grid (x, y)	16315 m, 33735m
Number of meshes (x, y)	251, 520
Spectral direction	Circle
Number of mesh in θspace	36
Discrete frequency (lowest, highest)	0.05 Hz, 1 Hz
Grid resolution in frequency-space	31
Bottom Grid (input file)	Regular and Uniform
Number of meshes (x, y)	251, 520
Δx, Δy (grid resolution)	65 m, 65 m
(b) Details of boundary condition	
Boundary spectral shape	JONSWAP as default
Gamma	3.3—peak enhancement parameter of JONSWAP spectrum
Peak	Peak Period. This is default
DSPR	Expressing the width of the directional distribution
Power	The directional width is expressed with the power m itself, this option is default
Initial Values for Stationary computation	Zero (0), The initial spectral densities are all 0
(c) Physical processes employed	
Wave-generation mode	GEN3 Komen
Whitecapping	Calculation based on Komen
Computation of quadruplet wave interaction	Iquad = 2 (default)
Breaking	Constant with alpha= 1 and gamma (breaker index) = 0.73 (default)
Bottom friction	Activated using JONSWAP formulation, Friction coefficient: constant (0.067 $m^2\ s^{-3}$)
Triad	To activate the triad wave-wave interactions
Wave limiter	Activated; upper threshold = 10.0;Threshold for fraction of breaking waves = 1.0

3. Results

3.1. Field Observation and Collected Data

Analyses of Water Sample

Measured mean, minimum and maximum values for Total, Organic, and Inorganic Suspended Solids (TSS, OSS and ISS, respectively), and chlorophyll-a concentrations (Chl-a) are presented in Table 4 along with the ratios of OSS/TSS and ISS/TSS in September 2015 (rainy season) and in February 2016 (dry season). Turbidity is also shown in February 2016 in FTU (Formazine Turbidity Unit).

Table 4. Mean, minimum and maximum values of water quality variables in (a) rainy season (September 2015) and (b) dry season (February 2016).

	TSS (mg/L)	OSS (mg/L)	ISS (mg/L)	Chl-a (µg/L)		OSS/TSS	ISS/TSS
(a) In Rainy Season (September 2015)							
Mean	5.96	2.09	3.87	5.34		0.43	0.57
Min.	0.95	0.55	0.15	0.81		0.21	0.16
Max.	25.70	7.30	18.40	13.77		0.84	0.79

	TSS(mg/L)	OSS (mg/L)	ISS (mg/L)	Chl-a (µg/L)	Turbidity (FTU)	OSS/TSS	ISS/TSS
(b) In Dry Season (February 2016)							
Mean	7.82	3.80	4.01	20.79	1.41	0.46	0.54
Min.	1.75	0.80	0.60	1.56	0.21	0.22	0.20
Max.	52.00	31.11	20.89	266.01	3.46	0.80	0.78

At the top of the lagoon with very shallow waters of around 0.5 to 1.5 m deep (Figure 1), Chl-a concentrations were higher than those at other stations in deep waters while TSS were low. In both rainy and dry seasons, the ratios of ISS/TSS were slightly higher than OSS/TSS on average and thus the contributions of inorganic substances, such as sand, silt or clay particles, and organic matters are nearly equal. This result revealed that the influence of algae was significant to spectral reflectance of the study area.

3.2. Turbidity Retrieval

Following the concept of Ritchie et al. (2003), linear regression models for turbidity were developed using the remote sensing reflectance extracted from Landsat 8 OLI on 14 February 2016 and observed turbidity in February 2016, as shown in Table 5. It can be seen that the lowest R^2 is 0.07 for the model No. 10 using the ratio of B5/B3 as the regression parameter while the highest R^2 is 0.84 for the model No. 1 adopting the single Band 4 (B4) (red region) (Figure 3). Selecting Band 4 as the regression parameter, the regression model for turbidity was determined as:

$$\text{Turbidity} = 380.32 R_{rs}(\lambda_{B4}) - 1.7826 \tag{9}$$

where λ_{B4} is the wavelength of Band 4 (B4).

Table 5. Regression models for turbidity retrieval (B1 = 433 nm, B2 = 482 nm, B3 = 562 nm, B4 = 655 nm and B5 = 865 nm).

No.	Regression Models	R^2	Significant
1	Turbidity = 380.32 × R_{rs} (B4) − 1.7826	$R^2 = 0.84$	$p < 0.05$
2	Turbidity = −8.1043 × R_{rs} (B5/B4) + 7.2697	$R^2 = 0.79$	$p < 0.05$
3	Turbidity = 297.86 × R_{rs} (B3) − 2.8208	$R^2 = 0.70$	$p < 0.05$
4	Turbidity = 604.54 × R_{rs} (B5) − 2.2241	$R^2 = 0.48$	$p < 0.05$
5	Turbidity = 424.54 × R_{rs} (B2) − 4.4504	$R^2 = 0.45$	$p < 0.05$
6	Turbidity = 504.31 × R_{rs} (B1) − 7.1769	$R^2 = 0.42$	$p < 0.05$
7	Turbidity = 12.895 × R_{rs} (B5/B1) − 3.158	$R^2 = 0.37$	$p < 0.05$
8	Turbidity = 6.3388 × R_{rs} (B4/B3) − 2.4028	$R^2 = 0.35$	$p < 0.05$
9	Turbidity = 5.5354 × R_{rs} (B5/B2) − 1.0947	$R^2 = 0.13$	$p < 0.05$
10	Turbidity = 3.5623 × R_{rs} (B5/B3) + 2.8059	$R^2 = 0.07$	$p < 0.05$

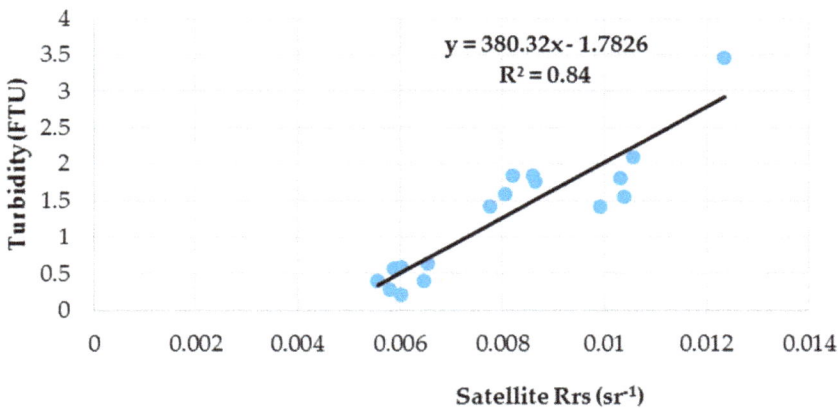

Figure 3. Correlation between remote sensing reflectance at 655 nm extracted from Landsat 8 OLI on 14 February 2016 and turbidity from in situ by using AAQ sensor.

A comparison of turbidities between computed using Equation (9) and observed at 17 stations (see Figure 1 and observed data missing at Station 11) is shown in Figure 4 and Table 6 along with squared residual, root mean square error (RMSE) and scatter index (SI), which shows a high correlation factor ($R^2 = 0.84$). The root mean square error (0.28) and scatter index (0.22) are low and acceptable [15,42]. The retrieved turbidities using Landsat 8 OLI are thus consistent with measured turbidities. These results indicate that the proposed model could be used to retrieve turbidity in highly turbid waters in the bay and lagoon.

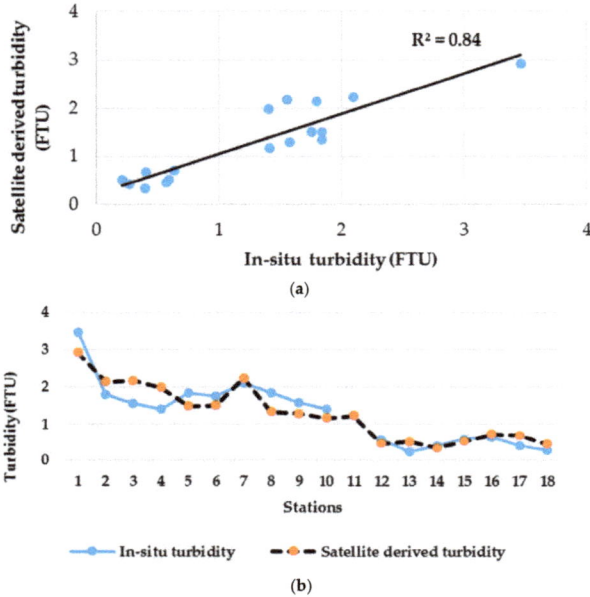

Figure 4. Scatter plots (**a**) and line graphs (**b**) comparing between satellite retrieved and observed turbidities at 18 stations shown in Figure 1 (observed turbidity missing at Station 11).

Table 6. Comparison of observed and retrieved turbidities with statistical analyses for squared residual, root mean square (RMSE) and scatter index (SI) from Station 1 to Station 18 except Station 11 in Figure 1.

Stations	Turbidity Observed	Turbidity Computed	Squared Residual	RMSE	SI
Station 1	3.46	2.92	0.30		
Station 2	1.80	2.14	0.12		
Station 3	1.56	2.17	0.37		
Station 4	1.41	1.99	0.33		
Station 5	1.84	1.49	0.12		
Station 6	1.76	1.51	0.06		
Station 7	2.10	2.23	0.02		
Station 8	1.85	1.33	0.26		
Station 9	1.58	1.28	0.09	0.28	0.22
Station 10	1.41	1.17	0.06		
Station 12	0.57	0.45	0.02		
Station 13	0.21	0.50	0.08		
Station 14	0.40	0.33	0.01		
Station 15	0.60	0.51	0.01		
Station 16	0.63	0.71	0.01		
Station 17	0.40	0.68	0.07		
Station 18	0.27	0.43	0.03		

3.3. Spatio-Temporal Variation in Turbidity

Using Equation (9), spatial variation in turbidity was retrieved for each of the collected Landsat 8 OLI images summarized in Table 2 from 29 June 2013 to 14 February 2016, and classified into rainy season, dry season and shortly after rainfall in dry season, respectively. By retrieving turbidities from pixels of the Landsat 8 OLI images, the relationship between turbidity and its frequency was obtained in rainy season, dry season and shortly after rainfall in dry season, as shown in Figure 5.

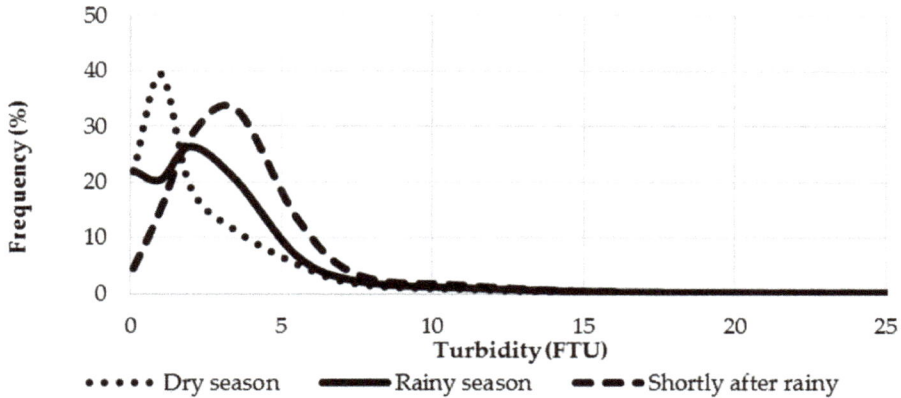

Figure 5. Frequency distribution of mean turbidity retrieved from Landsat 8 OLI images from 2013 to 2016 using the empirical model of Equation (9) in rainy season (four scenes), dry season (six scenes), and shortly after rainfall in dry season (eight scenes).

Turbidities ranging from 2.5 to 3.5 FTU were the most frequent values in rainy season (9 October 2015) with the highest frequency of 40%. Similarly, shortly after rainfall in dry season, turbidities around 3.5 FTU were the dominant values with the highest frequency of 55.8% (16 August 2013). By contrast, the most frequent values of turbidity in dry season were around 1.0 FTU with the frequency of 34% (26 January 2015).

3.4. Turbidity Distribution in Dry Season

Spatial variations in retrieved turbidity in dry season (on 8 February 2014, 2 July 2014, 26 January 2015, 15 March 2015, 3 June 2015 and 14 February 2016) are shown in Figure 6. Generally, turbidity is low in dry season except in shallow waters and areas adjacent to the coastlines. There are insignificant fluctuations of the mean turbidities ranging from 1.09 to 1.51 FTU during dry season. Small amount of precipitations less than cumulative 150 mm for ten-day period were recorded for these days. It is worth noting that rain rarely occurs during the dry season and almost no water flows in the rivers. High turbidity appearing near the river mouths and near shorelines might be due to resuspension of bed sediment.

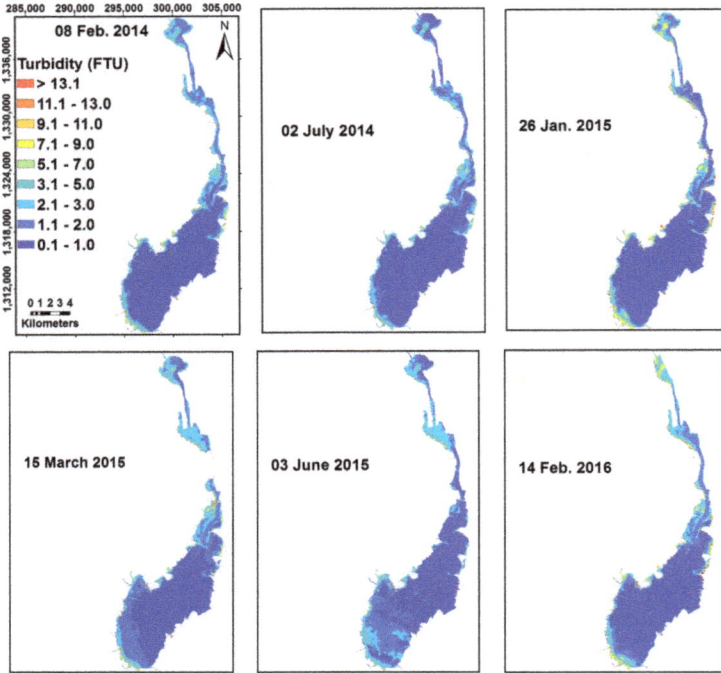

Figure 6. Turbidity distributions in dry season (cloud cover areas masked).

3.5. Turbidity Distribution in Rainy Season

Spatial variations in retrieved turbidity in rainy season (on 20 September 2014, 7 September 2015, 23 September 2015 and 9 October 2015) are shown in Figure 7. Compared with in dry season, the mean turbidities in rainy season were slightly higher (Table 6), notably in deep waters in Cam Ranh Bay. Both patterns in rainy and dry seasons show high turbidity in shallow water areas adjacent to coastlines. Turbidity in rainy season is slightly higher around the bay mouth than that in dry season. Depending on the amount of precipitation, the pattern as well as concentration of turbidity changed. The relationship between rainfall and turbidity will be further discussed in the section of discussion.

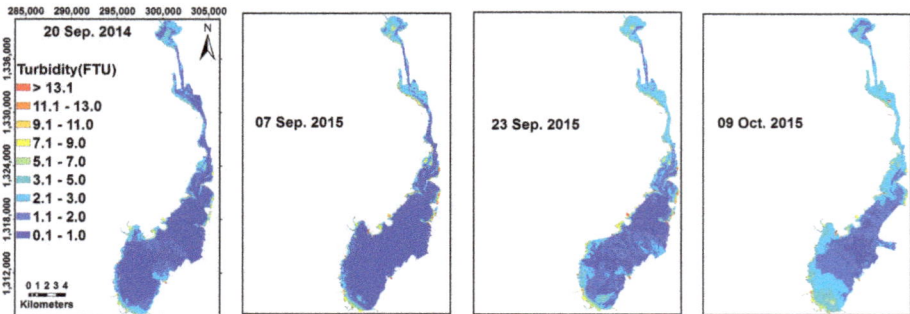

Figure 7. Turbidity distributions in rainy season (cloud cover areas masked).

3.6. Turbidity Distribution in Shortly after Raining in Dry Season

Spatial variations in retrieved turbidity shortly after rainfall in dry season (on 29 June 2013, 16 August 2013, 12 March 2014, 15 May 2014,16 April 2015, 21 July 2015, 6 August 2015 and 22 August 2015) are shown in Figure 8. River discharges into the bay and lagoon are controlled by dams in their up streams and thus after heavy rain, waters are impounded in the reservoirs first [38]. Turbidities in the river mouths therefore do not immediately become high after rainfall and strongly depend on the operation of the dams. Similar to the turbidity patterns in dry and rainy seasons, high turbidities appear in shallow water areas adjacent to the coastlines. This reason might be due to resuspension of bed sediment. After rainfall, high turbidities appear not only in shallow water areas in the river mouths but also in deep waters where turbid waters discharging from the rivers could penetrate far offshore in the bay (Figure 8).

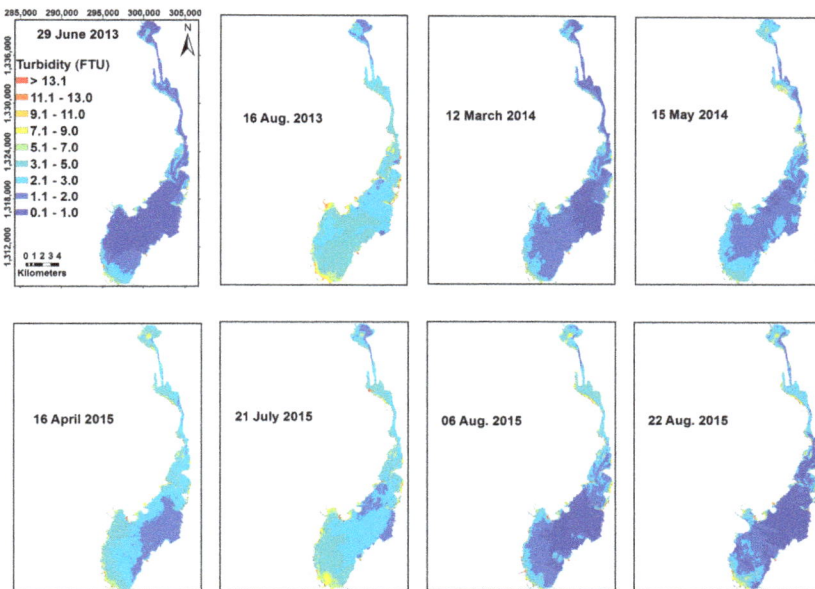

Figure 8. Turbidity distributions shortly after rainfall in dry season (cloud cover areas masked).

3.7. Statistics of Retrieved Turbidities

A statistical summary of turbidities in dry season, rainy season and shortly after rainfall in dry season retrieved from the 18 images is shown in Table 7. A wide range of turbidity was seen from 0.01 to 68.89 FTU. Turbidities in dry season ranged from 0.01 to 68.89 FTU with their mean values from 1.02 to 1.51 (see Table 7 and Figure 5). In rainy season, turbidities ranged from 0.01 to 46.64 FTU with their mean values from 1.18 to 2.35 FTU (see Table 7 and Figure 5). Shortly after rainfall in dry season, turbidities ranged from 0.01 to 44.2 FTU with their mean values from 1.29 to 3.68 FTU (see Table 7 and Figure 5).

Table 7. Statistics of seasonal variation in turbidities across the bay and lagoon retrieved from Landsat 8 OLI images from 2013 to 2016.

Seasons	Time	Max.	Min.	Median	Mean	Std.
Dry	8 February 2014	22.28	0.01	0.30	1.09	1.52
	2 July 2014	22.93	0.01	0.54	1.02	1.26
	26 January 2015	42.42	0.01	0.32	1.29	2.10
	15 March 2015	68.89	0.01	0.89	1.29	1.65
	3 June 2015	50.66	0.01	0.92	1.24	1.56
	14 February 2016	48.37	0.01	0.71	1.51	2.05
Rainy	20 September 2014	27.50	0.01	0.68	1.18	1.47
	7 September 2015	46.64	0.01	0.10	1.20	2.19
	23 September 2015	40.24	0.01	1.46	1.90	1.92
	9 October 2015	37.38	0.02	2.04	2.35	1.58
Shortly after rainfall during dry season	29 June 2013	19.69	0.01	0.83	1.29	1.39
	16 August 2013	23.17	0.10	3.27	3.68	1.62
	12 March 2014	21.92	0.01	1.39	1.76	1.26
	15 May 2014	19.16	0.01	1.91	2.29	1.32
	16 April 2015	39.78	0.10	2.64	2.96	1.50
	21 July 2015	42.34	0.01	2.93	3.35	1.73
	6 August 2015	42.47	0.01	1.49	2.06	1.97
	22 August 2015	44.20	0.01	1.37	1.88	2.11

Figure 9 shows the mean turbidities extracted from 18 Landsat 8 OLI scenes. It also reveals that the turbidity had significant temporal variability between 2013 and 2016. In dry season, mean turbidity ranged from 1.0 to 1.5 FTU and slightly higher in rainy season from 1.2 to 2.3 FTU. The highest fluctuation occurred shortly after rainfall from 1.3 to 3.7 FTU.

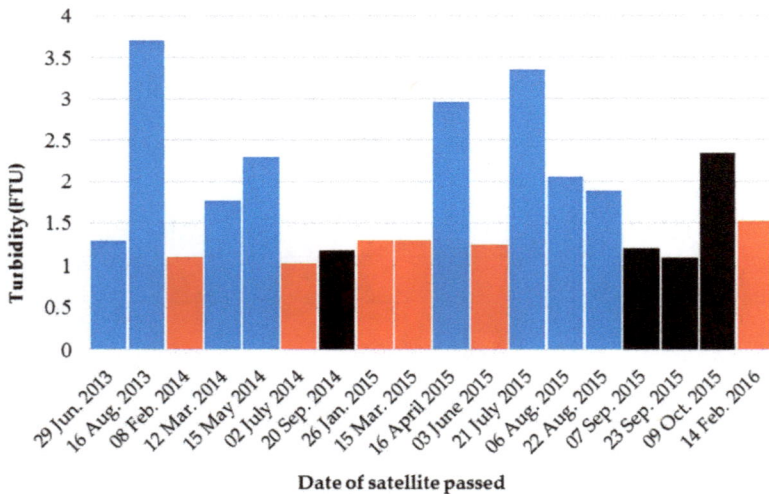

Figure 9. Landsat 8 OLI extracted mean turbidities in Cam Ranh Bay and Thuy Trieu Lagoon from 2013 to 2016 in both dry and rainy seasons (blue for shortly after rainfall in dry season, red and black for dry and rainy seasons, respectively).

3.8. Distribution of Wave Characteristics and Bed Shear Stress

In this study, wind was considered the most important input data for SWAN simulation because waves were mainly induced by wind. Due to the lack of measured significant wave height data, only observed wind was used to verify the input files. A comparison between observed wind provided by Institute of Oceanography and estimated data from NCEP FNL at 10 m height is presented in Figure 10.

The average observed and estimated wind speeds were 6.77 and 6.55, respectively. RMSE (2.39) and SI (0.35) values were small. These results showedhigh consistency between observed and estimated wind speed. Based on those, wind speed downloaded from NCEP FNL was appropriate and could be used in this study with high precision.

To estimate typical variations in significant wave heights and wave induced bed shear stresses across Cam Ranh Bay and Thuy Trieu Lagoon in dry season (northeast monsoon) and rainy season (southwest monsoon), respectively, SWAN model was forced by northeast (NE) winds and southwest (SW) winds with wind speeds at 1, 3, 5, 7, 9, 11 and 13 m/s which were chosen based on the statistics of wind fields in the study site extracted from the NCEP FNL dataset from 1999 to 2015. Figures 11 and 12 show computed significant wave heights and wave induced bed shear stresses in dry season and rainy season, respectively.

During rainy season (SW monsoon), the bed shear stress is higher in SW monsoon than in NE monsoon when wind speed was greater than 5 m/s in shallow waters, the top of Thuy Trieu Lagoon. However, in deep waters, Cam Ranh Bay, wind speed less than 9 m/s did not generate high bed shear stresses. The results at the central part of the study area were the most sensitive to change in wind speed. When wind speed was greater than 9 m/s, almost the entire study area had high bed shear stress.

The variations of significant wave height and bed shear stress to different wind speeds in NE monsoon (rainy season) were shown in Figure 12. Wind speeds greater than 5 m/s could generate a large area of high bed shear stresses leading to resuspension of sediments during NE monsoon. In deep waters, resuspension of sediments might occur when wind speeds were greater than 9 m/s. When wind speed reached 13 m/s, most of the study area showed high bed shear stresses including deep waters in Cam Ranh Bay (see Figures 11a and 12a).

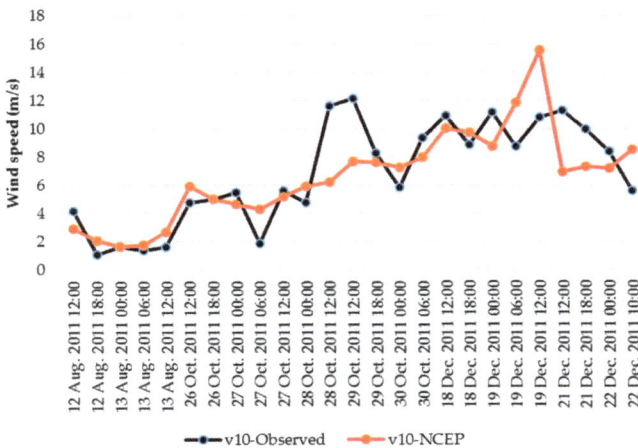

Figure 10. Comparison between observed and estimated (NCEP) wind speeds at 10 m height.

(a)

(b)

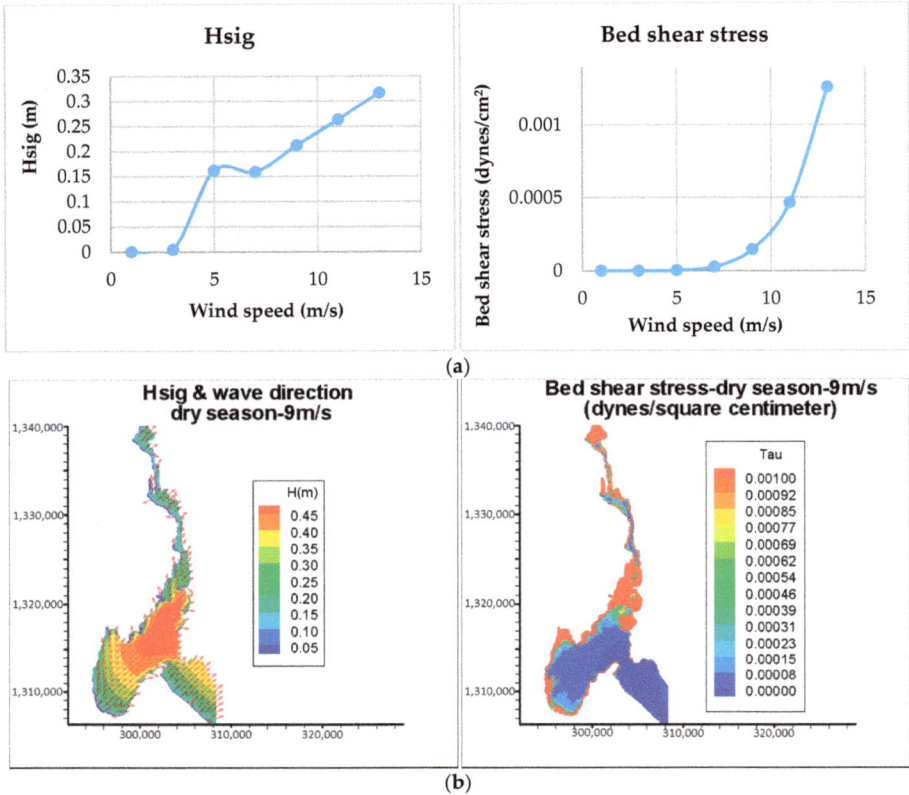

Figure 11. Typical variations of significant wave heights and directions as well as bed shear stresses in dry season (NE monsoon, −45 degree). Top panels (**a**) show relationships between wind speeds and significant wave heights and bed shear stresses, respectively, at 296,437.06 lat. 1,308,613.36 long. and bottom panels (**b**) show spatial distributions with the wind speed of 9 m/s.

Aside from the dependence on wind speed, the sensitivity of significant wave height and bed shear stress to different wind directions were analyzed. Wind speed ranging from 5 to 9 m/s was dominant in the period from 1999 to 2015 based on the wind analysis of NCEP FNL. Moreover, NE and SW are the main directions of wind in dry and rainy seasons, respectively. Based on those, wind speed at 7 m/s was used as a representative to assess the sensitivity of wave spectrum to wind direction for every 10 degree interval.

According to Figure 13, the amount of resuspension was influenced not only by wind speed, but also by wind direction. The simulation results showed that SW winds caused less resuspension than NE winds in deep waters, Cam Ranh Bay. In the upper part of Thuy Trieu Lagoon and the center of the study area, shallow waters had much influence on resuspension in both SW and NE monsoon. At the bay mouth, where the water depth is around 30 m, the bed shear stress was always small and approximately 0 in both NE and SW monsoon.

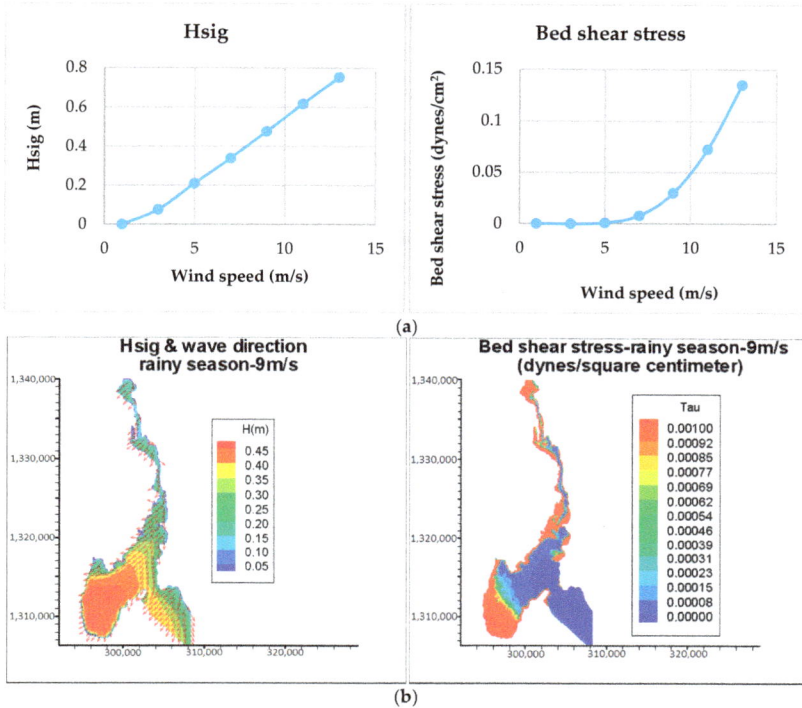

Figure 12. Typical variations of significant wave heights and directions as well as bed shear stresses in rainy season (SW monsoon, −225 degree). Top panels (**a**) show relationships between wind speeds and significant wave heights and bed shear stresses, respectively, at 296,437.06 lat. 1,308,613.36 long. and bottom panels (**b**) show spatial distributions with the wind speed of 9 m/s.

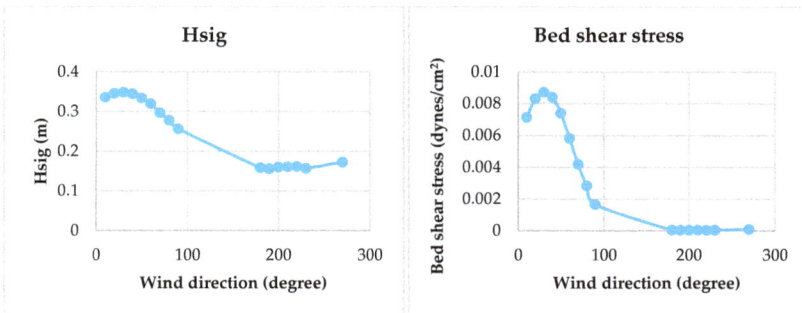

Figure 13. Variations of significant wave height and bed shear stress to different wind directions (at station 296,437.06 lat. 1,308,613.36 long.) in NE and SW monsoons.

4. Discussion

4.1. Turbidity Algorithm

Most of the existing methods interpreting turbidity propose site-specific empirical relationships between turbidity and reflectance at different satellite wavebands by fitting field turbidity measurements with either field- or satellite-derived reflectance [7]. Single band or ratio of two bands is routinely used to develop turbidity as well as other water component models [32,35]. In the present study, turbidity showed the highest correlation with Band 4 of Landsat 8 OLI, red region (655 nm) for a single season.

The best fit of the red band to characterize the turbidity of Cam Ranh waters is not surprising. Many researchers using remote sensing reflectance (R_{rs}) for turbidity as well as TSS retrieval have shown that there is high relationship between red band of satellite images and their concentrations. Güttler et al. [1] developed an algorithm for turbidity in the Northwestern Black Sea coastal zone based on the high correlation between turbidity and red band of Landsat TM. Lobo et al. used a time-series analysis of Landsat-MSS/TM/OLI images to assess the impacts of gold mining activities to Amazonian waters. This result indicated that water reflectance of red band of Landsat was the most appropriate for establishing a robust empirical model for TSS retrieval [43]. Vanhellemont and Ruddick used Landsat 8 OLI Band 4 to retrieve turbidity in the Southern North Sea where the water depth was less than 50 m. They also used MODIS red band, 645 nm, for testing turbidity retrieval algorithm. The results showed that red region was appropriate for interpret turbidity from satellite images [11]. In addition, Garaba and Zielinski were successful in using Landsat 8 OLI to interpret turbidity in estuary located in Wadden Sea where turbidity was always high. The result revealed that turbidity highly correlated with remote sensing reflectance Band 4, red region, of Landsat 8 OLI [44].

In low turbid waters where total suspended solids (TSS) are less than 50 g/m^3, there is high correlation between turbidity and remote sensing reflectance of red region [6]. However, for waters with very high turbidity or TSS exceeding 50 mg/m^3, near infrared (NIR) wavelengths are recommended to use for turbidity retrieval [6]. The shift to NIR wavelength for turbidity retrieval is due to the saturation of R_{rs} in visible wavelengths when TSS concentration exceeds 50 g/m^3 [6,45]. These results also indicated that, in very turbid waters where TSS exceeds 100 g/m^3, the R_{rs} signal in the short-wavelength infrared region (SWIR, 1000–3000 nm) could also be used to retrieve turbidity and inherent optical properties [6,46]. The sensitivity of remote-sensing reflectance to turbidity was inversely related to suspended sediment concentration [45]. This study also indicated that NIR wavelength should be used to retrieve suspended sediment matter for the very high turbid waters.

In our study, the TSS measured in two field trips were less than 20 mg/L except station 14 where algae blooming occurred (52 mg/L and slightly higher than 50 mg/L, the threshold). Therefore, the algorithm that was developed for turbidity retrieval was stable and reliable when using red band.

Notably, the algorithm developed from in situ R_{rs} data and then applied to Landsat-based on derived R_{rs} data with a spectral response function for satellite images can result in errors or uncertainty of the model [47]. The reason of this uncertainty is due to the differences of spectral reflectance between R_{rs} measured and calculated from the images. They cannot be exactly the same, even though satellite images are undergone rigorous calibrations. For adopting the spectra for Landsat sensors, the derived-turbidity model is the best precision with the highest determination coefficient ($R^2 = 0.84$; $p < 0.05$), using only one red band available from the entire Landsat satellite [47]. Furthermore, there is no difference in the spatial and spectral resolution of all 18 satellite images because only Landsat 8 OLI was used. In other words, the turbidity algorithm can be applied for all the Landsat 8 OLI images without uncertainty.

4.2. Factors and Processes Determining Turbidity

In order to discuss factors and processes determining turbidity, a correlation between measured sediment grain size and estimated turbidity has firstly been investigated as shown in Figure 14 representing a negative weak correlation between turbidity and median sediment diameter (d_{50}) in the bay and lagoon. This correlation is similar to that in a study by Richardson and Jowett (2002), which also showed a similar negative weak correlation ($R = -0.34$) [48]. These results indicate the easier occurrence of wave induced resuspension of finer sediment leading to higher turbidity while the correlation factor is not very high, which means turbidity is also influenced by other factors, including the relationship between bed shear stress and sediment properties and the effect of turbid water discharges and dispersion during floods as well as the influence of clear oceanic water intrusions through the bay entrance leading to lower turbidity.

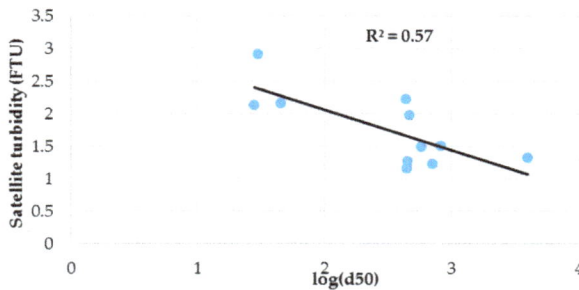

Figure 14. Correlation between Landsat 8 OLI extracted turbidities and log (d_{50}).

During the dry season, low turbidity waters appeared in most of the study areas except shallow waters and areas adjacent to coastlines. Turbidity in Thuy Trieu Lagoon was always slightly higher than in Cam Ranh Bay (see Figures 6–8), especially in the central part of the lagoon, while, around the bay mouth, turbidities were in the range from 0.1 to 1.0 FTU and lower than in the lagoon. These higher turbidities are considered to be mainly caused by wave induced resuspension of bed sediments in shallow waters, which is confirmed by the appearance of higher turbidity areas corresponding to shallower water areas, as bed shear stress tends to decrease with increasing water depth [32]. By extracting and averaging bed shear stresses (BSS) from each of Landsat scenes, correlations were taken between BSS and turbidities under small rainfall and heavy rainfall conditions, respectively, as shown in Figure 15.

Figure 15a indicates that the variation of turbidity was mainly governed by resuspension of bed sediments under small rainfall conditions. However, under heavy rainfall conditions, relationship between bed shear stress and turbidity became insignificant, as shown in Figure 15b, which will be discussed later. Under calm weather conditions, the amount of fresh water discharges into the bay and lagoon are basically small due to being controlled by the dams and thus no significant sediment supply occurs from the rivers [10]. Wind speed under calm weather conditions is not high enough to generate shear zones leading to high turbidity even in deep waters in Cam Ranh Bay. According to the results from SWAN simulation, winds with the speed of less than 9 m/s were not a major factor causing higher turbidity in deep waters (see Figure 12). This result may be in favor of that of a previous case study in China [47].

SWAN simulation results also showed that only wind speeds greater than or equal to 9 m/s could generate high bed shear stress areas leading to resuspension of bed sediments in deep waters of Cam Ranh Bay (see Figures 11 and 12). However, for the closed water areas like Cam Ranh Bay and Thuy Trieu Lagoon, wind speed usually less than 9 m/s. The combination of high wind speed with large amount of precipitation will result in high turbidity during NE monsoon. Besides, the wave heights are associated with the fetch dependent on the shape of the study area and the wind direction. Higher waves were found at the head of the lagoon during SW monsoon and at the southern part of the bay and areas adjacent to the coastlines in NE monsoon (Figures 11b and 12b), as these two areas are the downwind regions in SW and NE monsoons, respectively.

To see the relationship between turbidity and the amount of precipitation from 2013 to 2016, retrieved turbidities and the corresponding cumulative rainfalls during ten days before taking each of the images are plotted in the top panel of Figure 16a and their correlation is shown in Figure 16b. Among a variety of factors affecting turbidity, its correlation factor of $R^2 (= 0.54, p < 0.05, \text{t-test})$ is considerably high [15,49] and thus these rainfall events are considered as a major contributor for increasing in turbidity through a large amount of river discharges with high turbidity.

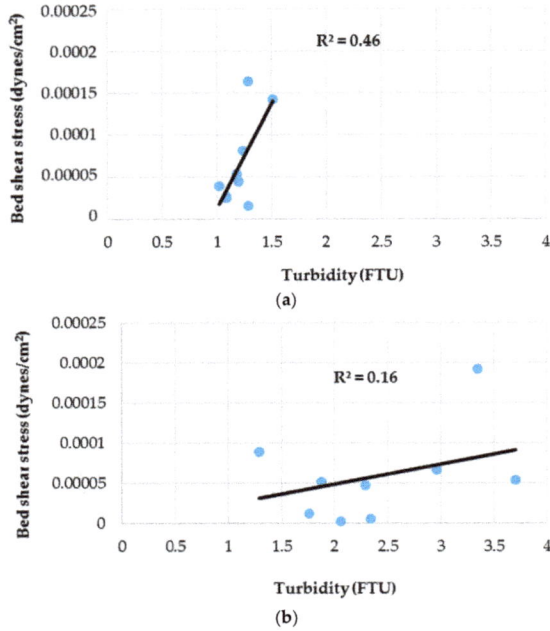

Figure 15. Correlation between the mean Landsat OLI extracted turbidities and computed bed shear stresses under (**a**) small rainfall condition and (**b**) heavy rainfall condition.

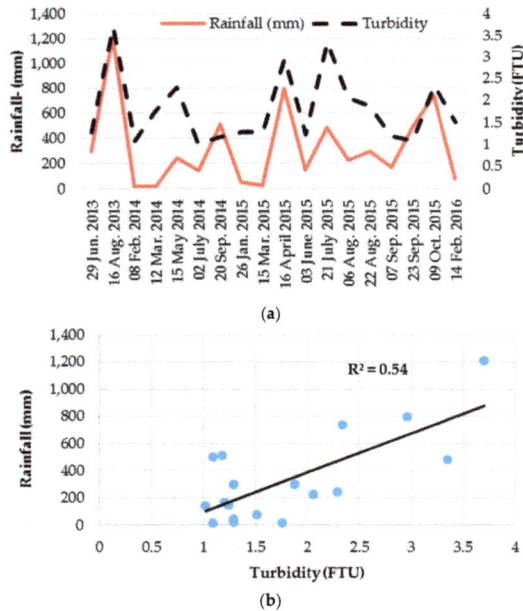

Figure 16. (**a**) Time series of measured rainfalls and the mean Landsat 8 OLI extracted turbidities and (**b**) their correlation.

According to Figure 16, turbidity in the bay contributed from the rivers was strongly influenced by a large amount of precipitation through the two main processes: (i) after rainfall, the river runoff bringing a large amount of sediment into the river mouths, which is expected to increase the concentration of suspended sediment before settlement; and (ii) the stronger current and tidal straining effect causing the significantly high resuspension after heavy rain [50].

5. Conclusions

To our knowledge, this is the first study on monitoring water quality, especially turbidity, in Cam Ranh Bay and Thuy Trieu Lagoon, using remote sensing. This study presented a method to estimate turbidity as well as the possibility of mapping water quality in a turbid deep water from remotely sensing data. A regional algorithm was developed and successfully applied to Landsat 8 OLI images to retrieve surface turbidity in the study area. The observed and computed turbidity by using Landsat 8 OLI were highly consistent, which indicates that the proposed model could be used to retrieve turbidity in high turbid waters like Cam Ranh Bay and Thuy Trieu Lagoon. Turbidity in Cam Ranh Bay and Thuy Trieu Lagoon was controlled by resuspension of bed sediment and influence of a large amount of precipitation, during both the rainy season and dry season. We demonstrated that resuspension of bed sediment is a major process controlling turbidity in shallow waters and near coastlines while rainfall is a key factor affecting turbidity in deep waters. The approach based on spatiotemporal scale is essential to determine and correctly interpret the differential effects engendered by the turbidity controlling processes in Cam Ranh Bay and Thuy Trieu Lagoon. The results can serve as a basis for future observations of turbidity pattern change in Cam Ranh Bay and Thuy Trieu Lagoon and can be applied for other similar waters. Some possible solutions proposed to decrease turbidity in the study area are minimizing the dredging works, and good control of the operation of dams and reservoirs inthe upstream and aquaculture activities in the study area. To reduce the effect of wind resulting in resuspension of bed sediment, planting trees around the entire area is encouraged.

Acknowledgments: The authors would like to thank researchers working in Institute of Oceanography, Tong Phuoc Hoang Son, Le Thi Vinh, Phan Minh Thu and Bui Hong Long, for their kind supports in this research. We also thank Hae Chong O at the University of Tokyo and Sylvain Ouillon at the University of Toulouse who advised and commented on the paper. This study was partially funded by JSPS KAKENHI Grant No. 25303016.

Author Contributions: Quang Nguyen Hao and Jun Sasaki conceived the study; Quang Nguyen Hao, Jun Sasaki and Higa Hiroto organized the field trip and designed the numerical experiments; Huan Nguyen Huu provided water quality data and bathymetric and conducted experiments; Quang Nguyen Hao performed the simulations and remote sensing data processing; Quang Nguyen Hao, Jun Sasaki and Higa Hiroto analyzed the data and simulations; Quang Nguyen Hao, Higa Hiroto and Huan Nguyen Huu wrote the paper; and Jun Sasaki edited the final manuscript.

Conflicts of Interest: The authors declare no conflict of interest.

References

1. Güttler, F.N.; Niculescu, S.; Gohin, F. Turbidity retrieval and monitoring of Danube Delta waters using multisensor optical remote sensing data: An integrated view from the delta plain lakes to the western–northwestern Black Sea coastal zone. *Remote Sens. Environ.* **2013**, *132*, 86–101. [CrossRef]
2. Zheng, G.J. *Hydrodynamics and Water Quality: Modeling Rivers, Lakes, and Estuary*; John Wiley & Sons, Inc.: Hoboken, NJ, USA, 2007; pp. 130–134.
3. Minnesota Pollution Control Agency. Turbidity: Description, Impact on Water Quality, Sources, Measures—A General Overview, USA. 2008. Available online: https://www.pca.state.mn.us/sites/default/files/wq-iw3-21.pdf (accessed on 18 February 2016).
4. Carson, A.B.; Benjamin, M.J.; Krista, K.B.; Daniel, B.Y.; Christian, E.Z. Reconstructing turbidity in a glacially influenced lake using the Landsat TM and ETM+ surface reflectance climate data record archive, Lake Clark, Alaska. *Remote Sens.* **2015**, *7*, 13692–13710. [CrossRef]
5. Aksnes, D.L.; Dupont, N.; Staby, A.; Fiksen, Ø.; Kaartvedt, S.; Aure, J. Coastal water darkening and implications for mesopelagic regime shifts in Norwegian fjords. *Mar. Ecol. Prog. Ser.* **2009**, *387*, 39–49. [CrossRef]

6. Gernez, P.; Barille, L.; Lerouxel, A.; Mazeran, C.; Lucas, A.; Doxaran, D. Remote sensing of suspended particulate matter in turbid oyster farming ecosystems. *J. Geophys. Res. Oceans* **2014**, *119*, 7277–7294. [CrossRef]

7. Dogliotti, A.I.; Ruddick, K.G.; Nechad, B.; Doxaran, D.; Knaeps, E. A single algorithm to retrieve turbidity from remotely-sensed data in all coastal and estuarine waters. *Remote Sens. Environ.* **2015**, *156*, 157–168. [CrossRef]

8. Allan, M.G.; Hamilton, D.P.; Hicks, B.J.; Brabyn, L. Landsat remote sensing of chlorophyll a concentrations in central North Island lakes of New Zealand. *Int. J. Remote Sens.* **2011**, *32*, 2037–2055. [CrossRef]

9. Phan, M.T.; Tong, P.H.S.; Teruhisa, K. Using remote sensing technique for analyzing temporal changes of seagrass beds by human impacts in waters of Cam Ranh Bay, Vietnam. In *Remote Sensing of the Marine Environment II, 85250T, Proceedings of the SPIE8525, San Diego, CA, USA, 11 December 2012*; Frouin, R.J., Ebuchi, N., Pan, D., Saino, T., Eds.; SPIE: Kyoto, Japan, 2012.

10. Phan, M.T.; Nguyen, H.H.; Bui, H.L. Study on environmental capacity in Cam Ranh Bay and Thuy Trieu Lagoon. *J. Mar. Sci. Technol.* **2013**, *13*, 371–381. [CrossRef]

11. Vanhellemont, Q.; Ruddick, K. Turbid wakes associated with offshore wind turbines observed with Landsat 8. *J. Remote Sens. Environ.* **2014**, *145*, 105–115. [CrossRef]

12. Kishino, M.; Tanaka, A.; Joji, I. Retrieval of Chlorophyll a, suspended solids, and colored dissolved organic matter in Tokyo Bay using ASTER data. *Remote Sens. Environ.* **2005**, *99*, 66–74. [CrossRef]

13. Lim, H.S.; MatJafri, M.Z.; Abdullah, K.; Asadpour, R. A Two-Band algorithm for total suspended solid concentration mapping using THEOS data. *J. Coast. Res.* **2012**, *29*, 624–630. [CrossRef]

14. Tebbs, E.J.; Remedios, J.J.; Harper, D.M. Remote sensing of chlorophyll-a as a measure of cyanobacterial biomass in Lake Bogoria, a hypertrophic, saline–alkaline, flamingo lake, using Landsat ETM+. *Remote Sens. Environ.* **2013**, *135*, 92–106. [CrossRef]

15. Zhang, Y.; Zhang, Y.; Shi, K.; Zha, Y.; Zhou, Y.; Liu, M. A Landsat 8 OLI-Based, semianalytical model for estimating the total suspended matter concentration in the slightly turbid Xin'anjiang reservoir (China). *IEEE J. Sel. Top. Appl. Earth Obs. Remote Sens.* **2016**, *9*, 398–413. [CrossRef]

16. Doxaran, D.; Froidefond, J.M.; Castaing, P.; Babin, M. Dynamics of the turbidity maximum zone in a macrotidal estuary (the Gironde, France): Observations from field and MODIS satellite data. *Estuar. Coast. Shelf Sci.* **2009**, *81*, 321–332. [CrossRef]

17. Petus, C.; Chust, G.; Gohin, F.; Doxaran, D.; Froidefond, J.M.; Sagarminaga, Y. Estimating turbidity and total suspended matter in the Adour River plume (South Bay of Biscay) using MODIS 250-m imagery. *Cont. Shelf Res.* **2010**, *30*, 379–392. [CrossRef]

18. Ouillon, S.; Douillet, P.; Petrenko, A.; Neveux, J.; Dupouy, C.; Froidefond, J.M.; Andréfouët, S.; Caravaca, A.M. Optical algorithms at satellite wavelengths for Total Suspended Matter in tropical coastal waters. *Sensor* **2008**, *8*, 4165–4185. [CrossRef] [PubMed]

19. Ali, P.Y.; Jie, D.; Sravanthi, N. Remote sensing of chlorophyll-a as a measure of red tide in Tokyo Bay using hotspot analysis. *J. Remote Sens. Appl. Soc. Environ.* **2015**, *2*, 11–25. [CrossRef]

20. Zhang, C.; Han, M. Mapping chlorophyll-a concentration in Laizhou Bay using Landsat 8 OLI data. In Proceedings of the E-proceedings of the 36th IAHR World Congress, The Hague, the Netherlands, 28 June–3 July 2015.

21. Tong, P.H.S.; Truong, M.C.; Hoang, C.T. Detecting chlorophyll-a concentration and bloom patterns at upwelling area in South central Vietnam by high resolution multi-satellite data. *J. Environ. Sci. Eng. A* **2015**, *4*, 215–224. [CrossRef]

22. Bui, H.L.; Tran, V.C.; Nguyen, H.H.; To, D.T. Self-cleaning ability by the tide of Cam Ranh Bay—Thuy Trieu Lagoon (Khanh Hoa). *J. Sci. Technol. Environ.* **2013**, *2*, 20–22.

23. Greenberg, A.E.; Clescert, L.S.; Eaton, A.D. *Standard Methods for the Examination of Water and Wastewater*, 18th ed.; Method 2540D; American Public Health Association: Washington, DC, USA, 1992; pp. 2–56.

24. Parsons, T.R.; Maita, Y.; Lalli, M.C. *A Manual of Chemical and Biological Methods for Seawater Analysis*; Pergamon Press: Oxford, UK, 1984; pp. 110–173.

25. Adam, T. Remote Sensing Models of Algal Blooms and Cyanobacteria in Lake Champlain. Master's Thesis, Department of Civil and Environmental Engineering of the University of Massachusetts Amherst, Amherst, MA, USA, February 2012.

26. Retsch GmbH Haan. Sieve Analysis Taking a Close Look at Quality. 2009. Available online: http://www.mep.net.au/wpmep/wpcontent/uploads/2013/07/MEP_expert_guide_sieving_en.pdf (accessed on 15 March 2016).

27. Wesley, J.M.; Anatoly, A.G.; Richard, L.P.; Daniela, G.; Donald, C.R.; Bryan, C.L.; Tadd, M.B.; Paul, B. Estimation of chlorophyll-a concentration in turbid productive waters using airborne hyperspectral data. *Water Res.* **2012**, *46*, 993–1004. [CrossRef]

28. Kaliraj, S.; Chandrasekar, N.; Mages, N.S. Multispectral image analysis of suspended sediment concentration along the Southern coast of Kanyakumari, Tamil Nadu, India. *J. Coast. Sci.* **2014**, *1*, 63–71.

29. Jorge, D.S.F.; Amore, D.J.; Barbossa, C.F. Efficiency estimation of four different atmospheric correction algorithms in a sediment-loaded tropic lake for Landsat 8 OLI sensor. In Proceedings of the Anais XVII Simpósio Brasileiro de Sensoriamento Remoto—SBSR, João Pessoa-PB, Brasil, 25–29 April 2015.

30. Watanabe, F.; Sayuri, Y.; Enner, A.; Thanan, W.; Pequeno, R.; Nilton, N.I.; Cláudio, C.F.B.; Luiz, H.S.R. Estimation of chlorophyll-a concentration and the trophic state of the Barra Bonita Hydroelectric Reservoir using OLI/Landsat 8 images. *Int. J. Environ. Res. Public Health* **2015**, *12*, 10391–10417. [CrossRef] [PubMed]

31. ENVI. *Atmospheric Correction Module: QUAC and FLAASH User's Guide*; EXELIS Visual Information Solutions: Boulder, CO, USA, 2009.

32. Ouillon, S.; Douillet, P.; Andrefouet, S. Coupling satellite data with in situ measurements and numerical modeling to study fine suspended-sediment transport: A study for the lagoon of New Caledonia. *Coral Reefs* **2004**, *23*, 109–122. [CrossRef]

33. Hellweger, F.L.; Schlossera, P.; Lalla, U.; Weissel, J.K. Use of satellite imagery for water quality studies in New York Harbor. *Estuar. Coast. Shelf Sci.* **2004**, *61*, 437–448. [CrossRef]

34. Lim, J.; Choi, M. Assessment of water quality based on Landsat 8 operational land imager associated with human activities in Korea. *J. Environ. Monit. Assess.* **2015**, *187*, 384. [CrossRef] [PubMed]

35. Ritchie, C.J.; Zimba, V.P.; Everitt, H.J. Remote sensing techniques to assess water quality. *Photogramm. Eng. Remote Sens.* **2003**, *69*, 695–704. [CrossRef]

36. Booij, N.; Ris, R.C.; Holthuijsen, L.H. A third-generation wave model for coastal regions. *J. Geophys. Res.* **1999**, *104*, 7667–7681. [CrossRef]

37. SWAN Team. Swan Scientific and Technical Documentation: SWAN Cycle III Version 41.01AB. Delft University of Technology, The Netherlands, 2015. Available online: http://swanmodel.sourceforge.net/download/zip/swantech.pdf (accessed on 20 February 2016).

38. Dalyander, P.S.; Butman, B.; Sherwood, R.C.; Signell, P.R.; Wilkin, L.J. Characterizing wave- and current-induced bottom shear stress: U.S. middle Atlantic continental shelf. *Cont. Shelf Res.* **2013**, *52*, 73–86. [CrossRef]

39. Phan, T.B. Process Simulation the Transmission of Pollutants under Impacts of Dynamic Factors in Cam Ranh Bay by Using Numerical Models. Master Thesis of Department of Meteorology, Hydrology and Oceanography, Ha Noi University of Science, Ha Noi, Vietnam, 2012. Available online: http://hus.vnu.edu.vn/files/ChuaPhanLoai/LuanVanThacSi-ChuaPhanLoai%20(423).pdf (accessed on 10 April 2016).

40. Rasmeemasmuang, T.; Sasaki, J. Modeling of mud accumulation and bed characteristics in Tokyo Bay. *Coast. Eng. J.* **2008**, *50*, 277–308. [CrossRef]

41. Sheng, P.Y.; Lick, W. The Transport and Resuspension of Sediments in a Shallow Lake. *J. Geophys. Res.* **1979**, *84*, 1809–1826. [CrossRef]

42. Bonansea, M.; Rodriguez, M.C.; Pinotti, L.; Ferrero, S. Using multi-temporal Landsat imagery and linear mixed models for assessing water quality parameters in Río Tercero reservoir (Argentina). *Remote Sens. Environ.* **2015**, *158*, 28–41. [CrossRef]

43. Lobo, F.L.; Costa, M.P.F.; Novo, E.M.L.M. Time-series analysis of Landsat-MSS/TM/OLI images over Amazonian waters impacted by gold mining activities. *Remote Sens. Environ.* **2013**, *157*, 170–184. [CrossRef]

44. Garaba, S.P.; Zielinski, O. An assessment of water quality monitoring tools in an estuarine system. *J. Remote Sens. Appl. Soc. Environ.* **2015**, *2*, 1–10. [CrossRef]

45. Fang, S.; Suhyb, S.M.H.D.; Zhou, Y.X.; Li, J.F.; Su, Z.; Kuang, D.B. Remote-sensing reflectance characteristics of highly turbid estuarine waters—A comparative experiment of the Yangtze River and the Yellow River. *Int. J. Remote Sens.* **2010**, *31*, 2639–2654.

46. Vanhellemont, Q.; Ruddick, K. Advantages of high quality SWIR bands for ocean colour processing: Examples from Landsat 8. *Remote Sens. Environ.* **2015**, *161*, 89–106. [CrossRef]

47. Zhubin, Z.; Yunmei, L.; Yulong, G.; Yifan, X.; Ge, L.; Chenggong, D. Landsat-based long-term monitoring of total suspended matter concentration pattern change in the wet season for Dongting Lake, China. *Remote Sens.* **2015**, *7*, 13975–13999. [CrossRef]

Water **2017**, *9*, 570

48. Richardson, J.; Jowett, I.G. Effects of sediment on fish communities in East Cape streams, North Island, New Zealand. *N. Z. J. Mar. Freshw. Res.* **2002**, *36*, 431–442. [CrossRef]
49. Goransson, G.; Larson, M.; Bendz, D. Variation in turbidity with precipitation and flow in a regulated river system—River GötaÄlv, SW Sweden. *Hydrol. Earth Syst. Sci.* **2013**, *17*, 2529–2542. [CrossRef]
50. Yu, Y.; Zhang, H.; Lemckert, C. Salinity and turbidity distributions in the Brisbane River estuary, Australia. *J. Hydrol.* **2014**, *519*, 3338–3352. [CrossRef]

Article

Application of the Support Vector Regression Method for Turbidity Assessment with MODIS on a Shallow Coral Reef Lagoon (Voh-Koné-Pouembout, New Caledonia)

Guillaume Wattelez [1,2,*], **Cécile Dupouy** [1,2], **Jérôme Lefèvre** [1,2,3], **Sylvain Ouillon** [3], **Jean-Michel Fernandez** [4] **and Farid Juillot** [2,5]

[1]   Aix Marseille Univ, Université de Toulon, CNRS, IRD, MIO UM 110, 13288 Marseille, France
[2]   Institut de Recherche pour le Développement (IRD), BP A5 98848 Nouméa CEDEX, New Caledonia; cecile.dupouy@ird.fr (C.D.); jerome.lefevre@ird.fr (J.L.); farid.juillot@ird.fr (F.J.)
[3]   LEGOS, Université de Toulouse, IRD, CNES, CNRS, UPS, 14 avenue Edouard Belin, 31400 Toulouse, France; sylvain.ouillon@legos.obs-mip.fr
[4]   Analytical and Environmental Laboratory (AEL), IRD-Nouméa, BP A5, 98800 Nouméa, New Caledonia; jmfernandez@ael-environnement.nc
[5]   Institut de Minéralogie, de Physique des Matériaux et de Cosmochimie (IMPMC), UMR IRD 206, UMR CNRS 7590, MNHN, Université Pierre et Marie Curie, Campus Jussieu, 75005 Paris, France
*   Correspondence: guillaume.wattelez@univ-nc.nc; Tel.: +687-290-591

Received: 23 June 2017; Accepted: 19 September 2017; Published: 27 September 2017

Abstract: Particle transport by erosion from ultramafic lands in pristine tropical lagoons is a crucial problem, especially for the benthic and pelagic biodiversity associated with coral reefs. Satellite imagery is useful for assessing particle transport from land to sea. However, in the oligotrophic and shallow waters of tropical lagoons, the bottom reflection of downwelling light usually hampers the use of classical optical algorithms. In order to address this issue, a Support Vector Regression (SVR) model was developed and tested. The proposed application concerns the lagoon of New Caledonia—the second longest continuous coral reef in the world—which is frequently exposed to river plumes from ultramafic watersheds. The SVR model is based on a large training sample of in-situ turbidity values representative of the annual variability in the Voh-Koné-Pouembout lagoon (Western Coast of New Caledonia) during the 2014–2015 period and on coincident satellite reflectance values from MODerate Resolution Imaging Spectroradiometer (MODIS). It was trained with reflectance and two other explanatory parameters—bathymetry and bottom colour. This approach significantly improved the model's capacity for retrieving the in-situ turbidity range from MODIS images, as compared with algorithms dedicated to deep oligotrophic or turbid waters, which were shown to be inadequate. This SVR model is applicable to the whole shallow lagoon waters from the Western Coast of New Caledonia and it is now ready to be tested over other oligotrophic shallow lagoon waters worldwide.

Keywords: turbidity; remote-sensing; MODerate Resolution Imaging Spectroradiometer (MODIS); Support Vector Regression (SVR); oligotrophic lagoon; bathymetry; reflectance; seabed colour; coral reef; New Caledonia

---

1. Introduction

In numerous tropical Pacific islands, the clarity of coastal lagoon waters is an essential parameter allowing the development of massive coral reefs and numerous benthic living species of prime importance for ecology and for fishing. This richness is essentially due to the oligotrophy of surrounding oceanic waters as it is found along the Great Barrier Reef and many other Pacific Islands. In these areas, human forcing on river-derived inputs of sediments, nutrients and organic matter to the coastal oceans can have negative effects on marine biogeochemical cycles and biodiversity [1–3]. It is therefore important to monitor these inputs at high spatial and time scales in order to estimate both their temporal and spatial fluctuations and to anticipate their possible influences on marine biota from ocean colour remote sensing [4].

With about one third of its terrestrial surface (8000 km^2) covered with ultramafic rocks, about 85% of endemic terrestrial plants and trees species, 24 tree species of the 70 identified in mangrove ecosystems worldwide, about 2800 species of marine molluscs and with the second longest continuous coral reef in the world [5–8], New Caledonia is one of the tropical and intertropical areas most concerned with the potential impacts of anthropogenic forcings at continental margins on marine biodiversity [9]. The main island (Grande Terre) of this small archipelago is characterized by a large occurrence of ultramafic rocks (i.e., peridotites) as the geological setting [10]. Strong weathering of these rocks upon tropical climate lead to deep lateritic covers that are enriched in trace metals like nickel or cobalt [11–14]. Natural geological and climatic events have then made the lateritic covers on ultramafic rocks a very important economical resource for New Caledonia [15]. However, these pedogeological formations are subject to acute erosion [16] and this natural process is significantly enhanced by mining activities [17–19]. Due to the shoreline location of these covers, eroded lateritic materials are directly transported to the coastal ecosystems, as evidenced by remote sensing or sea measurements [20–22].

Remote sensing provides efficient tools for monitoring sediment transport at high spatial and temporal scales since it offers a synoptic and instantaneous field view of the total suspended matter (TSM) concentration (e.g., [23–29]). Hu et al. [30] first determined a single band algorithm using the MODerate Resolution Imaging Spectroradiometer (MODIS) 645 nm-reflectance for mapping the turbidity in the Tampa Bay (Florida, USA). More recently, Nechad et al. [31] and Novoa et al. [32] proposed single band algorithms using channels 520 to 885 nm based on equations of the radiative transfer. Dogliotti et al. [33] developed a general algorithm designed to map turbidity concentrations from 2 to 1000 FNU, with a switching band algorithm that uses the red 645 nm band for low turbidity values (i.e., lower than 15 FNU) and the Near Infrared (NIR) 859 nm band for high turbidity values (up to 1000 FNU). Other attempts for estimating turbidity from MODIS NASA algorithms have been proposed on the basis of either MODIS-645 nm reflectance [34,35] or the ratios of the MODIS reflectance at 645 nm over 667 nm [25,36–38]. More recently, supervised methods based on classification of spectrally-enhanced quasi-true colour MODIS images have also been proposed by Álvarez-Romero et al. [34] for mapping river plumes in the Great Barrier Reef (Australia).

The only algorithm available at the moment for the oligotrophic waters of the New Caledonian lagoon is the one developed by Ouillon et al. [39]. However, this algorithm that relies on polynomial and exponential models using in-situ reflectance channels over deep waters or turbid waters is not suitable for the oligotrophic and shallow waters (shallower than 5 m) of the Western lagoon of New Caledonia. This is probably because the effect of bottom reflectance over the coral reefs ecosystems for the retrieval of water quality parameters such as chlorophyll-*a* concentration ([chl-*a*]) [40] or turbidity [41] is particularly strong in this context. Indeed, such a contribution of both bathymetry and bottom colour in oligotrophic waters has already been shown to impact the detection of [chl-*a*] from MERIS reflectance [42,43] and AVNIR2/MODIS reflectance [44]. In a similar context, analytical algorithms using 8 "pure" bottom end-members pointed to the same conclusion [45]. However, it has also been shown that the remote sensing reflectance signal shows no significant contamination (R_{rscorr} < 0.0005) from bottom reflectance for water depths larger than 17 m for MODIS images with the brightest reflectance (i.e., white sands and corals such as those found at some places along the Great Barrier Reef [46]).

For about two decades, supervised learning based on neural networks or support vector machines (SVM) has been largely used to estimate oceanic parameters [47–49]. Zhan et al. [50]

successfully retrieved oceanic chlorophyll concentration with data from the SeaBAM dataset. In this study, we propose a trained algorithm based on support vector regression [51] to get more accurate assessments of remote sensing turbidity [52] in the oligotrophic shallow waters of the Western lagoon of New Caledonia. A similar approach already gave interesting improvements for [chl-*a*] assessment in the lagoon and open ocean waters of New Caledonia [53]. Due to the possible strong influence of bathymetry and bottom colour, our support vector regression (SVR) model considers not only reflectance channels but also these two physical parameters as independent variables. Comparison of the results of our approach with published algorithms for estimation of turbidity from in-situ reflectance channels [39] or from MODIS images [33] emphasizes the potential of our model at retrieving the in-situ turbidity in shallow oligotrophic waters.

2. Materials and Methods

2.1. Study Area

New Caledonia is a South Pacific archipelago located between longitudes 162° and 169° E and latitudes 19° and 23° S. The study area—the Voh-Koné-Pouembout (VKP) lagoon in the Northern Province of New Caledonia—extends from 164.5° to 164.9° E and from 20.89° to 21.22° S (Figure 1). This lagoon is particularly concerned by the enhanced inputs of sediments due to mining activities since the Koniambo Nickel SAS (KNS) company started mining nickel at the Koniambo regolith in 2013. During the arrangement of the Koniambo regolith for mining vehicle access, as well as the construction of the nickel pyrometallurgical plant including large dredging in the lagoon, the whole area was monitored in order to assess the possible environmental impacts, especially on fish, coral reefs and marine vegetation [8]. Although it can be controlled by both river discharge and resuspension [18], turbidity was defined as an indicator for assessing water quality in the lagoon [54].

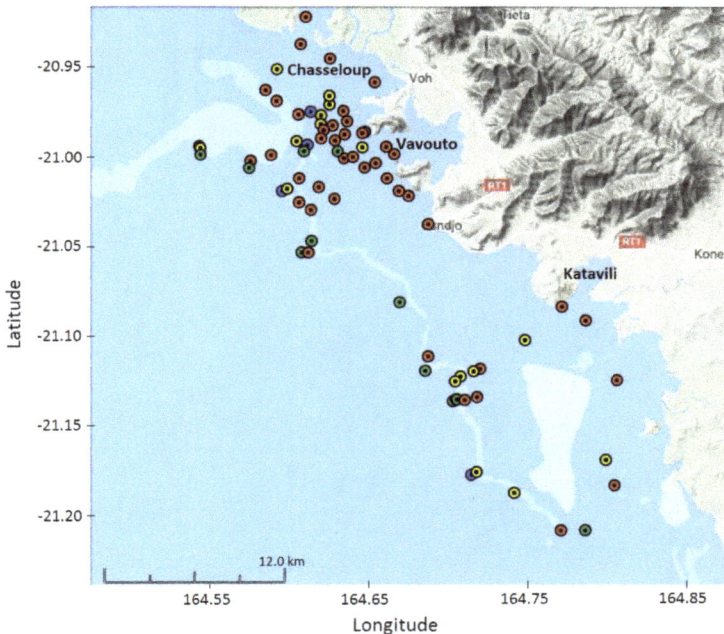

Figure 1. Visited stations at the Voh-Koné-Pouembout (VKP) lagoon (Google Maps Terrain overlay). Points colours correspond to bathymetry, i.e., ◉: 0–10 m depth; ◉: 10–20 m depth; ◉: 20–30 m depth; ◉: > 30 m depth.

The three main bays of the VKP lagoon are the Chasseloup Bay, the Vavouto Bay and the Katavili Bay (Figure 1). The watersheds contributing to water discharge in these bays are principally drained by the Voh, the Taléa/Coco, the Pandanus, the Confiance and the Koné rivers.

The major fraction of the shoreline at this area is made of mangrove, a very productive ecosystem that protects the coast from erosion, acts as a refuge for marine biodiversity and potentially contributes to CO_2 fixation [55]. The sea bottom is made of mud (red to grey), sand (white to grey), fringing or reticulated coral reefs (white) or vegetation as sea grass or algae (grey) and it is delimited by a barrier reef [8].

The bathymetry of the VKP lagoon (Figure 2a) was extracted from the official database of New Caledonia administration [56]. A double-check was achieved by determining at each station the maximal depth recorded by the Conductivity Temperature Depth (CTD) probe (between 5 and 20 profiles per station performed in 2014 and 2015, Figure 2b). Following this protocol, the maximal water depth is 63 m. About 75% of the stations show a depth lower than 13 m, and more than 90% show a depth lower than 30 m (Figure 2b). The deepest waters are located in channels at and around passes to the open ocean while 70% of the lagoon shows a water depth lower than 5 m (Figure 2a).

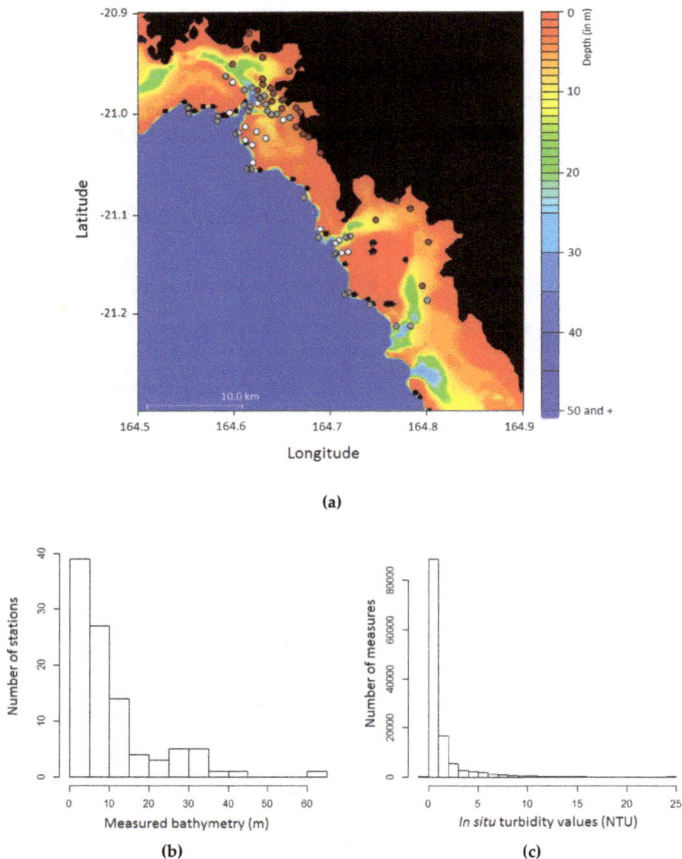

(a)

(b)

(c)

Figure 2. (**a**) Map of the bathymetry (in m) at the Voh-Koné-Pouembout (VKP) lagoon. Points colours correspond to bottom colour, i.e., ○: white bottom; ◑: grey bottom; ●: brown bottom; black areas correspond to land and those near the barrier reef are emerged reefs. (**b**) Histogram of the measured bathymetry on the visited stations. (**c**) Histogram of the in-situ turbidity values measured along CTD profiles.

2.2. Data

2.2.1. Field Measurements

In-situ turbidity values were collected by Analytical Environmental Laboratory (AEL) during a two-year survey (2014–2015) performed for the KNS company in an environmental monitoring context. During this survey, 76 stations were monitored with a SeaBird 19+ CTD probe (Bellevue, WA, USA) that provided measurements at several depths for different parameters including turbidity (in NTU) and fluorescence measured with an ECOFLNTU (from WetLabs, Philomath, OR, USA), pH, oxygen concentration, and salinity according to protocols described in [18]. The bottom colour at each station was estimated from visual in-situ observations [57].

In-situ turbidity values along profiles were generally low, but could exceptionally exceed 10.0 NTU offshore (an exceptional value of 24.0 NTU was recorded on 21 June 2014). More than 70% of turbidity values were below 1.0 NTU and more than 85% were below 2.0 NTU (Figure 2c). These turbidity values are typical of the New Caledonia lagoon [18,39], as well as of the Great Barrier Reef, depending on rain intensity [26,34]. In order to provide in-situ turbidity values representative of the CTD profile, we used the median value of all filtered values over a 10 m depth rather than taking the median values of the 3 first meters as in [39]. Such a calculation aimed at taking into account the variations of turbidity along the water column [18,39].

2.2.2. Satellite Data

MODIS Aqua images were processed from the level 1A to level 2 by creating 250 m resolution data as in Bailey and Werdell [58] for all MODIS data over New Caledonia [59]. Atmospheric corrections were made by default (SeaDAS, but a specific flag was applied composed of 6 SeaDAS flags Land, Cloud, High Sun Glint, Stray Light, High TOA Radiance and Atmospheric Correction Failure adapted to shallow coastal lagoons [59]). As a result, only a few pixels with negative reflectance values in the NIR (1240 nm) were found and subsequently eliminated from the coincidence research. Moreover, the match-ups with an R_{rs} (1240) value above 0.001 Sr^{-1} were discarded because such values in infrared channels were considered as indicative of wrong atmospheric corrections or of the presence of emerged reefs within the pixel.

The product of this MODIS database is marine remote sensing reflectance (R_{rs}) available at 14 channels: 412, 443, 469, 488, 531, 547, 555, 645, 667, 678, 748, 859, 869 and 1240 nm. The processing provides also non-phytoplankton absorption coefficients (a_{dg}), particulate backscattering coefficients (b_{bp}) at 7 channels: 412, 433, 488, 531, 547, 555 and 667 nm. Turbidity assessment according to Ouillon et al. [39] (see Equation (1) below) and Dogliotti et al. [33] were calculated as outputs of the Level2-imagery processing, as well as [chl-*a*] by Wattelez et al. [53].

$$\text{TURB3} = \begin{cases} 90.647 \left(R_{rs}(620) \times \dfrac{R_{rs}(681)}{R_{rs}(510)} \right)^{0.594} & \text{if Turb} < 1 \text{ FTU} \\ \text{Turb if Turb} \geq 1 \text{ FTU} \end{cases}$$

(1)

$$\text{with Turb} = -6204217\, R_{rs}(681)^3 + 179652\, R_{rs}(681)^2 + 36.49\, R_{rs}(681) + 0.452$$

2.2.3. Match-Ups

494 match-ups from MODIS Aqua images were selected using a 0.01° square (about 1×1 km^2) centred on the visited station and in a 2-day temporal window [53,58]. Table 1 summarizes numbers of match-ups according to the campaigns periods and lists the corresponding MODIS files.

Table 1. Campaigns periods, number of match-ups per period and corresponding MODIS files.

Period	Number of CTD Stations	Number of Match-Ups	MODIS Files
22–23/4/2014	31	30	A2014111025500 A2014113024000
19–29/5/2014	59	53	A2014138023500 A2014139032000 A2014141030500 A2014145024000 A2014147023000
24–27/6/2014	43	40	A2014173030500 A2014175025500 A2014177024000
28–30/7/2014	35	29	A2014206021000 A2014207025500 A2014209024000 A2014211023000 A2014229021500
18–20/8/2014	8	7	A2014227023000 A2014229021500
28–30/10/2014	28	27	A2014301030500 A2014302021000
21–23/1/2015	34	34	A2015021032500 A2015024021500
19–27/3/2015	33	32	A2015079022000 A2015085032500 A2015086023000
28–30/4/2015	24	19	A2015116024000 A2015117032500 A2015121030000
6–7/5/2015	18	17	A2015128030500
22–30/6/2015	36	35	A2015173023500 A2015174032000 A2015175022500 A2015176030500 A2015180024000
22–27/7/2015	17	17	A2015205023500 A2015207022000
17–26/8/2015	35	26	A2015228024000 A2015234020500 A2015236033000
28/9/2015–1/10/2015	47	45	A2015272030500 A2015273021000 A2015274025000
26–30/10/2015	41	40	A2015299024500 A2015300033000 A2015302031500
16–20/11/2015	39	39	A2015325032500
22/12/2015	6	4	A2015357032500

Satellite values were assigned according to three different methods as preconized for the research of coincident pixels [58], i.e., with the closest neighbour method (CL), the weighted mean method (WMM) and the filtered mean method (FMM). This approach has already been successfully used for lagoon waters of New Caledonia in Dupouy et al. [60] and Wattelez et al. [53].

2.3. Creation of the Support Vector Regression (SVR) Model

2.3.1. Sampling

Support Vector Regression (SVR) Models are built with a learning sample and then tested with a randomly selected test sample. In our study, the learning sample was constructed with 70% of the data and the test sample contained the remaining 30% of the data. This method is necessary to check the algorithm effectiveness without an overtraining effect. Each model was created and tested ten times (by using ten randomly selected samples). This process allowed selecting a model well fitted on average (and not on a particular random selection).

2.3.2. Indicators

Several indices were computed in order to compare the different models. These indices were the mean normalized bias (MNB), the mean normalized absolute error (MNAE), the mean absolute error (MAE) and the root mean square error (RMSE) with the following respective mathematical expressions:

$$MNB\,(y) = \frac{1}{n} \sum_{i=1}^{n} \frac{x_i - y_i}{x_i} \tag{2}$$

$$MNAE\,(y) = \frac{1}{n} \sum_{i=1}^{n} \frac{|x_i - y_i|}{x_i} \tag{3}$$

$$MAE\,(y) = \frac{1}{n} \sum_{i=1}^{n} |x_i - y_i| \tag{4}$$

$$RMSE\,(y) = \sqrt{\frac{1}{n} \sum_{i=1}^{n} (x_i - y_i)^2} \tag{5}$$

where n is the number of observations, x_i is the ith in-situ observation, y_i is the ith remote sensing assessment.

As a model was created and tested ten times, each index was computed ten times. Indicators of differences of two models were compared thanks to a paired Student's *t*-test. Different models were also compared using the values range, the coefficient of determination (R^2) in both linear and log-regression modes.

2.3.3. Support Vector Regression

In this study, the SVR was built with the following parameters: in-situ turbidity as the explained variable, and remote sensing parameters as the explanatory variables. We first performed tests with all remote sensing parameters (i.e., R_{rs} in the visible spectra from 412 to 678 nm, and a_{dg} and b_{bp}) as explanatory variables. Considering the low turbidity values of our in-situ dataset, we decided in a first approach not to select the MODIS NIR channels (i.e., 748, 859, 869 and 1240 nm) available in the products data set. This option was chosen on the basis of previous studies [31,33,39] which suggested that these channels should not bring information in the low turbidity values range of our study. The first SVR model was therefore deliberately based on visible channels, and the bathymetry and bottom colour were added as explanatory variables to test if these physical parameters bring some significant information. However, to check this assumption, we re-integrated the NIR channels in a second step in order to check the capacity of these channels at improving our SVR model in the case of oligotrophic shallow waters.

The SVR was implemented by using the "svm" function of the R package "e1071" [61]. This function uses an epsilon-regression with a radial kernel whose γ parameter is equal to $\frac{1}{m}$ where m is the number of explanatory variables, $\varepsilon = 0.1$ for the insensitive-loss function and a cost parameter $\mathcal{C} = 1$ is used in the Lagrange formulation.

2.3.4. Algorithm Steps

The forward stepwise approach was used by successively adding the optical parameters, one by one. First, each optical parameter was tested as an explanatory variable in the model containing only one explanatory variable. The one giving the best results according to the aforementioned indicators was retained. Then, the selected model was expanded with a second optical parameter as an explanatory variable, selected according to values of the indicators, and so on.

At each step and for each optical parameter, 10 models were built with 10 different random learning samples and then tested with the 10 corresponding test samples, giving 10 different values for the indicators. If the added parameter did not bring information statistically significant, the last model was kept as the best model.

The last steps were the following: testing the bathymetry and the bottom colour significance, one by one, then removing some parameters in the model and testing again if results were significantly different. For a model, 10 values of an indicator were available (recall that there are 6 indicators, i.e., MNB, MNAE, MAE, RMSE, R^2 and log R^2). Then, comparison of the different models one to another by successive paired t-tests on the series of indicators enabled checking the significance of the indicators of differences. During this latter procedure, the H_0 hypothesis was "There is no significant difference between the two tested models", whereas the H_1 hypothesis was "Indicators computed from the model using an additional parameter are better than the others". We considered that the final model was the one providing the best indicators results and using the lowest number of parameters.

2.4. Interpolated Maps for In-situ Values

The resulting model of this study must be compared to other usual models and to in-situ values. Maps are a useful tool to clearly perceive spatial structures induced by models. But in-situ data are punctual, that is why they were interpolated before mapping.

In an aim of building a map of the interpolated data and comparing with a model applied on a MODIS image, ordinary kriging was implemented on the corresponding MODIS 250m-satellite grid. Turbidity values on all the stations were used to get an empirical variogram from which a variogram model (exponential, Gaussian or else according to the spatial variation structure) was designed with the "fit.variogram" function of the gstat R package. Then, the "krige" function was applied on data with the fitted variogram model. Finally, the kriging output was mapped with the colour scale used for the satellite data mapping.

3. Results

3.1. Evaluation of the SVR Model at Visible Wavelengths

At each step of our SVR approach, R_{rs} values generally provided better results than R_{rs} ratios. Other optical parameters such as a_{dg} and b_{bp} did not add information so they were discarded from the explanatory variables.

Our approach converged with a 3-parameters model that includes R_{rs} (555), R_{rs} (645) and R_{rs} (667) as optical parameters (Optical Model, O.M.), with bathymetry (B) and bottom colour (C) added as explanatory variables. The indicators values computed with the corresponding assessments are shown in Table 2 C.B.O.M. part. The first t-test (t-test 1) aimed at checking the significance of adding (B) to the (O.M.) to yield a (B.O.M.), whereas the second one (t-test 2) aimed at checking the significance of adding (C) to the (B.O.M.) to yield a (C.B.O.M.).

Table 2. Minimum, mean and maximum values for the series of indicators computed on each test sample with C.B.O.M. and C.B.NIR.M. (see Section 3.3). The *t*-tests *p*-values indicate the statistical significance of improvements between models. The *t*-test 1: O.M. vs. B.O.M.; the *t*-test 2: B.O.M. vs. C.B.O.M.; *t*-test 3: C.B.O.M. vs. C.B.NIR.M.

Indicators	C.B.O.M.					C.B.NIR.M.			
	Min.	Mean	Max.	*p*-Value (*t*-Test 1)	*p*-Value (*t*-Test 2)	Min.	Mean	Max.	*p*-Value (*t*-Test 3)
MNB	−0.104	−0.050	0.033	0.0371 #	0.1659 *	−0.144	−0.064	−0.010	0.0439 *
MNAE	0.211	0.233	0.262	<0.001 *	<0.001 *	0.219	0.234	0.269	0.5484 *
MAE	0.094	0.139	0.172	<0.001 *	<0.001 *	0.109	0.136	0.174	0.1691 *
RMSE	0.126	0.235	0.330	<0.001 *	<0.001 *	0.146	0.220	0.333	0.0156 *
R^2	0.426	0.494	0.602	<0.001 *	0.0012 *	0.402	0.553	0.633	0.0312 *
R^2 (log)	0.431	0.539	0.639	<0.001 *	<0.001 *	0.478	0.590	0.684	0.0141 *
In-situ values (NTU)	0.216	0.549	2.417			0.216	0.549	2.417	
Assessment values (NTU)	0.230	0.513	1.291			0.221	0.525	1.235	

* *t*-test conditions are verified in this case; # *t*-test conditions were not verified in this case so a Wilcoxon paired rank test was applied.

The *t*-test 1 results clearly showed that (B) brings significant information in remotely-sensed assessment of turbidity in the VKP lagoon. Indeed, MNAE, MAE and RMSE computed with the (B.O.M.) were significantly lower than those computed with the (O.M.) and all the *p*-values were below 0.05 with a H_1 hypothesis being "The indicator is significantly below for the (B.O.M.)" (Table 2). Similarly, the R^2 and the log-R^2 computed with the (B.O.M.) were significantly larger than those computed with the (O.M.), the H_1 hypothesis being "The indicator is significantly above for the (B.O.M.)". Similar results were obtained for the *t*-test 2, which indicated that adding (C) as an explanatory variable to the (B.O.M.) significantly improved the remotely-sensed assessments (Table 2).

Comparison of the in-situ turbidity with the remote-sensed turbidity assessed by our different SVR models showed that adding (B) and then (C) to the (O.M.) significantly improved the quality of the model (Figure 3), especially above brown bottoms. Despite this improvement, the highest turbidity values (i.e., around 2 NTU in this study), remain underestimated with our SVR model, even if both bathymetry (B.O.M., Figure 3b) and bottom colour (C.B.O.M., Figure 3c) parameters are used.

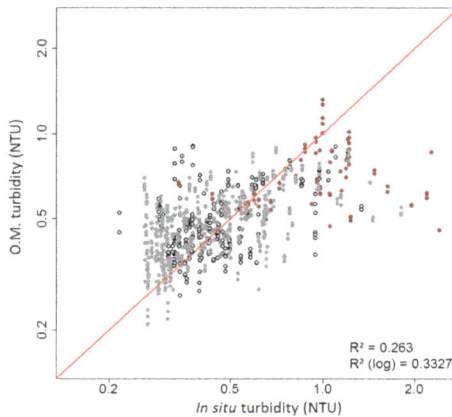

R² = 0.263
R² (log) = 0.3327

(a)

Figure 3. *Cont.*

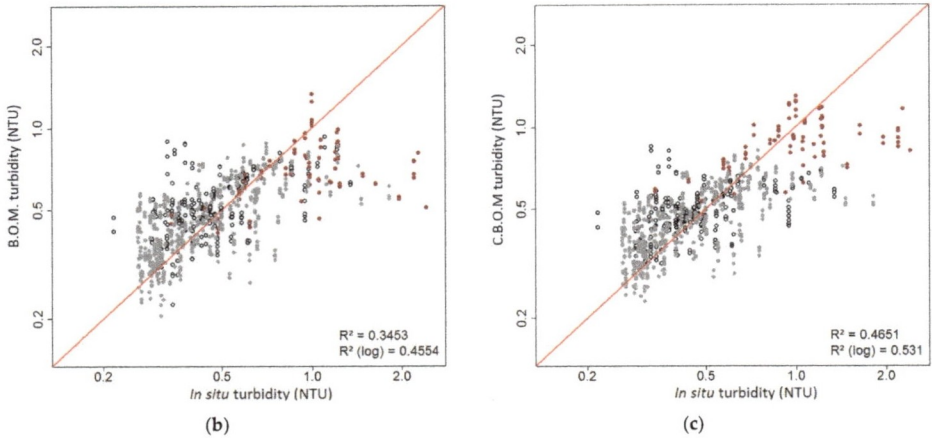

Figure 3. Log-linear regressions between in-situ turbidity and remote sensing turbidity assessed by the different optical SVR models. The red line is the first bisector. (a) O.M. Optical Model; (b) B.O.M. Optical Model + Bathymetry; (c) C.B.O.M. Optical Model + Bathymetry + bottom Colour. Points colours correspond to bottom colour, i.e., ○: white bottom; ◗: grey bottom; ●: brown bottom.

3.2. Comparison with Other Models

This SVR model was then compared with the already existing algorithms of Ouillon et al. [39] (hereafter referred as O2008, set for New Caledonia waters, based on in-situ reflectance data) and Dogliotti et al. [33] (hereafter referred as D2015, based on MODIS images), that have not yet been tested on satellite data over oligotrophic shallow waters. Figure 4 shows that the density of errors is close to 0 with C.B.O.M., which is not the case for the O2008 and D2015 models. This comparison shows then that C.B.O.M. is more suited than a general model in the lagoon of the VKP area.

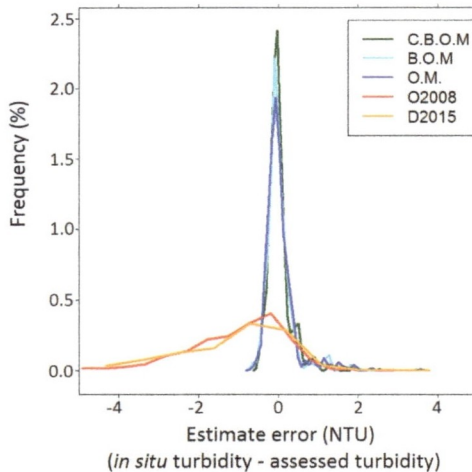

Figure 4. Error density distribution on the 10 test samples obtained with the different SVR models (i.e., C.B.O.M., B.O.M., and O.M.) and with the O2008 and D2015 models.

3.3. Using NIR Channels as Explanatory Variables

Testing the use of NIR channels (from 700 to 869 nm) through the three R_{rs} (859), R_{rs} (488)/R_{rs} (555) and R_{rs} (667)/R_{rs} (678) parameters for another SVR model including both bathymetry and bottom colour yields a new model (i.e., C.B.NIR.M.) that improved the quality of turbidity assessment (Figure 5). This improvement can be evaluated by comparison of Figures 3b and 5, which shows that the highest turbidity values retrieved with C.B.NIR.M. are above those retrieved with C.B.O.M. and then slightly better fit the in-situ values.

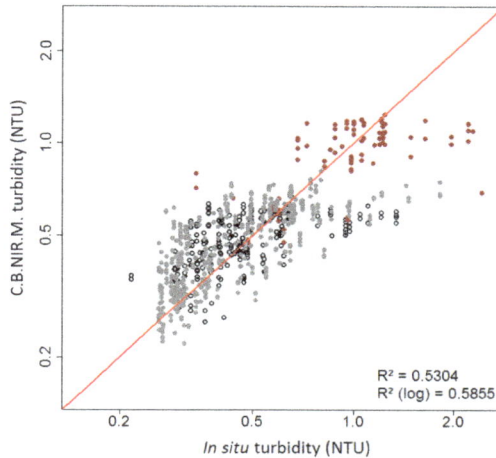

Figure 5. Log-linear regression between in-situ turbidity and remote sensing turbidity retrieved with C.B.NIR.M. The red line is the first bisector. Only the high turbidity values (corresponding to brown-bottom stations) are not well retrieved by the SVR model. Points colours correspond to bottom colour, i.e., ○: white bottom; ●: grey bottom; ●: brown bottom.

The statistical improvement of C.B.NIR.M. compared to C.B.O.M. is confirmed in Table 2 that compares the various indicators for both models with *t*-test 3 which aimed at checking a significant improvement in using C.B.NIR.M. instead of C.B.O.M. Table 2 indicates that RMSE (0.220 for C.B.NIR.M. and 0.235 for C.B.O.M. in mean), R^2 (0.553 for C.B.NIR.M. against 0.494 for C.B.OM. in mean) and log-R^2 (0.590 for C.B.NIR.M. against 0.539 for C.B.O.M. in mean) are significantly improved with C.B.NIR.M. with *t*-tests *p*-values < 0.05. Considering the NIR channels as explanatory variables enables in particular better retrieval of the remote sensing turbidity values on brown bottom, the in-situ values of which are high (between 1.5 and 2 NTU).

Despite this improvement, the highest in-situ turbidity values are not well fitted. With a maximum value of 1.23 NTU, C.B.NIR.M. seems unable to assess high turbidity values compared to C.B.O.M., which provides a maximum value of 1.29 NTU. Both models yield slightly underestimated values.

3.4. Application to MODIS Images

The B.O.M. was applied on MODIS images despite the efficiency of C.B.O.M. at retrieving the in-situ turbidity as the full C.B.O.M. model could not be applied on MODIS images since the bottom colour (C) is not known at all the image pixels. Figure 6a–c shows respectively the in-situ and the remotely-sensed turbidity assessed by O2008 and the B.O.M. on the VKP lagoon area for 21 April 2014 (low turbidity period). The in-situ map (Figure 6a) was made from a kriging interpolation based on a Gaussian variogram model. This map shows a coastal enhancement of turbidity up to 1.5 NTU (164.8° lon, −21.1° lat), while stations in the middle part of the lagoon (164.6° lon, −21.0° lat) and over barrier reefs show a moderate turbidity of 0.3 NTU. High turbidity values shown near the barrier reef

(164.55° lon, −21.00° lat) are probably due to localised coral resuspension. Figure 6b (O2008) shows that high turbidity values were located in shallow waters near the coast (164.78° lon, −21.12° lat) as well as on shallow reef flats (164.75° lon, −21.15° lat) due to bottom effect. Figure 5c shows that with the B.O.M., some pixels with turbidity above the mean level (164.75° lon, −21.15° lat) were retrieved in shallow waters though generally high measured turbidity values were not retrieved everywhere.

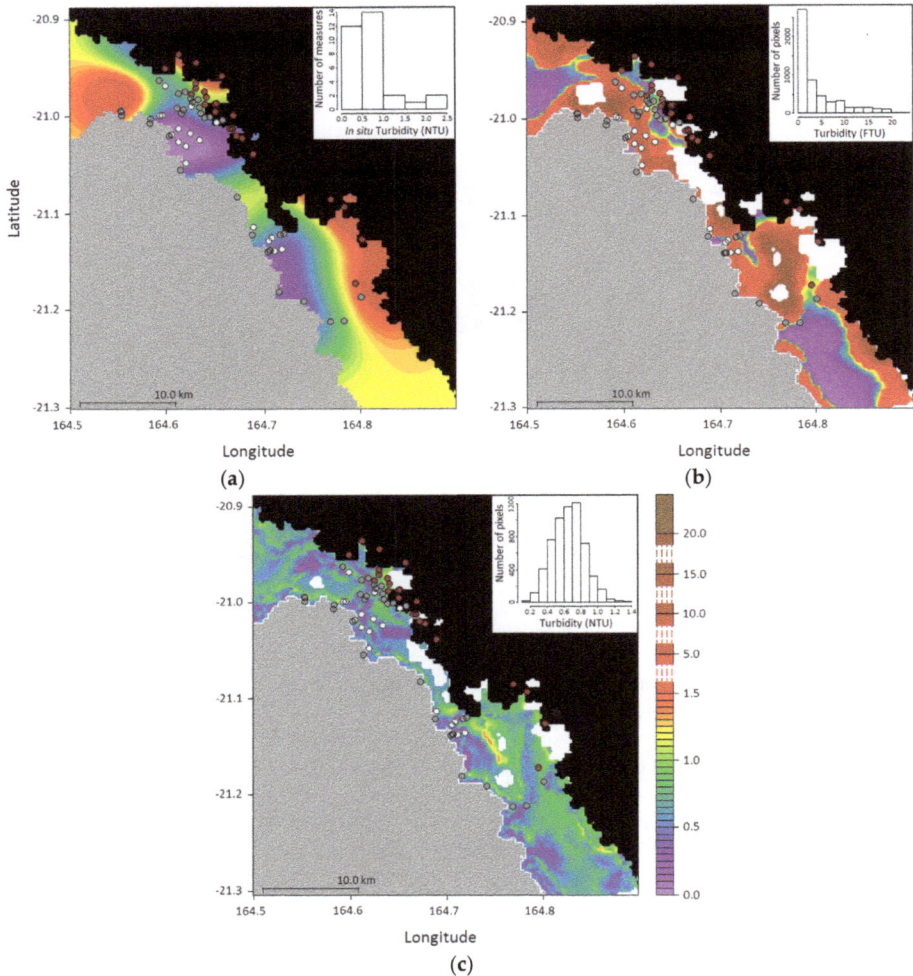

Figure 6. Turbidity in the VKP lagoon area on 21 April 2014. (**a**) in-situ turbidity values (in NTU) interpolated by ordinary kriging and their histogram, as measured with the CTD; (**b**) Map and histogram of turbidity (in FTU) retrieved from the MODIS image with the O2008 model; (**c**) Map and histogram of the turbidity values (in NTU) retrieved from the MODIS image with our B.O.M. Black areas correspond to MODIS land mask and grey areas correspond to deep ocean. Points colours correspond to bottom colour, i.e., ○: white bottom; ◔: grey bottom; ●: brown bottom. On maps (**b**) and (**c**) the white areas correspond to flagged pixels.

Similarly, Figure 7a–c show in-situ and retrieved turbidity values with O2008 and B.O.M. on the VKP area for the MODIS image captured on 24 June 2014. The in-situ map (Figure 7a) was made from a kriging interpolation based on an exponential variogram model. For this day, in-situ values highlight a coastal enhancement too (164.68° lon, −20.98° lat). With B.O.M., assessed values around the barrier reef were quite low (<1.5 NTU, Figure 6c) whereas values assessed by O2008 were usually high (up to 5 NTU).

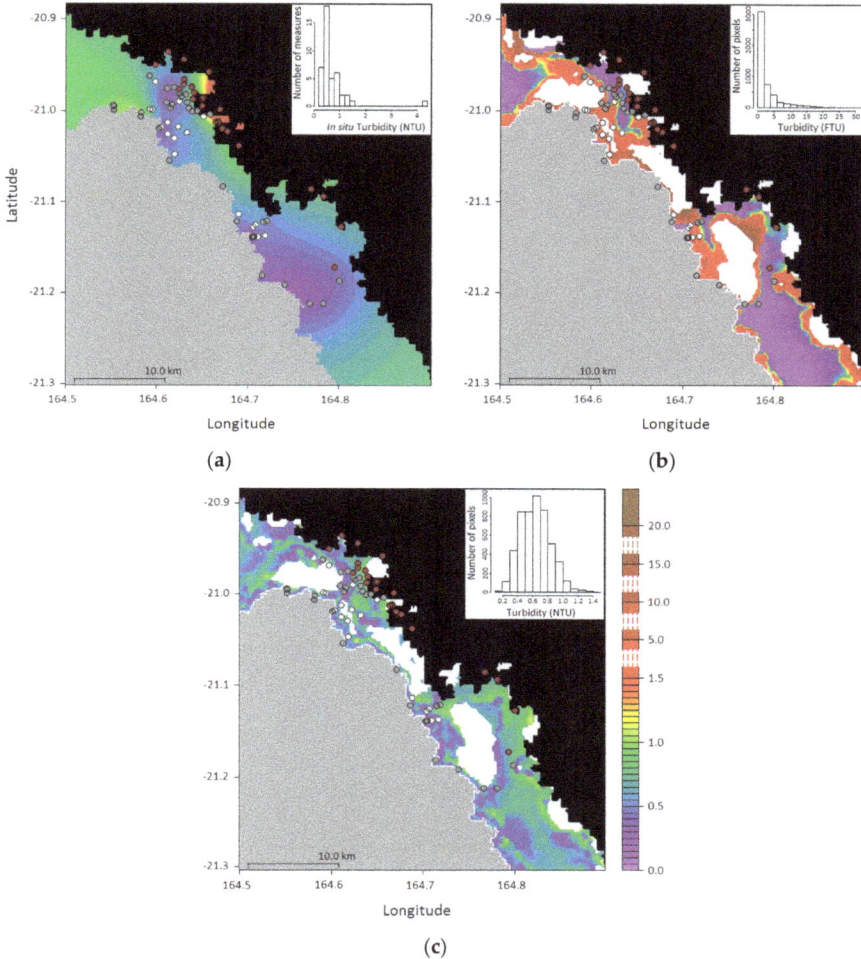

(a)

(b)

(c)

Figure 7. Turbidity in the VKP lagoon area on 24 June 2014. (**a**) in-situ turbidity values (in NTU) interpolated by ordinary kriging and their histogram, as measured with the CTD; (**b**) Map and histogram of turbidity (in FTU) retrieved from the MODIS image with the O2008 model; (**c**) Map and histogram of the turbidity values (in NTU) retrieved from the MODIS image with our B.O.M. Black areas correspond to MODIS land mask and grey areas correspond to deep ocean. Points colours correspond to bottom colour, i.e., ○: white bottom; ◉: grey bottom; ●: brown bottom. On maps (**b**) and (**c**) the white areas correspond to flagged pixels.

Finally, both Figures 6c and 7c on the two selected MODIS images of April and June confirm the capacity of the B.O.M. at retrieving the in-situ turbidity of the oligotrophic shallow waters of the West Coast of New Caledonia as high values do not appear on reefs.

Tables 3 and 4 exhibit the main quantile values for the turbidity assessments by different models for the days of 21 April 2014 and 24 June 2014 respectively. As expected, the ranges of turbidity values retrieved by B.O.M. are strongly reduced and their range (from 0.1 to 1.40 NTU, Figure 3b) more accurate than those retrieved by the O2008 model (Figures 6 and 7). These results indicate that, despite its difficulty at assessing the highest turbidity values, the present SVR model is more suited than the O2008 model that overestimated turbidity with more than 25% for pixels with a turbidity value above 5 FTU.

Table 3. Main quantile values of turbidity estimations by the O2008 model, the B.O.M. and the B.NIR.M. (see Sections 3.1 and 3.3) in the Voh-Koné-Pouembout lagoon area on 21 April 2014.

Model	Min.	1st Decile	1st Quartile	Median	3rd Quartile	9th Decile	Max.
O2008 (FTU)	0.0500	0.1360	0.4418	1.8898	5.6038	11.4643	23.4488
B.O.M. (NTU)	0.1036	0.4084	0.5195	0.6615	0.7768	0.8885	1.3938
B.NIR.M. (NTU)	0.2294	0.3880	0.4503	0.6029	0.7264	0.8191	1.3116

Table 4. Main quantile values of turbidity estimations by the O2008 model, the B.O.M. and the B.NIR.M. (see Sections 3.1 and 3.3) in the Voh-Koné-Pouembout lagoon area on 24 June 2014.

Model	Min.	1st Decile	1st Quartile	Median	3rd Quartile	9th Decile	Max.
O2008 (FTU)	0.0500	0.05	0.1209	0.7162	3.2655	6.8045	30.0500
B.O.M. (NTU)	0.1190	0.3935	0.4862	0.6373	0.7611	0.8964	1.4139
B.NIR.M (NTU)	0.2232	0.4104	0.5125	0.6465	0.7914	0.8904	1.3017

4. Discussion

4.1. Validity of the SVR for Turbidity or SPM Estimation, and Comparison to Previous Algorithms

The SVR model was tested using in-situ turbidity with a SeaBird 19+ CTD calibrated during the period 2014–2015. A relationship can be inferred between turbidity and Suspended Particulate Matter (SPM) [62,63] but this relation is highly dependent on the area explored and on the season considered as well [64]. Consequently, the present SVR model can be used to estimate suspended matter concentration, but only on the VKP region where this regression was set.

The SVR optical model uses 3 channels in the visible, i.e., R_{rs} (555), R_{rs} (645) and R_{rs} (667). Many algorithms have used the 667 nm wavelength (see for instance [26,33]). It has been shown that the sensitivity of one-band algorithms depends on both wavelength and turbidity range, with reflectance at shorter wavelengths more sensitive to low turbidity and reflectance at longer wavelengths more sensitive to high turbidity [31,39,65]. Even if these proposed algorithms show performance with low mean relative errors on a large turbidity range (e.g., with a 20% RMSE for turbidity ranging from 1 to 1000 FNU [33]), they were not developed for oligotrophic shallow tropical waters.

The O2008 algorithm developed over the Southern Lagoon of New Caledonia is designed for oligotrophic waters. However, it is restricted to water depth > 14 m or to water with turbidity > 1 FTU and depth > 10.5 m. In its present state, it can then not be applied successfully to the shallow parts of the lagoon where the bottom influence is not negligible. Indeed, in that context, the upwelling light emerging from the sea surface is affected by the bottom reflectance that is in fact composed of two terms (R_{rs}-water and R_{rs}-bottom). The measured R_{rs} can thus no more be considered to infer the inversion algorithm because its value over shallow waters is higher than the R_{rs}-water value. This mismatch will yield an overestimation of the turbidity retrieved from the O2008 or D2015 (for the same reason) algorithms compared to that retrieved using the SVR method.

As shown on Figure 8, the strongest differences between the O2008 model and our B.O.M. are on white bottom stations and on very shallow brown bottom stations. This figure also shows that the difference in retrieved turbidity between a generic algorithm established for oligotrophic and deep waters (O2008 model) and a SVR model (B.O.M.) decreases with increasing water depth (Figure 8).

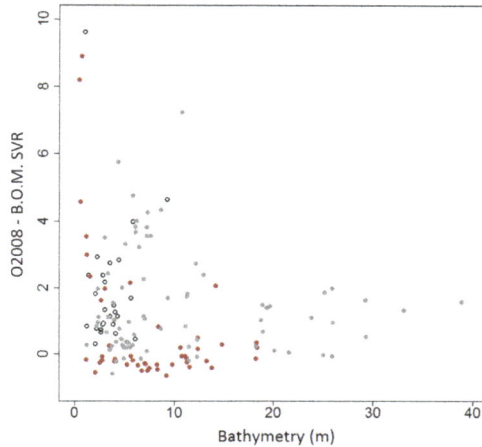

Figure 8. Differences in turbidity estimates between the O2008 model and B.O.M. according to bathymetry and bottom colour. Points colours correspond to bottom colour, i.e., ○: white bottom; ◕: grey bottom; ●: brown bottom.

The addition of a NIR channel (C.B.NIR.M.) slightly improved the model performance, in particular over the brown bottoms. The smaller penetration depth of near infra-red light as compared to the visible bands (due to the high absorption of light by water molecules, see for example References [66–68]) makes the upwelling radiance much more dependent on water and dissolved or suspended matter properties than on bottom colour. That is likely the reason why the performance of the C.B.NIR.M model did not depend anymore or very little on the bottom colour (see Figure 5).

Petus et al. [25] found that the MODIS-Aqua band at 859 nm was not sensitive enough to detect turbidity variations between 0.01 and 10 NTU in the Adour River. Nevertheless, as this study context is widely different from ours (oligotrophic waters), good results provided by C.B.NIR.M. may encourage us to test in the future another model using the visible and near-infrared bands but without considering anymore the bottom colour as a possible explanatory variable. However, the C.B.NIR.M. failed more than the C.B.O.M. at retrieving the highest values of turbidity with underestimation of the highest values by 49% and 46.5%, respectively (Table 2). This performance can be explained by reasons that are discussed in the following part, such as a temporal window of 48 hours between satellite overpass and field measurements, and that can be considered in the next applications of this method. Another reason of discrepancy may be considered in future tests as the penetration depth of light is very low in the NIR. Considering turbidity averaged over 10 m for model training could be reduced to a smaller depth below the surface in next applications.

4.2. Other Possible Improvements of Our Modelling Approach

4.2.1. Vertical Heterogeneity of In-situ Turbidity Profiles

Examination of the in-situ turbidity vertical profiles indicates that the coastal VKP lagoon area is rather mixed as is the South-Western lagoon [18,69]. We then decided to link the ocean colour on each pixel to the median value of the in-situ turbidity value from 0 to 10 m depth. However, the ocean colour on a pixel actually depends on turbidity values in the water column weighted by an exponential function of the depth of measurement [66–68] and, in shallow waters, on the bottom colour [70–72]. It could then be interesting to depict more precisely the function that links the surface value of in-situ turbidity to the vertical profile and to the bottom colour by considering separately the SPM and the seabed contributions to this parameter.

4.2.2. Turbidity Values Distribution

Figures 3 and 5 highlighted that both C.B.O.M. and C.B.NIR.M. underestimated high in-situ turbidity values. This limitation is considered to be linked to the asymmetric distributions of in-situ turbidity values in our learning data set (Figures 6a and 7a) and it should then be improved when a sufficient number of high turbidity values is available for the training of the SVR model (i.e., along a greater sampling period including significant climatic events). Since it is usually difficult to obtain images with plumes and high turbidity right after a significant climatic event because of the cloud coverage, an alternative way would be to train the SVR model with in-situ reflectance values rather than with satellite values.

Moreover, customizing the parameters of our SVR model, including the kernel function and the C cost parameter, could improve the results by providing a distribution of retrieved values closer to the distribution of the in-situ values.

4.2.3. Match-Up Research Procedure

A limitation of our approach during the match-up process in the lagoon waters from the VKP area in New Caledonia is the use of methods mainly developed for the open ocean [58], where sea floor and bathymetry do not influence the ocean colour and where spatial and temporal changes in biogeochemical parameters are quite low. However, in lagoon waters, many localized and transient phenomena, such as upwelling, river inputs and resuspension due to wind bursts, can influence in-situ turbidity values at short distance and time scales. A future improvement of our approach could then consist in reducing both the temporal and spatial windows that are used during the match-up process. However, such a reduction would imply a reduced number of coincidences, which would raise the issue of data representativeness and significance to create a robust model.

4.2.4. Model Conception

Generic models designed to remotely assess turbidity values are generally customized with in-situ turbidity values and in-situ reflectance values. By this way, resulting models have only to be fitted according to remote reflectance sensors. Since the current SVR model was customized with in-situ turbidity values and remote reflectance values from Aqua-MODIS, it is highly dependent from the MODIS sensors. Nevertheless, the methodology developed with this SVR model should be easily used with other sensors providing a sufficient training dataset in coincidence with in-situ data is available.

An alternative could be to develop a SVR model from in-situ turbidity and R_{rs} for any sensor obtained from in-situ hyperspectral R_{rs} values and the spectral sensitivity of the given sensor. Such an opportunity would overcome the need of many sets of match-up for different sensors.

4.2.5. Spectral Classification of MODIS Pixels in the VKP Lagoon

Figure 9 shows the MODIS reflectance spectra obtained on all coincident pixels of the VKP lagoon area for the 2014–2015 period that was used to construct our SVR model. Grey bottom pixels' reflectance spectra show little variability according to turbidity range. The highest R_{rs} values were linked to the brown bottom pixels, and to white bottom pixels with turbidity values below 1 NTU. On white bottom pixels, the higher reflectance values for lower turbidity values clearly show the prevalence of the bottom effect in the optical signal. Although the current data set was too small to strongly support this statement, this effect seems to be opposite on brown bottom pixels. This latter point illustrates the difficulty to remotely assess turbidity in oligotrophic shallow waters and it then emphasizes the interest of the SVR method.

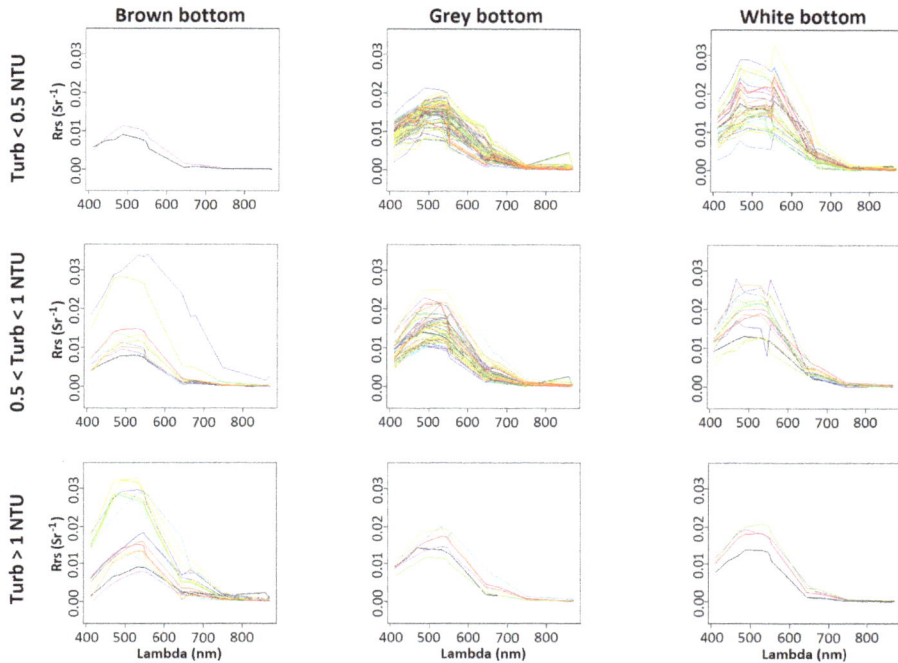

Figure 9. Spectra of MODIS reflectance on all pixels used for the construction of our SVR model and classified by both turbidity values (i.e., Upper row: Turb < 0.5 NTU, Middle row: 0.5 < Turb < 1, Lower row: Turb > 1) and bottom colour (i.e., Left: Brown bottom, Middle: Grey bottom, Right: White bottom).

4.2.6. Including the Bottom Colour

In the present SVR model, the optical signal of the bottom is integrated in the model conception and it is weighted by the bathymetry and the bottom colour. Some bathymetric patterns slightly appear on the derived turbidity maps (e.g., Figures 6c and 7c as compared to Figure 2) but a generalized integration of the bottom colour in the application should bring more accurate results. The main remaining challenge of this approach lies in constructing a SVR model that could include the bottom colour at each pixel. To reach this goal, a first approach would consist in collecting this variable by in-situ observations at the largest possible number of stations. However, this option appears difficult to apply at the lagoon scale. Another option would be to extrapolate a bottom colour from geological maps. An alternative approach could be to retrieve the bottom colour from an extremely clear image, where the water column is considered null, by using the Lyzenga's method from ocean colour reflectance [70,73,74], and then operate a spectral classification to associate each pixel of the image to a bottom colour as successfully applied in other areas of New Caledonia [42,43]. Such a method will be applied on the VKP lagoon area in order to evaluate its improvements toward the C.B.O.M. for turbidity retrieving in oligotrophic shallow waters. Figures 5 and 6 show an application of the B.O.M. SVR on two particular days for MODIS. We may also use the Sentinel 2 or Sentinel 3 data in order to produce synoptic maps for studying the temporal variations on the whole area.

5. Conclusions

This paper introduced an empirical algorithm—based on the SVR method—for assessing turbidity values in oligotrophic shallow waters from MODIS images. This algorithm was tested on the oligotrophic shallow waters of the West Coast of New Caledonia, but it may be applicable to other similar areas. The optical explanatory variables included in this SVR model were selected according to statistical considerations, and improving results of turbidity assessments given by generic algorithms which are not adapted to shallow oligotrophic waters. Since bathymetry and bottom colour showed to widely influence the remotely-sensed optical signal, both parameters were introduced as explanatory parameters in the SVR model. Despite the complex optical character of the waters at the VKP lagoon area studied, this latter approach significantly improved the capacity of our SVR model at retrieving the in-situ turbidity data in these oligotrophic shallow waters.

Since the SVR model introduced in this paper is based on a limited set of in-situ data with low turbidity values, extending the range of these data should improve its accuracy for higher turbidity values. Considering that this method is widely applicable on hyperspectral and multispectral remote sensors, such as Sentinels, it should allow a better monitoring of coastal shallow waters in coral reefs.

Acknowledgments: This work was financially supported by the Centre de Recherche pour le Nickel et son environment (www.cnrt.nc/) under the project DYNAMINE, "Dynamique des métaux de la mine au lagon en Nouvelle-Calédonie" project (CSF N° 3PS2013-CNRT.IRD/DYNAMINE) and by French Institute of Research for the Development (IRD—www.ird.fr/). The PC-cluster used for the ocean colour applications was funded through an IRD/IFREMER collaboration and by the IRD SPIRALES Valhysat program. We gratefully acknowledge the NASA Ocean Biology Processing Group (OBPG) for making MODIS ocean-colour imagery and products available. We thank Koniambo Nickel SAS (KNS) for allowing access to the set of in-situ turbidity measurements. Finally, we especially thank Morgan Mangeas (IRD, UMR 228 ESPACE-DEV) for his advices about SVR modelling.

Author Contributions: G.W. and C.D. designed the study. J.-M.F. performed the in-situ measurements of turbidity. J.L. developed the MODIS database Valhysat software. G.W. developed the SVR model and adapted it to the site. G.W., C.D. and F.J. performed the statistics. C.D., G.W., F.J. and S.O. analysed the results and wrote the paper.

Conflicts of Interest: The authors declare no conflict of interest.

References

1. Morrison, R.J.; Denton, G.; Tamata, U.B.; Grignon, J. Anthropogenic biogeochemical impacts on coral reefs in the Pacific Islands—An overview. *Deep Sea Res. II* **2013**, *96*, 5–12. [CrossRef]
2. Fabricius, K.E.; Logan, M.; Weeks, S.; Brodie, J. The effects of river run-off on water clarity across the central Great Barrier Reef. *Mar. Pollut. Bull.* **2014**, *84*, 191–200. [CrossRef] [PubMed]
3. Heinz, T.; Haapkylä, J.; Gilbert, A. Coral health on reefs near mining sites in New Caledonia. *Dis. Aquat. Org.* **2015**, *115*, 165–173. [CrossRef] [PubMed]
4. Chen, Z.; Muller-Karger, F.; Hu, C. Remote sensing of water clarity in Tampa Bay. *Remote Sens. Environ.* **2007**, *109*, 249–259. [CrossRef]
5. Adjeroud, M.; Fernandez, J.-M.; Carroll, A.G.; Harrison, P.L.; Penin, L. Spatial patterns and recruitment process of coral assemblages among contrasting environmental conditions in the southwestern lagoon of New Caledonia. *Mar. Pollut. Bull.* **2010**, *61*, 375–386. [CrossRef] [PubMed]
6. Myers, N.; Mittermeier, R.A.; Mittermeier, C.G.; da Fonseca, G.A.B.; Kent, J. Biodiversity hotspots for conservation priorities. *Nature* **2000**, *403*, 853–858. [CrossRef] [PubMed]
7. Alongi, D.M. Present state and future of the world's mangrove forests. *Environ. Conserv.* **2002**, *29*, 331–349. [CrossRef]
8. Adjeroud, M.; Gilbert, A.; Facon, M.; Foglia, M.; Moreton, B.; Heintz, T. Localised and limited impact of a dredging operation on coral cover in the northwestern lagoon of New Caledonia. *Mar. Pollut. Bull.* **2016**, *105*, 208–214. [CrossRef] [PubMed]
9. Ceccarelli, D.M.; McKinnon, A.D.; Andréfouët, S.; Allain, V.; Young, J.; Gledhill, D.C.; Flynn, A.; Bax, N.J.; Beaman, R.; Borsa, P.; et al. The coral sea: Physical environment, ecosystem status and biodiversity assets. *Adv. Mar. Biol.* **2013**, *66*, 213–290. [PubMed]
10. Cluzel, D.; Aitchison, J.C.; Picard, C. Tectonic accretion and underplating of mafic terranes in the Late Eocene intraoceanic fore-arc of New Caledonia (Southwest Pacific): Geodynamic implications. *Tectonophysics* **2001**, *340*, 23–59. [CrossRef]
11. Perrier, N.; Ambrosi, J.P.; Colin, F.; Gilkes, R.J. Biogeochemistry of a Regolith: The New Caledonian Koniambo Ultramafic Massif. *J. Geochem. Explor.* **2006**, *88*, 54–58. [CrossRef]

12. Fandeur, D.; Juillot, F.; Morin, G.; Olivi, L.; Cognigni, A.; Webb, S.M.; Brown, G.E. XANES evidence for oxidation of Cr(III)) to Cr(VI) by Mn-oxides in a lateritic regolith developed on serpentinized ultramafic rocks of New Caledonia. *Environ. Sci. Technol.* **2009**, *43*, 7384–7390. [CrossRef] [PubMed]

13. Dublet, G.; Juillot, F.; Morin, G.; Fritsch, E.; Fandeur, D.; Ona-Nguema, G.; Brown, G.E. Ni speciation in a New Caledonian lateritic regolith: A quantitative X-ray absorption spectroscopy investigation. *Geochim. Cosmochim. Acta* **2012**, *95*, 119–133. [CrossRef]

14. Dublet, G.; Juillot, F.; Morin, G.; Fritsch, E.; Fandeur, D.; Brown, G.E., Jr. Goethite aging explains Ni depletion in upper units of ultramafic lateritic ores from New Caledonia. *Geochim. Cosmochim. Acta* **2015**, *160*, 1–15. [CrossRef]

15. Lagadec, G.; Perret, C.; Pitoiset, A. *Nickel et Développement en Nouvelle-Calédonie, Perspectives de Développement Pour la Nouvelle-Calédonie*; Perret, C., Ed.; PUG: Grenoble, France, 2002; Chapter 1, pp. 21–42.

16. Join, J.L.; Robineau, B.; Ambrosi, J.P.; Costis, C.; Colin, F. Système hydrogéologique d'un massif minier ultrabasique de Nouvelle-Calédonie. *C. R. Geosci.* **2005**, *337*, 1500–1508. [CrossRef]

17. Fernandez, J.-M.; Ouillon, S.; Chevillon, C.; Douillet, P.; Fichez, R.; Le Gendre, R. A combined modelling and geochemical study of the fate of terrigenous inputs from mixed natural and mining sources in a coral reef lagoon (New Caledonia). *Mar. Poll. Bull.* **2006**, *52*, 320–331. [CrossRef] [PubMed]

18. Ouillon, S.; Douillet, P.; Lefebvre, J.P.; Le Gendre, R.; Jouon, A.; Bonneton, P.; Fernandez, J.M.; Chevillon, C.; Magand, O.; Lefèvre, J.; et al. Circulation and suspended sediment transport in a coral reef lagoon: The South-West lagoon of New Caledonia. *Mar. Pollut. Bull.* **2010**, *61*, 269–296. [CrossRef] [PubMed]

19. Fernandez, J.M.; Meunier, J.D.; Ouillon, S.; Moreton, B.; Douillet, P.; Grauby, O. Dynamics of Suspended Sediments during a Dry Season and Their Consequences on Metal Transportation in a Coral Reef Lagoon Impacted by Mining Activities, New Caledonia. *Water* **2017**, *9*, 338. [CrossRef]

20. Andrefouët, S.; Mumby, P.J.; McField, M.; Hu, C.; Muller-Karger, F.E. Revisiting coral reef connectivity. *Coral Reefs* **2002**, *21*, 43–48. [CrossRef]

21. Dupouy, C.; Minghelli-Roman, A.; Despinoy, M.; Röttgers, R.; Neveux, J.; Pinazo, C.; Petit, M. MODIS/Aqua chlorophyll monitoring of the New Caledonia lagoon during the 2008 La Nina event. In Proceedings of the Remote Sensing of Inland, Coastal, and Oceanic Waters, Noumea, New Caledonia, 19 December 2008; Frouin, R.J., Andrefouët, S., Kawamura, H., Lynch, M.J., Pan, T., Platt, T., Eds.; SPIE: Bellingham, WA, USA, 2008; Volume 7150, pp. 1–8.

22. Dupouy, C.; Röttgers, R.; Tedetti, M.; Martias, C.; Murakami, H.; Doxaran, D.; Lantoine, F.; Rodier, M.; Favareto, L.; Kampel, M.; et al. Influence of CDOM and Particle Composition on Ocean Colour of the Eastern New Caledonia Lagoon during the CALIOPE Cruises. *Proc. SPIE* **2014**, *9261*, 92610M. [CrossRef]

23. Wang, Y.J.; Yan, F.; Zhang, P.Q.; Dong, W.J. Experimental research on quantitative inversion model of suspended sediment concentration using remote sensing technology. *Chin. Geogr. Sci.* **2007**, *17*, 243–249. [CrossRef]

24. Gohin, F. Annual cycles of chlorophyll-*a*, non-algal suspended particulate matter, and Turbidity observed from space and in-situ in coastal waters. *Ocean Sci.* **2011**, *7*, 705–732. [CrossRef]

25. Petus, C.; Chust, G.; Gohin, F.; Doxaran, D.; Froidefond, J.M.; Sagarminaga, Y. Estimating turbidity and total suspended matter in the Adour River plume (South Bay of Biscay) using MODIS 250-m imagery. *Cont. Shelf Res.* **2010**, *30*, 379–392. [CrossRef]

26. Petus, C.; da Silva, E.T.; Devlin, M.; Wenger, A.S.; Álvarez-Romero, J.G. Using MODIS data for mapping of water types within river plumes in the Great Barrier Reef, Australia: Towards the production of river plume risk maps for reef and seagrass ecosystems. *J. Environ. Manag.* **2014**, *137*, 163–177. [CrossRef] [PubMed]

27. Han, B.; Loisel, H.; Vantrepotte, V.; Mériaux, X.; Bryère, P.; Ouillon, S.; Dessailly, D.; Xing, Q.; Zhu, J. Development of a semi-analytical algorithm for the retrieval of Suspended Particulate Matter from remote sensing over clear to very turbid waters. *Remote Sens.* **2016**, *8*, 211. [CrossRef]

28. Kabiri, K.; Moradi, M. Landsat-8 imagery to estimate clarity in near-shore coastal waters: Feasibility study—Chabahar Bay, Iran. *Cont. Shelf Res.* **2016**, *125*, 44–53. [CrossRef]

29. Constantin, S.; Doxaran, D.; Constantinescu, S. Estimation of water turbidity and analysis of its spatio-temporal variability in the Danube River plume (Black Sea) using MODIS satellite data. *Cont. Shelf Res.* **2016**, *112*, 14–30. [CrossRef]

30. Hu, C.; Chen, Z.; Clayton, T.D.; Swarzenski, P.; Brock, J.C.; Muller-Karger, F.E. Assessment of estuarine water-quality indicators using MODIS medium-resolution bands: Initial results from Tampa Bay, FL. *Remote Sens. Environ.* **2004**, *93*, 423–441. [CrossRef]

31. Nechad, B.; Ruddick, K.G.; Park, Y. Calibration and validation of a generic multisensor algorithm for mapping of total suspended matter in turbid waters. *Remote Sens. Environ.* **2010**, *114*, 854–866. [CrossRef]

32. Novoa, S.; Doxaran, D.; Ody, A.; Vanhellemont, Q.; Lafon, V.; Lubac, B.; Gernez, P. Atmospheric Corrections and Multi-Conditional Algorithm for Multi-Sensor Remote Sensing of Suspended Particulate Matter in Low-to-High Turbidity Levels Coastal Waters. *Remote Sens.* **2017**, *9*, 61. [CrossRef]

33. Dogliotti, A.I.; Ruddick, K.G.; Nechad, B.; Doxaran, D.; Knaeps, E.A. single algorithm to retrieve turbidity from remotely-sensed data in all coastal and estuarine waters. *Remote Sens. Environ.* **2015**, *156*, 157–168. [CrossRef]

34. Álvarez-Romero, J.G.; Devlin, M.J.; Teixeira da Silva, E.; Petus, C.; Ban, N.; Pressey, R.J.; Kool, J.; Roberts, S.; Cerdeira, W.A.; Brodie, J. A novel approach to model exposure of coastal-marine ecosystems to riverine flood plumes based on remote sensing techniques. *J. Environ. Manag.* **2013**, *119*, 194–207. [CrossRef] [PubMed]

35. Devlin, M.; McKinna, L.W.; Álvarez-Romero, J.G.; Petus, C.; Abott, B.; Harkness, P.; Brodie, J. Mapping the pollutants in surface riverine flood plume waters in the Great Barrier Reef, Australia. *Mar. Pollut. Bull.* **2012**, *65*, 224–235. [CrossRef] [PubMed]

36. Miller, R.L.; McKee, B.A. Using MODIS Terra 250 m imagery to map concentrations of total suspended matter in coastal waters. *Remote Sens. Environ.* **2004**, *93*, 259–266. [CrossRef]

37. Doxaran, D.; Froidefond, J.M.; Castaing, P.; Babin, M. Dynamics of the turbidity maximum zone in a macrotidal estuary (the Gironde, France): Observations from field and MODIS satellite data. *Estuar. Coast. Shelf Sci.* **2009**, *81*, 321–332. [CrossRef]

38. Lahet, F.; Stramski, D. MODIS imagery of turbid plumes in San Diego coastal waters during rainstorm events. *Remote Sens. Environ.* **2010**, *114*, 332–344. [CrossRef]

39. Ouillon, S.; Douillet, P.; Petrenko, A.; Neveux, J.; Dupouy, C.; Froidefond, J.-M.; Andréfouët, S.; Muñoz-Caravaca, A. Optical Algorithms at satellite wavelengths for Total Suspended Matter in Tropical Coastal Waters. *Sensors* **2008**, *8*, 4165–4185. [CrossRef] [PubMed]

40. Dupouy, C.; Neveux, J.; Ouillon, S.; Frouin, R.; Murakami, H.; Hochard, S.; Dirberg, G. Inherent optical properties and satellite retrieval of chlorophyll concentration in the lagoon and open waters of New Caledonia. *Mar. Pollut. Bull.* **2010**, *61*, 503–518. [CrossRef] [PubMed]

41. Hochberg, E.J.; Atkinson, M. Capabilities of remote sensors to classify coral, algae, and sand as pure and mixed spectra. *Remote Sens. Environ.* **2003**, *85*, 174–189. [CrossRef]

42. Minghelli-Roman, A.; Dupouy, C. Influence of water column chlorophyll concentration on bathymetric estimations in the lagoon of New Caledonia using several MERIS images. *IEEE J. Sel. Top. Appl. Earth Obs. Remote Sens.* **2013**, *77*, 1–7. [CrossRef]

43. Minghelli-Roman, A.; Dupouy, C. Correction of the water column attenuation: Application to the seabed mapping of the lagoon of New Caledonia using MERIS images. *IEEE J. Sel. Top. Appl. Earth Obs. Remote Sens.* **2014**, *7*, 2617–2629. [CrossRef]

44. Murakami, H.; Dupouy, C. Atmospheric correction and inherent optical property estimation in the southwest New Caledonia lagoon using AVNIR-2 high-resolution data. *Appl. Opt.* **2013**, *52*, 182–198. [CrossRef] [PubMed]

45. McKinna, L.I.W.; Fearns, P.R.C.; Weeks, S.J.; Werdell, P.J.; Reichstetter, M.; Franz, B.A.; Shea, D.M.; Feldman, G.C. A semianalytical ocean colour inversion algorithm with explicit water column depth and substrate reflectance parameterization. *J. Geophys. Res. Oceans* **2015**, *120*, 1741–1770. [CrossRef]

46. Reichstetter, M.; Fearns, P.R.C.S.; Weeks, S.J.; McKinna, L.I.W.; Roelfsema, C.; Furnas, M. Bottom reflectance in Ocean Colour Satellite Remote Sensing for Coral Reef Environments. *Remote Sens.* **2015**, *7*, 16756–16777. [CrossRef]

47. Keiner, L.E.; Yan, X.H. A neural network model for estimating sea surface chlorophyll and sediments from Thematic Mapper imagery. *Remote Sens. Environ.* **1998**, *66*, 153–165. [CrossRef]

48. Zhan, H. Application of Support Vector Machines in inverse problems in ocean colour remote sensing. *Support Vector Mach. Theory Appl.* **2005**, *177*, 387–398.

49. Chen, J.; Quan, W.T.; Cui, T.W.; Song, Q.J. Estimation of total suspended matter concentration from MODIS data using a neural network model in the China eastern coastal zone. *Est. Coast. Shelf Sci.* **2015**, *155*, 104–113. [CrossRef]

50. Zhan, H.; Shi, P.; Chen, C. Retrieval of oceanic chlorophyll concentration using support vector machines. *IEEE Trans. Geosci. Remote Sens.* **2003**, *41*, 2947–2951. [CrossRef]

51. Drucker, H.; Burges, C.J.C.; Kaufman, L.; Smola, A.; Vapnik, V. Support Vector Regression Machines. *Adv. Neural Inf. Process. Syst.* **1996**, *9*, 155–161.

52. Camps-Valls, G.; Bruzzone, L.; Rojo-Alvarez, J.L.; Melgeni, F. Robust Support Vector Regression for biophysical variable estimation from remotely sensed images. *IEEE Geosci. Remote Sens. Lett.* **2006**, *3*, 1–5. [CrossRef]

53. Wattelez, G.; Dupouy, C.; Mangeas, M.; Lefèvre, J.; Touraïvane; Frouin, R. A statistical algorithm for estimating Chlorophyll Concentration in the New Caledonian lagoon. *Remote Sens.* **2016**, *8*, 45. [CrossRef]

54. Touraïvane; Allenbach, M.; Mangeas, M.; Bonte, C. Monitoring the turbidity associated with the dredging in Vavouto Bay in New Caledonia. In Proceedings of the 19th International Congress on Modelling and Simulation, Perth, Australia, 12–16 December 2011.

55. Leopold, A.; Marchand, C.; Renchon, A.; Deborde, J.; Quiniou, T.; Allenbach, M. Net ecosystem CO_2 in the "Coeur de Voh" mangrove, New Caledonia: Effects of water stress on mangrove productivity in a semi-arid climate. *Agric. For. Meteorol.* **2016**, *223*, 217–232. [CrossRef]

56. Banque des Données Bathymétriques de la Nouvelle-Calédonie (BDBNC). *Banque des Données Bathymétriques de la Nouvelle-Calédonie*; Atlas de la Direction des Technologies et Systèmes d'Information, Direction des Technologies et des Service de l'Information (DTSI): Noumea, France, 2009.

57. Kumar-Roiné, S.; Achard, R.; Kaplan, H.; Haddad, L.; Laurent, A.; Drouzy, M.; Hubert, M.; Pluchino, S.; Fernandez, J.M. *Suivi Environnemental du Milieu Marin de la Zone VKP*; Volet 4: Surveillance Physicochimique. Période: Septembre 2015—Août 2016 et Novembre 2016. Rapport AEL 131121-KS-02; AEL/LEA: Noumea, France, 2017.

58. Bailey, S.W.; Werdell, P.J. A multi-sensor approach for the on-orbit validation of ocean colour satellite data products. *Remote Sens. Environ.* **2006**, *102*, 12–23. [CrossRef]

59. Lefèvre, J. *The VALHYSAT Project: MODIS-DB Database: Description Guide of the Database*; Valhysat Report 1. Noumea: IRD Internal Report; IRD: Noumea, New Caledonia, 2010.

60. Dupouy, C.; Savranski, T.; Lefèvre, J.; Despinoy, M.; Mangeas, M.; Fuchs, R.; Faure, V.; Ouillon, S.; Petit, M. Monitoring optical properties of the Southwest Tropical Pacific. In Proceedings of the Remote Sensing of the Coastal Ocean, Land, and Atmosphere Environment, Incheon, Korea, 4 November 2010; Frouin, R.J., Rhyong Yoo, H., Won, J.-S., Feng, A., Eds.; SPIE: Bellingham, WA, USA, 2010; Volume 7858, p. 13. [CrossRef]

61. R Core Team. *R: A Language and Environment for Statistical Computing*; R Foundation for Statistical Computing: Vienna, Austria, 2014.

62. Jouon, A.; Ouillon, S.; Douillet, P.; Lefebvre, J.P.; Fernandez, J.-M.; Mari, X.; Froidefond, J.M. Spatio-temporal variability in suspended particulate matter concentration and the role of aggregation on size distribution in a coral reef lagoon. *Mar. Geol.* **2008**, *256*, 36–48. [CrossRef]

63. Kaplan, H.; Laurent, A.; Hubert, M.; Moreton, B.; Kumar-Roiné, S.; Fernandez, J.M. *Suivi de la Qualité Physico-Chimique de l'eau de mer de la Zone sud du Lagon de Nouvelle-Calédonie: 2ème Memester 2016*; Contrat AEL/Vale-NC n°3052-Avenant n°1; AEL/LEA: Noumea, France, 2016.

64. Jafar-Sidik, M.B.J.; Gohin, F.; Bowers, D.G.; Howarth, J.; Hull, T. The relationship between Suspended Particulate Matter and Turbidity at a mooring station in a coastal environment: Consequences for satellite-derived products. *Oceanologia* **2017**, in press. [CrossRef]

65. Shen, F.; Verhoef, W.; Zhou, Y.X.; Salama, M.S.; Liu, X.L. Satellite estimates of wide-range suspended sediment concentrations in Changjiang (Yangtze) estuary using MERIS data. *Estuaries Coasts* **2010**, *33*, 1420–1429. [CrossRef]

66. Gordon, H.R.; Clarke, D.K. Remote sensing optical properties of a stratified ocean: An improved interpretation. *Appl. Opt.* **1980**, *19*, 3428–3430. [CrossRef] [PubMed]

67. Nanu, L.; Robertson, C. The effect of suspended sediment depth distribution on coastal water spectral reflectance: Theoretical simulation. *Int. J. Remote Sens.* **1993**, *14*, 225–239. [CrossRef]

68. Ouillon, S. An inversion method for reflectance in stratified turbid waters. *Int. J. Remote Sens.* **2003**, *24*, 535–548. [CrossRef]

69. Fichez, R.; Chifflet, S.; Douillet, P.; Gérard, P.; Gutierrez, F.; Jouon, A.; Ouillon, S.; Grenz, C. Biogeochemical typology and temporal variability of lagoon waters in a coral reef ecosystem subject to terrigeneous and anthropogenic inputs (New Caledonia). *Mar. Poll. Bull.* **2010**, *61*, 309–322. [CrossRef] [PubMed]

70. Lyzenga, D. Passive remote sensing techniques for mapping water depth and bottom features. *Appl. Opt.* **1978**, *17*, 379–383. [CrossRef] [PubMed]

71. Tolk, B.L.; Han, L.; Rundquist, D.C. The impact of bottom brightness on spectral reflectance of suspended sediments. *Int. J. Remote Sens.* **2000**, *21*, 2259–2268. [CrossRef]

72. Mobley, C.D.; Sundman, L.K. Effects of optically shallow bottoms on upwelling radiances: Inhomogeneous and slopping bottoms. *Limnol. Oceanogr.* **2003**, *48*, 329–336. [CrossRef]

73. Lyzenga, D. Remote sensing of bottom reflectance and Water attenuation parameters in shallow water using aircraft and Landsat data. *Int. J. Remote Sens.* **1981**, *2*, 71–82. [CrossRef]

74. Lyzenga, D.; Malinas, N.; Tanis, F. Multispectral Bathymetry Using a Simple Physically Based Algorithm. *IEEE Trans. Geosci. Remote Sens.* **2006**, *44*, 2251–2259. [CrossRef]

![water logo] *water*

MDPI

Article

Modelling Hydrology and Sediment Transport in a Semi-Arid and Anthropized Catchment Using the SWAT Model: The Case of the Tafna River (Northwest Algeria)

Amin Zettam [1,2,*], **Amina Taleb** [1], **Sabine Sauvage** [2], **Laurie Boithias** [3], **Nouria Belaidi** [1] and **José Miguel Sánchez-Pérez** [2]

[1] Laboratoire d'Écologie et Gastion des Ecosystmes Naturels (LECGEN), University of Tlemcen, 13000 Tlemcen, Algeria; taleb_14@hotmail.com (A.T.); belaidi_nr@yahoo.fr (N.B.)
[2] Laboratoire Ecologie Fonctionnelle et Environnement (EcoLab), Université de Toulouse, CNRS, INPT, UPS, 31400 Toulouse, France; sabine.sauvage@univ-tlse3.fr (S.S.); jose-miguel.sanchez-perez@univ-tlse3.fr (J.M.S.-P.)
[3] Géosciences Environnement Toulouse, Université de Toulouse, CNES, CNRS, IRD, UPS, 31400 Toulouse, France; laurie.boithias@ird.fr
* Correspondence: zettam.amine@gmail.com; Tel.: +213-551-71-5309

Academic Editor: Sylvain Ouillon
Received: 15 December 2016; Accepted: 7 March 2017; Published: 14 March 2017

Abstract: Sediment deposits in North African catchments contribute to around 2%–5% of the yearly loss in the water storage capacity of dams. Despite its semi-arid climate, the Tafna River plays an important role in Algeria's water self-sufficiency. There is continuous pressure on the Tafna's dams to respond to the demand for water. The Soil and Water Assessment Tool (SWAT) was used to evaluate the contribution of different compartments in the basin to surface water and the dams' impact on water and sediment storage and its flux to the sea in order to develop reservoir management. The hydrological modelling fitted well with the observed data (Nash varying between 0.42 and 0.75 and R^2 varying between 0.25 and 0.84). A large proportion of the surface water came from surface runoff (59%) and lateral flow (40%), while the contribution of groundwater was insignificant (1%). SWAT was used to predict sediments in all the gauging stations. Tafna River carries an average annual quantity of 2942 t·yr^{-1} to the Mediterranean Sea. A large amount of water was stored in reservoirs (49%), which affected the irrigated agricultural zone downstream of the basin. As the dams contain a large amount of sediment, in excess of 27,000 t·yr^{-1} (90% of the sediment transported by Tafna), storage of sediment reduces the lifetime of reservoirs.

Keywords: soil erosion; SWAT; water scarcity; sediment transport modelling; Tafna catchment; North Africa

1. Introduction

As in most semi-arid and arid regions, which cover over 40% of the world's land surface, water resource management in the Middle East and North Africa is more complex than it is in humid zones due to the lack of perennial rivers and other readily available water sources [1]. The population of the Middle East and North Africa was 432 million in 2007, and is projected to reach nearly 700 million by 2050 [2]. This alone would lead to a 40% drop in per capita water availability in the region by 2050 [3]. In Maghreb (Northwest Africa), which has only scarce water resources, most damage is associated with the loss of alluvial sediments from the catchment and subsequent dam siltation [4]. The study of semi-arid North African environments is problematic for several reasons. These include data gaps and considerable anthropic pressures coupled with increasingly intense dry seasons [5].

As in all North African countries, water in Algeria is one of its most valuable resources because it is one of the poorest countries in the region in terms of water potential [6]. Algeria's rivers transport a large quantity of sediments [7,8]. The sediment deposited in Algerian dams is estimated to be 20×10^6 m^3·yr^{-1} [9]. Competition for water between agriculture, industry, and drinking water supply—accentuated by a drought in Algeria—has shown the need for greater attention to be paid to water [10] and for it to be managed at the large basin scale [11]. Surface water resources in Algeria are evaluated to be approximately 8376 billion m^3 for an average year [12]. These water resources in Algeria are characterized by wide variability—the resources for the last nine years have been significantly below this average [13]. In this context, several dams were built in Algeria to ensure water resources for the supply of drinking water to all its cities and allowed approximately 12,350 km^2 of irrigated land to be developed [12–14]. However, dam reservoirs lose about 20×10^6 to 30×10^6 m^3 of water storage every year [15,16].

Despite its semi-arid climate, the Tafna catchment plays an important role in water self-sufficiency in northwest Algeria [17]. There is always huge pressure on Tafna dams, which have a capacity of 398×10^6 m^3, in order to meet the demand expressed specifically and continuously by the largest cities of northwest Algeria (Oran, which is Algeria's second largest city with 10,000 m^3·day^{-1}; Sidi Bel Abbes, 20,000 m^3·day^{-1}; Ain Temouchent, 15,000 m^3·day^{-1}; and Tlemcen, 37,000 m^3·day^{-1}) [12,17].

The deposits of sediment in Maghreb contribute about 2%–5% of the yearly loss in the dams' water storage capacity. In Algeria, the intercepted runoff in dams and weirs hold about 5.2 billion m^3, which makes up 42% of total runoff [18]. The construction of dams has raised questions about their hydrological impacts on water resources at basin scale, especially where there are conflicts between upstream and downstream water users [19,20].

Hydrological models serve many purposes [21]. The accuracy and skill of flow prediction models can have a direct impact on decisions with regard to water resources management. Various statistical and conceptual streamflow prediction models have been developed to help urban planners, administrators, and policy makers make better and more informed decisions [22]. Hydrological models including distributed physically-based model—such as SHE [23], TOPMODEL [24], HEC [25], VIC [26], IHDM [27], and WATFLOOD [28]—are capable of simulating temporal-spatial variations in hydrological processes and assist in the understanding of mechanisms of influence behind land use impacts [29].

Out of the distributed physically-based models, the Soil and Water Assessment Tool (SWAT) [30] has been used widely to assess agricultural management practices [31], help identify pollution sources and contaminant fate [32,33], evaluate the impacts of climate change [34], and assess the hydrology and sediment transfer in various catchments [35,36]. Many authors have applied SWAT in semi-arid areas, such as southeast Africa [37], southern Australia [38], in the Mediterranean coastal basin in Spain [39], and in North Africa [40,41]. Some authors have focused on the impact of dams on water balance using SWAT because of its reservoir module [42,43], as shown in China [44] and Pakistan [45].

By applying the SWAT model, which is not widely used in Algeria, to a semi-arid anthropized catchment, the objectives of this study were: (1) to evaluate the contribution of the different compartments of the basin to surface water; (2) to evaluate the impact of the construction of dams in semi-arid catchments on water and sediment storage and (3) on suspended sediment flux to the sea, in order to facilitate, plan, and assess the management of these important reservoirs, which are a crucial part of water self-sufficiency in semi-arid regions.

2. Materials and Methods

To achieve these objectives, the study was divided into two parts. First a model with dams was considered in which hydrology and sediment flux were calibrated on all the gauging stations. Then a model without dams was considered, retaining the same calibration parameters as the first model, which revealed the impact of the installation of this infrastructure on hydrology and sediment flux in this basin. To verify this method, the flows of the two projects were verified on the basis of the literature published by the Algerian National Agency of Hydrologic Resources [46–48].

2.1. Study Site

The Tafna watershed covers much of western Algeria (Figure 1). The Tafna Wadi is the main stream with a drainage area of 7245 km^2 and elevation varying from sea level to 1100 m.a.s.l. After a 170-km course, the river reaches the Mediterranean Sea near the town of Beni-Saf. It is located between 34°11' N, 35°19' N latitude and 0°50' W, 2°20' W longitude. The catchment area of the Tafna is divided into two zones that are of a different geological nature: the upstream sector where the river runs in a canyon through Jurassic rocks rich in limestone and dolomite, and the downstream sector where it runs in a tertiary basin characterized by marls covered by recent alluvium [49].

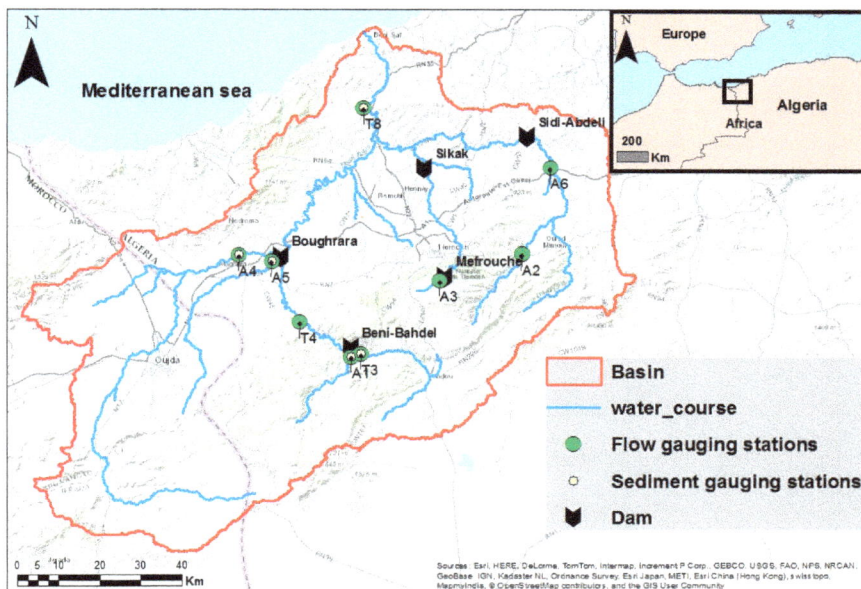

Figure 1. Location of the Tafna River catchment and its dams and gauging stations (A: located in tributaries; T: located in the main watercourse).

The climate is Mediterranean with two main seasons: a long, dry, hot summer-autumn and a winter-spring with abrupt and frequent heavy rainfall. During the summer, most of the streams, especially in their downstream parts, become mostly dry between June and October. The annual average water temperature varies from 11° in winter to 28° in summer [5]. Annual rainfall is between 240 and 688 mm·yr^{-1} [50]. The flow at the watershed outlet ranges from 0 to 108 m^3·s^{-1} [51]. The Tafna River has several tributaries. The most important tributary is the Mouillah Wadi, situated in Maghnia region, which is an industrial area. This tributary is polluted by domestic sewage and industrial effluent from the Moroccan cities of Oujda and El Abbes and the Ouerdeffou Wadi. Another important tributary is the Isser Wadi, but its water supply to the Tafna has decreased significantly since the construction of the Al Izdahar dam, which retains most of the water during the rainy season for irrigation purposes [52]. Five dams have been constructed in the catchment of the Tafna: Beni Bahdel, Meffrouch, Hammam Boughrara, Al Izdahar (Sidi-Abdeli), and Sikkak. Their capacities vary between 15 and 177 million m^3 (Table 1).

According to the Algerian Ministry of Agriculture, agriculture occupies an important place in the catchment of the Tafna, with cereal covering 1699 km^2 (23.6% of the total area), horticulture 342 km^2 (4.75% of the total area) and arboriculture 263 km^2 (3.65% of the total area). The basin has about 1,450,000 inhabitants [53]. The more densely populated areas are the cities of Oujda (Morocco) with 548,280 inhabitants, followed by the city of Tlemcen (Algeria), which has 140,158 inhabitants.

Table 1. Characteristics of the dams built in the Tafna catchment.

Dams	Capacity (Mm^3)	Construction Date	Used for
Beni Bahdel	65.5	1952	Drinking water/irrigation
Hammame Boughrara	177	1998	Drinking water/irrigation
Mefrouche	15	1963	Drinking water
Sikkak	30	2005	Drinking water/irrigation
Al Izdahar (Sidi Abdeli)	110	1988	Drinking water/irrigation

2.2. Discharge and Sediment Monitoring

Tafna's daily discharge and monthly sediment measurement has been monitored since 2003 by the National Agency of Hydrologic Resources (ANRH) at nine hydrometric stations (Figure 1). River discharge was obtained from the water level, which is continuously measured by a limnimetric ladder and float water level recorder using a rating curve. Suspended sediments are defined as the portion of total solids retained by a fiberglass membrane (Whatman GF/F) of 0.6 μm porosity. The sediment collected was weighed after being dried at 105 °C for 24 h. The difference in the weight of the filter before and after filtration allowed the calculation of the suspended sediment concentration based on the volume of water filtered (C, in g·L^{-1}).

2.3. Modelling Approach

2.3.1. The SWAT Model

SWAT was developed at the USDA Agricultural Research Service [30]. It was designed for application in catchments ranging from a few hundred to several thousand square kilometres. The model is semi-distributed: the catchment is first divided into sub-catchments and then into hydrologic response units (HRUs), which represent homogeneous combinations of soil type, land use type, and slope. Any identical combination of these three features is assumed to produce a similar agro-hydrologic response [54].

The Hydrological Component in SWAT

SWAT uses a modified SCS curve number method (USDA Soil Conservation Service, 1972) to compute the surface runoff volume for each HRU. The peak runoff rate is estimated using a modification of the rational method [55]. Daily climatic data are required for calculations. Flow is routed through the channel using a variable storage coefficient method [56].

The hydrologic cycle as simulated by SWAT is based on the water balance equation:

$$SW_t = SW_0 + \sum_{i=1}^{i} \left(R_{day} - Q_{surf} - E_a - W_{seep} - Q_{gw} \right)$$

where SW_t is the final soil water content on day i (mm), SW_0 is the initial soil water content on day i (mm), t is the time (days), R is the amount of precipitation on day i (mm), Q_{surf} is the amount of surface runoff on day i (mm), E_a is the amount of evapotranspiration on day i (mm), W_{seep} is the amount of water entering the vadose zone from the soil profile on day i (mm), and Q_{gw} is the amount of return flow to the stream on day i (mm) [57].

The water balance of dams is given by the following equation:

$$V = V_{stored} + V_{flowin} - V_{flowout} + V_{pcp} - V_{evap} - V_{seep}$$

where V is the volume of water in the impoundment at the end of the day (m^3 H_2O), V_{stored} is the volume of water stored in the water body at the beginning of the day (m^3 H_2O), V_{flowin} is the volume of water entering the water body during the day (m^3 H_2O), $V_{flowout}$ is the volume of water flowing out the water body during the day (m^3 H_2O), V_{pcp} is the volume of the precipitation falling on the water body during the day (m^3 H_2O), V_{evap} is the volume of water removed from the water body by evaporation during the day (m^3 H_2O), and V_{seep} is the volume of water lost from the water body by seepage during the day (m^3 H_2O) [57].

Flow is routed through the channel using a variable storage coefficient method [56] or the Muskingum routing method [57].

Suspended Sediment Modelling Component in SWAT

The sediment from sheet erosion for each HRU is calculated using the modified universal soil loss equation (MUSLE) [58]. Details of the MUSLE equation factors can be found in theoretical documentation of SWAT [59]. Sediment was routed through stream channels using a modification of Bagnold's sediment transport equation [60]. The deposition or erosion of sediment within the channel depends on the transport capacity of the flow in the channel.

2.4. SWAT Data Inputs

The following spatialized data were used in this study: (i) a digital elevation model with a 30 m × 30 m resolution from the US Geological Survey (Figure 2a); (ii) a soil map [61] (Figure 2b); (iii) a land-use map [62] (Figure 2c); (iv) daily climate data between 2000 and 2013 from eight meteorological stations (Figure 2) provided by the Algerian National Office of Meteorology that were used to simulate the reference evapotranspiration in the model using the Hargreaves method because it is the best in semi-arid regions [63]; and (v) daily discharge outflow data for the five Tafna dams provided by the Algerian National Agency for Dams and Transfers (ANBT). Version 2012 of ArcSWAT (Texas Agrilife Research, Usda Agricultural Research Service, Temple, TX, USA) was used to set up the model. The catchment was discretized into 107 sub-basins with a minimum area of 7020 km² (Figure 2d) and 1067 HRUs. To measure the impact of the dams, two SWAT projects were undertaken with and without dams, retaining the same parameter values.

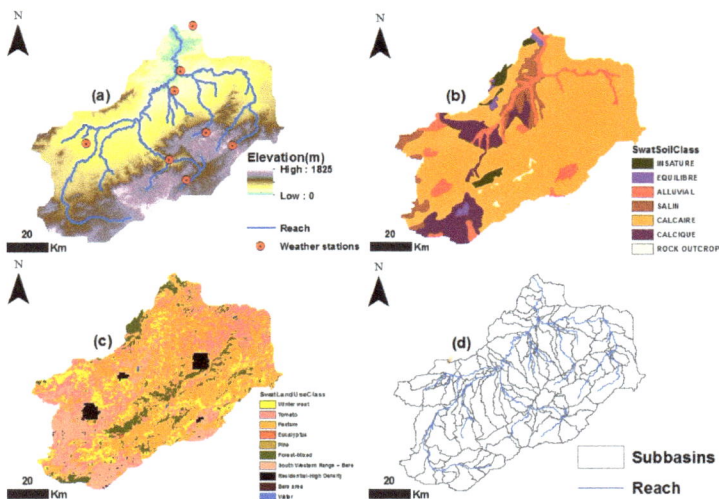

Figure 2. (**a**) 30 m digital elevation model; (**b**) main soils; (**c**) main land uses; and (**d**) SWAT DEM delineated sub-basins of the Tafna catchment

2.5. Model Calibration

In this study, the SUFI-2 (sequential uncertainty fitting, ver. 2) algorithm [64] was used for calibration and sensitivity analysis for flow and sediment output. This program is currently linked to SWAT in the calibration package SWAT-CUP (SWAT calibration uncertainty procedures) (EAWAG, Zurich, Switzerland.). The whole simulation was performed daily from January 2000 to December 2013 (excluding a three-year warm-up from 2000 to 2003). Stream flow was calibrated at a monthly time-step because of the lack of good observed daily data from January 2003 to August 2011, while sediments were calibrated at a daily time-step from January 2003 to December 2006 except for station A5, which was from January 2003 to December 2005. 150 simulations were performed for each gauging station by SWAT-CUP.

2.6. Model Evaluation

The monthly discharge performance of the model was evaluated using the Nash-Sutcliffe efficiency (NSE) index [65] and the coefficient of determination (R^2):

$$\text{NSE} = 1 - \frac{\sum_{i=1}^{n}(O_i - S_i)^2}{\sum_{i=1}^{n}(O_i - \overline{O})^2}$$

$$R^2 = \left\{ \frac{\sum_{i=1}^{n}(O_i - \overline{O})(S_i - \overline{S})}{[\sum_{i=1}^{n}(O_i - \overline{O})^2]^{0.5}[\sum_{i=1}^{n}(S_i - \overline{S})^2]^{0.5}} \right\}$$

where O_i and S_i are the observed and simulated values, n is the total number of paired values, $\overline{O}$ is the mean observed value, and $\overline{S}$ is the mean simulated value.

In this study, monthly NSE was deemed satisfactory at >0.5 [66] and daily and monthly R^2 satisfactory at >0.5 [66].

3. Results

3.1. Discharge and Sediment Calibration

For discharge and sediment calibration, the following parameters, presented in (Table 2), were calibrated.

Table 2. Calibrated parameter values with a ranking of the most sensitive parameters (Rank 1 = most sensitive).

Parameter		Definition	Units	Initial Range	Calibrated Range	Rank
	CN2.mgt	SCS runoff curve number for moisture condition II		[35; 98]	[38.5; 94]	3
	SOL_Z.sol	Depth from soil surface to bottom of layer	(mm)	[0; 4500]	[1500; 3500]	16
	SOL_AWC.sol	Soil available water storage capacity	(mm H$_2$O/mm soil)	[0; 1]	[0.116; 0.169]	7
	SOL_K.sol	Soil conductivity	(mm·h^{-1})	[0; 2000]	[4.71; 180]	14
Parameters related to flow	ALPHA_BF.gw	Base flow alpha factor characterizes the groundwater recession curve	(days)	[0; 1]	[0.055; 0.975]	4
	GW_DELAY.gw	Groundwater delay: time required for water leaving the bottom of the root zone to reach the shallow aquifer	(days)	[0; 500]	[89.223; 176.363]	13
	GW_REVAP.gw	Groundwater "revap" coefficient: controls the amount of water which evaporates from the shallow aquifer		[0.02; 0.2]	[0.069; 0.191]	9
	REVAPMN.gw	Threshold depth of water in the shallow aquifer for "revap" to occur	(mm)	[0; 1000]	[185; 892.294]	12
	RCHRG_DP.gw	Deep aquifer percolation fraction		[0; 1]	[0.176; 0.673]	2

Table 2. *Cont.*

Parameter		Definition	Units	Initial Range	Calibrated Range	Rank
Parameters related to flow	ESCO.hru	Soil evaporation compensation coefficient		[0; 1]	[0.50; 0.86]	5
	OV_N.hru	Manning's "n" value for overland flow		[0.01; 30]	[0.177; 0.823]	11
	CH_N2.rte	Manning's "n" value for the main channel		[−0.01; 0.3]	[0.01; 0.2]	10
	CH_K2.rte	Effective hydraulic conductivity of main channel	$(\text{mm} \cdot \text{h}^{-1})$	[−0.01; 500]	[58; 406]	8
	EVRCH.bsn	Reach evaporation adjustment factor		[0.5; 1]	0.669	6
	TRNSRCH.bsn	Fraction of transmission losses from main channel that enter deep aquifer		[0; 1]	0.211	1
	SURLAG.bsn	Surface runoff lag coefficient		[0; 1]	2.15	15
Parameters related to sediment	USLE-K.sol	USLE soil erodibility factor	$0.013 \, (\text{t} \cdot \text{m}^2 \cdot \text{hr})/ (\text{m}^3 \cdot \text{t} \cdot \text{cm}))$	[0; 0.65]	0.005	1
	USLE-P.mgt	USLE equation support practice factor		[0; 1]	[0.003; 0.8]	2
	PRF.bsn	Peak rate adjustment factor for sediment routing in the main channel		[0; 1]	0.18	3

For the calibrated parameter set, the average annual rainfall of the total simulation period over the area of the catchment is 364 $\text{mm} \cdot \text{yr}^{-1}$. The model predicted the potential evapotranspiration to be 1301.4 $\text{mm} \cdot \text{yr}^{-1}$, and runoff as 26.16 $\text{mm} \cdot \text{yr}^{-1}$.

In this study, only the monthly calibration of flow without validation was performed because there were several difficulties with calibration due to the poor measurement of daily water flow in the gauging stations (renovations of the limnimetric scales and maintenance of the stations need to be undertaken). The flow was calibrated at nine gauging stations. Monthly simulated discharges were satisfactorily correlated to observations for the calibration periods, except for the A4 station (Figure 3, Table 3). It should be noted that the values of the performance of the model evaluation parameters were more satisfactory in the upstream portion, with NSE varying between 0.5 and 0.75 and R^2 between 0.49 and 0.84, while in the downstream part NSE between 0.42 and 0.59 and R^2 between 0.25 and 0.62 were found.

The hygrogram of the Tafna River modelled by SWAT (Figure 4) showed that a large proportion of surface water came from surface runoff (59%) and lateral flow (40%), while the contribution of groundwater was insignificant (1%).

Table 3. Model performance for the simulation of runoff.

Stations	NSE	R^2
A1	0.67	0.7
A2	0.67	0.7
A3	0.53	0.58
A4	0.42	0.25
A5	0.59	0.62
A6	0.5	0.49
T3	0.75	0.84
T4	0.66	0.73
T8	0.51	0.53

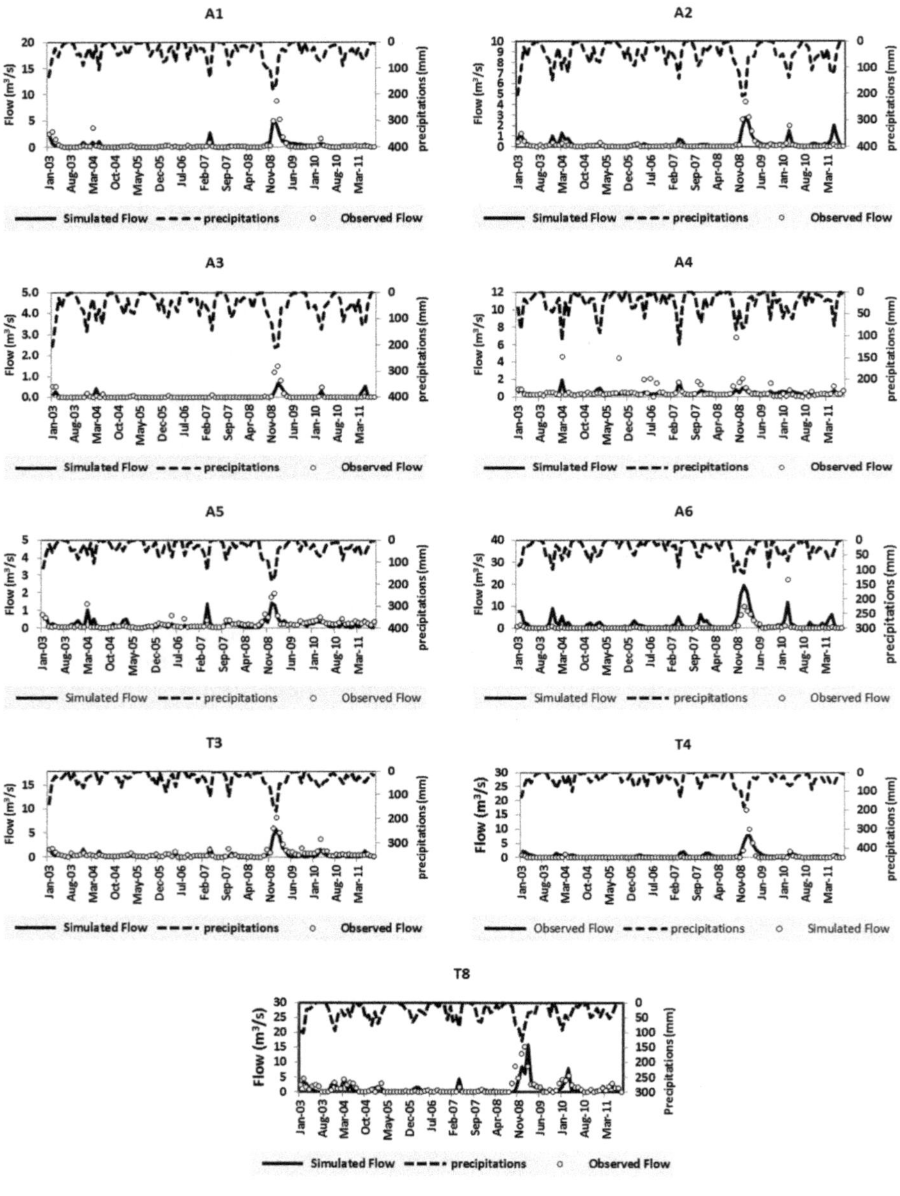

Figure 3. Monthly simulated and observed discharge ($m^3 \cdot s^{-1}$) at the gauging stations (calibration period: January 2003–August 2011).

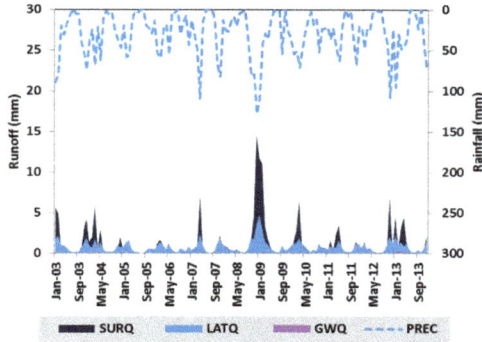

Figure 4. Hygrogram of the Tafna at the outlet modelled by SWAT (SURQ: surface runoff/LATQ: lateral flow/GWQ: groundwater flow/PREC: rainfall).

The results of sediment calibration are shown in Figure 5.

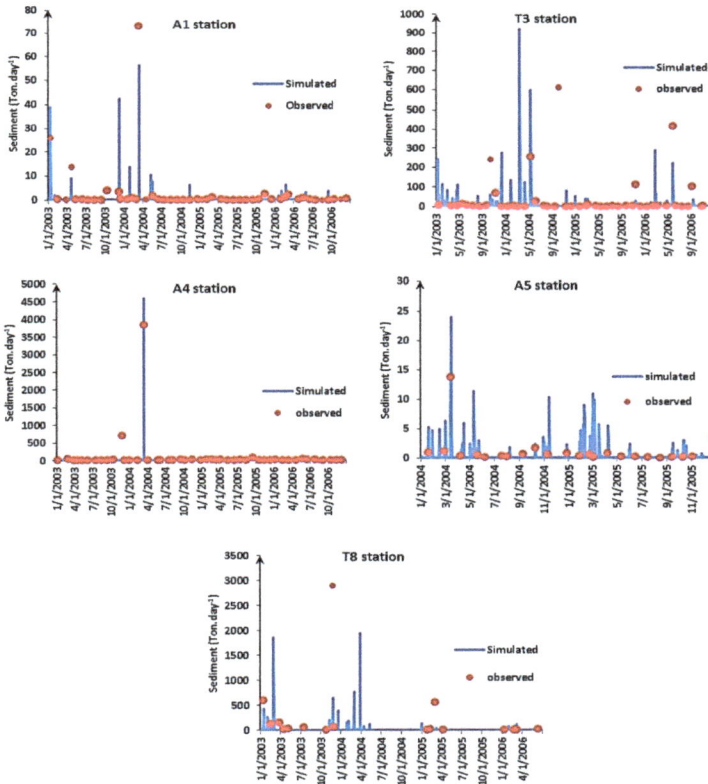

Figure 5. Daily simulated and observed sediment (t·day^{-1}) at the gauging sediment stations. (There is no observed data after 2006).

Figure 5 compares graphically measured and simulated daily sediment yield values for the calibration. Although the observed data were limited, sediment estimation by the model showed that simulated and measured sediment yields were in a similar range for the calibration period.

The annual simulations for each sub-catchment (Figure 6) show that rainfall (Figure 6A) varied between 270 and 550 mm·yr^{-1}, and the largest quantity (450–550 mm·yr^{-1}) was in the Tlemcen Mountains sub-basins. The potential evapotranspiration (Figure 6B) was between 993 and 1300 mm in the downstream portion, while it was between 1300 and 1500 mm in the entire basin. Surface runoff (Figure 6C) varied between 0 and 120 mm·yr^{-1} (semi-arid catchment). The highest values were located in the upstream sub-basin (between 10 and 30 mm), while the lowest were downstream (between 0 and 10 mm). The rate of soil erosion ranged from 0 to 0.2 t·ha^{-1}·yr^{-1} (Figure 6D), and the eastern upstream basins were identified as areas with high soil erosion in the Tafna.

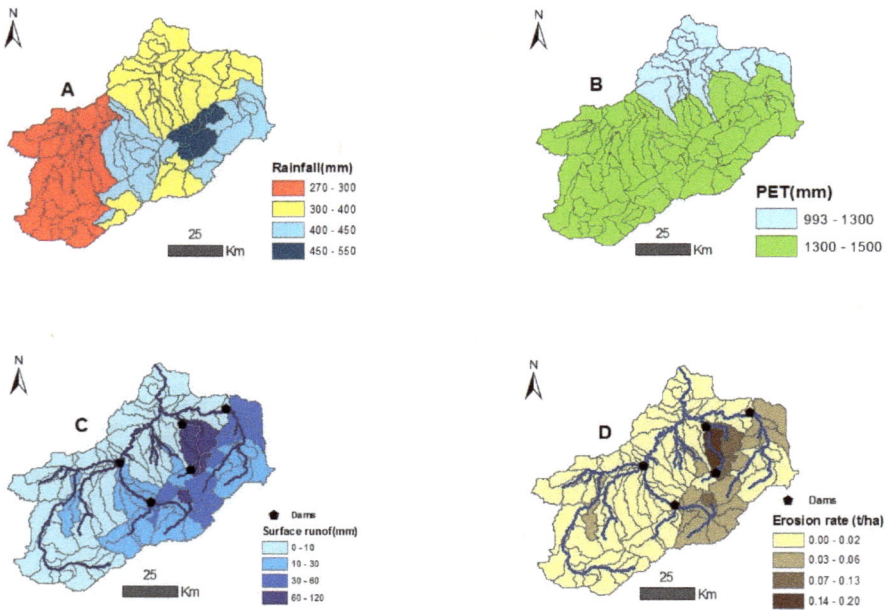

Figure 6. Inter-annual averages for each sub-catchment between 2000 and 2013. (**A**) rainfall (mm·yr^{-1}); (**B**) Hargreaves potential evapotranspiration (mm·yr^{-1}); (**C**) simulated surface runoff loads (mm·yr^{-1}); (**D**) simulated sediment yield (t·ha^{-1}·yr^{-1}).

3.2. *Impact of Dams on Water Balance and Sediment Loading*

To assess the impact of dam construction on water balance and sediments, the SWAT model was run both with and without dams.

3.2.1. Impact of Dams on Water Balance

Figure 7 shows the average annual basin value for water balance, calculated as a relative percentage of average annual rainfall.

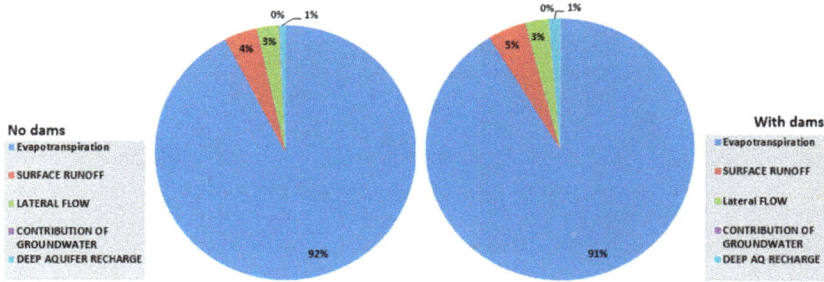

Figure 7. Impact of dams on average annual water balance as a relative percentage to precipitation.

The results of the simulation between 2000 and 2013 (Figure 7) show that the construction of the dams did not disturb the Tafna water balance.

The simulation results (Figure 8) reveal that dams greatly reduced the quantity of water arriving at the outlet of the Tafna between 2003 and 2013. A large amount of water was stored in five reservoirs (49%). This difference is significant according to ANOVA ($p = 0.006 < 0.05$).

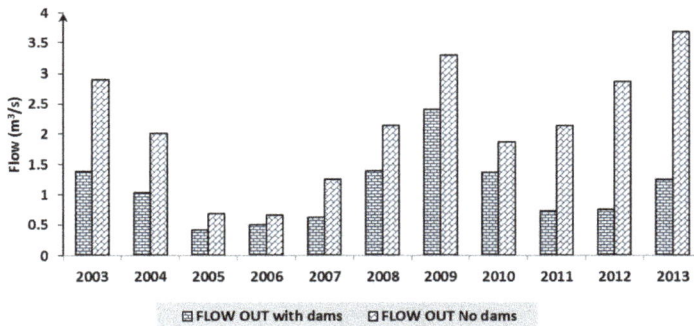

Figure 8. Impact of dams on flow at the outlet of the basin.

3.2.2. Impact of Dams on Sediment

The cumulative annual sediment load was also compared at the outlet of the basin (Figure 9) to quantify the amount of sediment stored in Tafna's dams.

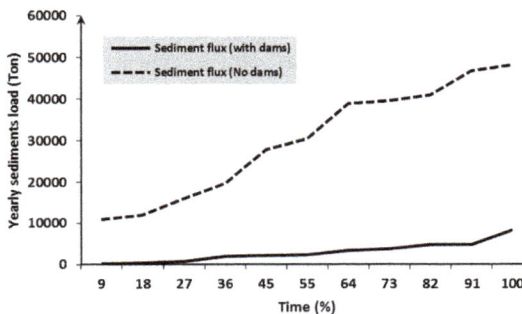

Figure 9. Cumulative annual sediment load at the outlet of the basin between 2003 and 2013.

The SWAT simulation showed that the reservoirs in these basins stocked a large amount of sediment, in excess of 27,000 t·yr^{-1} (90%). Large quantities were retained mainly during flood events, representing 87%–95% of the annual sediment export.

4. Discussion

Simulation results between 2003 and 2011 showed that the model adequately predicted the watershed hydrology of the Tafna River. These values remained consistent with the values published by ANRH , which were based on a series of observations between September 1965 and August 2002 [46,47,50] concerning runoff (SWAT value = 26.16 mm·yr^{-1}; ANRH (10–100 mm·yr^{-1})), potential evapotranspiration (SWAT value = 1301.4 mm·yr^{-1}; ANRH (900–1400 mm·yr^{-1})), rainfall (SWAT value = 364 mm·yr^{-1}; ANRH (250–550 mm·yr^{-1})) and limited groundwater resources [48]. The Tafna was characterized by very irregular flow with frequent dry summers, indicating very limited permanent reserves [67]. This study confirmed those results, affirming the good performance of SWAT in Mediterranean Karstic semi-arid watersheds [68].

The fraction of transmission losses from the main channel that enter the deep aquifer, the deep aquifer percolation fraction, and the curve number were the most sensitive parameters for stream flow. The value of the first parameter was 0.211, and the second parameter varied between 0.176 and 0.673. The Tafna sub-basins are essentially formed by semi-permeable and permeable formations that cover the whole surface of the basin, thus increasing the infiltration of surface water [69]. It was also evident that the curve number was the third most sensitive factor, varying relatively between −0.5 and 0.09. In a study of the Hathab river in Tunisia, the curve number ranged relatively between −0.5 and +0.5 [70]. In another study in Hamadan–Bahar watershed in Iran, this parameter ranged relatively between −0.32 and 1.02 [71]. These results confirmed the effect of land use spatial heterogeneity on runoff spatial heterogeneity in semi-arid catchments [72].

The results were analyzed by computing the coefficients of efficiency and determination on a monthly basis for nine water flow gauges. Multi-gauge calibration is an important step in developing a reliable watershed model in semi-arid watersheds, because the single outlet calibration of the watershed in arid and semi-arid regions can be misleading and thus requires spatial calibration to capture the spatial heterogeneity and discontinuities in the watershed [73]. Goodness-of-fit indices were satisfactory for discharge for the monthly calibration period. The Nash-Sutcliffe efficiencies at the nine flow gauges ranged from 0.42 to 0.75 and the coefficient R^2 was varying from 0.25 to 0.84 for the calibrated monthly flow for sub-basins for the period from January 2003 to August 2011. It should be noted that the upstream basin stations (A1, T3, A2, A6, A3) that had a less anthropic influence had higher index efficiencies (Nash varying between 0.50 and 0.75 and R^2 varying between 0.49 and 0.84) than the downstream stations (A4, A5, T8), which had a lower index efficiency (Nash varying between 0.42 and 0.59 and R^2 varying between 0.25 and 0.62). These are influenced by domestic and industrial waste from the major cities of Tlemcen and Maghnia in Algeria and Oujda in Morocco. In a study of the Medjerda River basin in Tunisia, the authors found a range of Nash-Sutcliffe efficiencies between 0.31 and 0.65, and range of coefficient R^2 between 0.62 and 0.8 [40]. In another study on the semi-arid river of the Hamadan-Bahar watershed in Iran, the authors found a Nash-Sutcliffe efficiency range of between 0.33 and 0.77, and a range of coefficient R^2 of between 0.38 and 0.83 [68]. The calibrated SWAT model can be used successfully to predict the volume inflow to the dams and facilitate the storage and release of water [44].

Gaps between observed and simulated flow values were partly explained by errors in observed and simulated values [31]. Uncertainty in the observed discharge values came from the precision of the sensor and the use of a rating curve. Errors in simulated values could be attributed to actual local rainfall storms that were not well represented by the SWAT rainfall data interpolation [31]. The model could not capture the small peaks. Aside from the uncertainty attributed to the precipitation input, the SCS curve number method, which works on daily rainfall depths, does not consider the duration and intensity of precipitation. Representing this precipitation characteristic is necessary for semi-arid watersheds, where high-intensity short-duration precipitation occurs [74].

There were few observed suspended sediment data because, ANRH measures sediment once a month (sometimes there are no measurements) and sampling during flood periods is problematic. It is very difficult to assess the quality of the model performance as suspended matter sampling was not systematically performed for all storms, while the major losses of suspended matter occur during a small number of intensive rain events [39]. However, it was noticed that the simulated values were in the same range as the observed values. A similar observation was made in a study in Tunisia where the predicted concentrations of suspended matter were in the order of magnitude of the measured concentrations [39]. However, the low sampling frequency and lack of detailed land use and land management data did not allow an in-depth evaluation of the SWAT performance. It is recommended that future studies collect data at a greater frequency and spread along the river stretch [39]. Despite these shortcomings, the results from this study were still useful for representing the measured data [75]. The USLE soil erodibility factor, USLE equation support practice factor and peak rate adjustment factor for sediment routing in the main channel were the three sensitive parameters for sediment calibration. The value of the first parameter was 0.005, the second varied between 0.003 and 0.8, and the value of the third was 0.18.

A quantification of changes in water balance is necessary, especially after the construction of dams, for integrated watershed management in order to identify their effects on the basin [18]. The simulation results showed a considerable reduction in the quantity of water arriving at the outlet of the Tafna between 2003 and 2013, with a large amount of water stored in five reservoirs (49%). This decrease in flow downstream, which represents 18% of the basin surface, can affect the irrigated agricultural zone, especially as most of this land depends on Tafna water.

The comparison between the cumulative annual sediment load at the outlet with and without dams shows that reservoirs stock a large quantity of sediment, in excess of 27,000 $t \cdot yr^{-1}$ (90%). Large quantities are retained mainly during flood events, representing 87%–95% of the annual sediment export. In the Koiliaris river in Greece, flood events account for 63%–70% of the annual sediment export in a wet or dry year [68]. Between 37% and 98% of sediment settles in North African reservoirs [76]. These deposits contribute around 2%–5% of the yearly loss of water storage capacity [77]. The mean annual suspended sediment flux in North African rivers was estimated to be 254 million tons [78]. This storage reduces the lifetime of dams. The results of this study showed the need to implement a water resources management strategy to reduce reservoir sediment deposition, as in Tunisia where there are contour ridges for water harvesting in semi-arid catchments. The result was checked using the SWAT model in the Merguellil catchment (central Tunisia) and the contour ridges for water harvesting retained a large proportion of the entrained sediment (26%) [18].

5. Conclusions

In the present study, the hydrological SWAT model was applied to the Tafna River, which is a semi-arid basin. The model reproduced water flow and sediment in all gauging stations. The model's weakness at simulating runoff for some months was probably due to errors in the observed values and to the poor representation of small peaks. The weakness of the model at simulating sediment was due to the improper runoff simulation and the nature and accuracy of the measured sediment data.

Prediction of runoff and soil loss is important for assessing soil erosion hazards and determining suitable land uses and soil conservation measures for a catchment [75]. In turn, this can help to derive the optimum benefit from the use of the land while minimizing the negative impacts of land degradation and other environmental problems. As there are limited data available from the study region, the model developed here could help assess different land management options [75].

The application of the model enabled an evaluation of the contribution of the different compartments of the basin to surface water. SWAT has shown that a large proportion of surface water comes from surface runoff and lateral flow, while the contribution of groundwater was insignificant.

It was also noted that the application of the model gave a general idea of the impact of dam building on water balance and sediment in the Tafna semi-arid watershed. It highlighted that a large amount of water (49%) was stored in five reservoirs, decreasing the water flow in the downstream part of the basin, and could affect the irrigated agricultural zone, especially as most of this land depends on Tafna water. The dams of the Tafna have been built for the supply of drinking water and for irrigation. However, according to statistics of the National Agency of Basin and Transfer (ANBT) from January 2003 to July 2011, the largest dam basin (Hammame Boughrara) is devoted exclusively to drinking water.

These hydraulic structures were observed to stock a large quantity of sediment—in excess of 27,000 $t \cdot yr^{-1}$ (90%). Large quantities were retained mainly during flood events, representing 87%–95% of the annual sediment export. The results of this study showed the need for the implementation of a better water resources management strategy such as reforestation, and contour ridges for water harvesting upstream of the reservoirs to reduce the amount of sediment transported by the river. This is particularly important with dams in semi-arid and arid regions where water resources are limited and vary greatly with more intense low flow episodes and where rivers transport a high quantity of sediments, in order to reduce the siltation of dams and increase their lifetime. In fact, hydrological models such as SWAT demonstrate that it is a useful tool for understanding hydrological processes, even when the amount of measured data available is poor. It can be useful for identifying the most appropriate location for reservoirs and optimizing them to reduce their impact on water resources [18].

However, a general problem in watershed modelling that still needs to be addressed is the limited availability of data, especially in terms of measured water quality for calibrating and validating these models. The lack of a long time series of sediment with a daily time step and high spatial resolution limited this study's ability to evaluate the simulations [18].

Finally, the results obtained were very encouraging. SWAT allows the dynamics of water and sediment on the Tafna to be correctly represented. This model can be useful for understanding the impact of sediment transport on the water storage capacity of dams in a semi-arid region.

Acknowledgments: The authors thank the National Agency of Hydrologic Resources (ANRH) Oran, the Algerian agency in charge of stream gauging, which provided all water flow and nitrate concentration data, especially Sidi Mohammed Boudaliya and Belkacem Sardi. The National Agency for Dams and Transfers (ANBT) Ain youcef-Tlemcen, which provided all the data on dam management, especially Bensmaine. Ayoub Bouazzaoui, a PhD student at the University of Tlemcen-Algeria, in the SVTU faculty, agroforestry department, for his help during the realization of soil map.

Author Contributions: Amin Zettam, Amina Taleb, Sabine Sauvage, Nouria Belaidi and José Miguel Sánchez-Pérez conceived and designed the experiments; Amin Zettam performed the experiments; Amin Zettam, Sabine Sauvage and José Miguel Sánchez-Pérez analyzed the data; Laurie Boithias contributed analysis tools; Amin Zettam, Amina Taleb, Sabine Sauvage, Laurie Boithias and José Miguel Sánchez-Pérez wrote the paper.

Conflicts of Interest: The authors declare no conflict of interest.

References

1. Souza, J.O.P.; Correa, A.C.B.; Brierley, G.J. An approach to assess the impact of landscape connectivity and effective catchment area upon bedload sediment flux in Saco CreekWatershed, Semiarid Brazil. *Catena* **2016**, *138*, 13–29. [CrossRef]

2. Roudi-Fahimi, F.; Kent, M. Challenges and opportunities—The population of the Middle East and North Africa. *Popul. Bull.* **2007**, *62*, 1–20.

3. Terink, W.; Immerzeel, W.; Droogers, P. Climate change projections of precipitation and reference evapotranspiration for the Middle East and Northern Africa until 2050. *Int. J. Climatol.* **2013**, *33*, 3055–3072. [CrossRef]

4. Megnounif, A.; Terfous, A.; Ouillon, S. A graphical method to study suspended sediment dynamics during flood events in the Wadi Sebdou, NW Algeria (1973–2004). *J. Hydrol.* **2013**, *497*, 24–36. [CrossRef]

5. Taleb, A.; Belaidi, N.; Sanchez-Perez, J.M.; Vervier, P.; Sauvage, S.; Gagneur, J. The role of the hyporheic zone of a semi-arid gravel bed stream located downstream of a heavily polluted reservoir (Tafnawadi, Algeria). *River Res. Appl.* **2008**, *24*, 183–196. [CrossRef]

6. Touati, B. Les Barrages et la Politique Hydraulique en Algérie:état, Diagnostic et Perspectives d'un Aménagement Durable. Ph.D. Thèse, Université de Constantine, Constantine, Algérie, 2010.

7. Bourouba, M. Contribution a l'etude de l'erosion et des transports solides de l'Oued Medjerda superieur (Algerie orientale). *Bull. Reseau Erosion* **1998**, *18*, 76–97. (In French)

8. Colombani, J.; Olivry, J.C.; Kallel, R. Phénomènes exceptionnels d'érosion et de transport solide en Afrique aride et semi-aride. In *Challenges in African Hydrology and Water, Ressources*; IAHS Publication: Oxfordshire, UK, 1984; Volume 144, pp. 295–300. (In French)

9. Kettab, A. Les ressources en eau en Algérie: Stratégies, enjeux et vision. *Desalination* **2001**, *136*, 25–33. (In French) [CrossRef]

10. Remini, B. *La Problématique de l'eau en Algérie*; Office des publications Universitaires: Alger, Algérie, 2005; p. 162. (In French)

11. Boithias, L.; Acuña, V.; Vergoñós, L.; Ziv, G.; Marcé, R.; Sabater, S. Assessment of the water supply:demand ratios in a Mediterranean basin under different global change scenarios and mitigation alternatives. *Sci. Total Environ.* **2014**, *470–471*, 567–577. [CrossRef] [PubMed]

12. Surface Water Resources Mobilization; Document of Algerian Ministry of Water Resources. Available online: http://www.mree.gov.dz (accessed on 13 March 2017).

13. Hamiche, A.; Boudghene Stambouli, A.; Flazi, S. A review on the water and energy sectors in Algeria: Current forecasts, scenario and sustainability issues. *Renew. Sustain. Energy Rev.* **2015**, *41*, 261–276. [CrossRef]

14. Bouzid-Lagha, S.; Djelita, B. Study of eutrophication in the Hamman Boughrara Reservoir (Wilaya de Tlemcen, Algeria). *Hydrol. Sci. J.* **2012**, *57*, 186–201. [CrossRef]

15. Bessenasse, M.; Kettab, A.; Paquier, A.; Galeas, G.; et Ramez, P. Simulation numérique de la sédimentation dans les retenues de barrages : Cas de la retenue de Zardezas, Algérie. *J. Water Sci.* **2003**, *16*, 103–122. (In French) [CrossRef]

16. Remini, B.; Avenard, J.-M.; Kettab, A. Évolution dans le temps de l'envasement dans une retenue de barrage dans laquelle est pratiquée la technique du soutirage. *La Houille Blanche* **1997**, *6*, 4–8. (In French)

17. Benyahia, M.; Bechlaghem, N.; Habi, M.; Kerfouf, A. Importance des ressources hydriques de la wilaya de Tlemcen dans le cadre de l'Oranie (Algérie Nord Occidentale) et Perspectives de développement durables. In Proceedings of the Vème Colloque International-Energie, Changement Climatiques et Développement Durable, Hammamet, Tunisia, 15–17 Juin 2009. (In French)

18. Ounissi, M.; Bouchareb, N. Nutrient distribution and fluxes from three Mediterranean coastal rivers (NE Algeria) under large damming. *C. R. Geosci.* **2013**, *345*, 81–92. [CrossRef]

19. Abouabdillah, A.; White, M.; Arnold, J.G.; De Girolamo, A.M.; Oueslati, O.; Maataoui, A.; Lo porto, A. Evaluation of soil and water conservation measures in a semi-arid river basin in Tunisia using SWAT. *Soil Use Manag.* **2014**, *30*, 539–549. [CrossRef]

20. Le Goulven, P.; Leduc, C.; Bachta, M.S.; Poissin, J.C. Sharing scarce resources in a Mediterranean river basin: Wadi Merguellil in Central Tunisia. In *River Basin Trajectories: Societies, Environments and Development*; Molle, F., Wester, P., Eds.; MPG Books Group: Bodmin, UK, 2009; pp. 147–170, ISBN-13: 978-1-84593-538-2.

21. Todini, E. Hydrological catchment modelling: Past, present and future. *Hydrol. Earth Syst. Sci.* **2007**, *11*, 468–482. [CrossRef]

22. Noori, N.; Kalin, L. Coupling SWAT and ANN models for enhanced daily stream flow Prediction. *J. Hydrol.* **2016**, *533*, 141–151. [CrossRef]

23. Abbott, M.B.; Bathurst, J.C.; Cunge, J.A.; O'Connell, P.E.; Rasmussen, J. An introduction to the European Hydrological System—Système Hydrologique Européen, "SHE", 1: History and Philosophy of a physically-based, distributed modelling system. *J. Hydrol.* **1986**, *87*, 45–59. [CrossRef]

24. Wendling, J. Théorie de TOPMODEL. Extrait de thèse de Doctorat, Université de Grenoble, Grenoble, France, 1992. (In French)

25. Feldman, A.D. *Hydrologic Modeling System HEC-HMS*; Technical Reference Manual; U.S. Army Corps of Engineers, Hydrologic Engineering Center (HEC): Davis, CA, USA, 2000.

26. Liang, X.; Lettenmaier, D.P.; Wood, E.F.; Burges, J. A simple hydrologically based model of land surface water and energy fluxes for general circulation models. *J. Geophys. Res.* **1994**, *99*, 14415–14428. [CrossRef]

27. Beven, K.; Calver, A.; Morris, E.M. *Institute of Hydrology Distributed Model, Report No. 8*; Institute of Hydrology: Wallingford, UK, 1987.

28. Kouwen, N.; Soulis, E.D.; Pietroniro, A.; Donald, J.; Harrington, R.A. Grouping Response Units for Distributed Hydrologic Modelling. *ASCE J. Water Resour. Plan. Manag.* **1993**, *119*, 289–305. [CrossRef]

29. Lin, B.; Chen, X.; Yao, H.; Chen, Y.; Liu, M.; Gao, L.; James, A. Analyses of land use change impacts on catchment runoff using different time indicators based on SWAT model. *Ecol. Indic.* **2015**, *58*, 55–63. [CrossRef]

30. Arnold, J.G.; Srinivasan, R.; Muttiah, R.S.; Williams, J.R. Large-area hydrologic modeling and assessment: Part, I. Model development. *J. Am. Water Res. Assoc.* **1998**, *34*, 73–89. [CrossRef]

31. Moriasi, D.N.; Steiner, J.L.; Arnold, J.G. Sediment measurement and transport modeling: Impact of riparian and filter strip buffers. *J. Environ. Qual.* **2011**, *40*, 807–814. [CrossRef] [PubMed]

32. Boithias, L.; Sauvage, S.; Taghavi, L.; Merlina, G.; Probst, J.L.; Sanchez Perez, J.M. Occurrence of metolachlor and trifluralin losses in the Save river agricultural catchment during floods. *J. Hazard. Mater.* **2011**, *196*, 210–219. [CrossRef] [PubMed]

33. Oeurng, C.; Sauvage, S.; Sánchez-Pérez, J.M. Temporal variability of nitrate transport through hydrological response during flood events within a large agricultural catchment in south-west France. *Sci. Total Environ.* **2010**, *409*, 140–149. [CrossRef] [PubMed]

34. Singh, A.; Gosain, A.K. Climate-change impact assessment using GIS based hydrological modelling. *Water Int.* **2011**, *36*, 386–397. [CrossRef]

35. Arnold, J.G.; Moriasi, D.N.; Gassman, P.W.; Abbaspour, K.C.; White, M.J.; Srinivasan, R.; Santhi, C.; Harmel, R.D.; Van Griensven, A.; Van Liew, M.W.; et al. SWAT: Model use, calibration, and validation. *Trans. ASABE* **2012**, *55*, 1491–1508. [CrossRef]

36. Gassman, P.W.; Sadeghi, A.M.; Srinivasan, R. Applications of the SWAT Model Special Section: Overview and Insights. *J. Environ. Qual.* **2014**, *43*, 1–8. [CrossRef] [PubMed]

37. Baker, T.J.; Miller, S.N. Using the Soil and Water Assessment Tool (SWAT) to assess land use impact on water resources in an East African watershed. *J. Hydrol.* **2013**, *486*, 100–111. [CrossRef]

38. Shrestha, M.K.; Recknagela, F.; Frizenschafb, J.; Meyer, W. Assessing SWAT models based on single and multi-site calibration for the simulation of flow and nutrient loads in the semi-arid Onkaparinga catchment in South Australia. *Agric. Water Manag.* **2017**, in press. [CrossRef]

39. Molina-Navarro, E.; Trolle, D.; Martínez-Pérez, S.; Sastre-Merlín, A.; Jeppesen, E. Hydrological and water quality impact assessment of a Mediterranean limno-reservoir under climate change and land use management scenarios. *J. Hydrol.* **2014**, *509*, 354–366. [CrossRef]

40. Bouraoui, F.; Benabdallah, S.; Jrad, A.; Bidoglio, G. Application of the SWAT model on the Medjerda river basin (Tunisia). *Phys. Chem. Earth* **2005**, *30*, 497–507. [CrossRef]

41. Sellami, H.; Benabdallah, S.; La Jeunesse, I.; Vanclooster, M. Quantifying hydrological responses of small Mediterranean catchments under climate change projections. *Sci. Total Environ.* **2016**, *543*, 924–936. [CrossRef] [PubMed]

42. Neitsch, S.L.; Arnold, J.G.; Kiniry, J.R.; Srinivasan, R.; Williams, J.R. *Soil and Water Assessment Tool User's Manual, Version 2000*; Texas Water Resources Institute: College Station, TX, USA, 2002; p. 412.

43. Neitsch, S.L.; Arnold, J.G.; Kiniry, J.R.; Williams, J.R.; King, K.W. *Soil and Water Assessment Tool Theoretical Documentation, Version 2000*; Texas Water Resources Institute: College Station, TX, USA, 2002; p. 458.

44. Wang, G.; Xia, J. Improvement of SWAT 2000 modelling to assess the impact of dams and sluices on streamflow in the Huai River basin of China. *Hydrol. Process.* **2010**, *24*, 1455–1471. [CrossRef]

45. Ghoraba, S.M. Hydrological modeling of the Simly Dam watershed (Pakistan) using GIS and SWAT model. *Alex. Eng. J.* **2015**, *54*, 583–594. [CrossRef]

46. National Agency of Hydrologic Resources (ANRH). *Map of Potential Evapotranspiration in the North of Algeria*; ANRH: Alger, Algeria, 2003.

47. National Agency of Hydrologic Resources (ANRH). *Map of Average Annual Runoff in the North of Algeria*; ANRH: Alger, Algeria, 2003.

48. National Agency of Hydrologic Resources (ANRH). *Map of Groundwater Resources in the North of Algeria*; ANRH: Alger, Algeria, 2003.

49. Guardia, P. Géodynamique de la Marge Alpine du Continent Africain. D'après l'Etude de l'Oranie Nord Occidentale. Relations Structurales et Paléogéographiques Entre le tell Extrême et L'avant Pays Atlassique+ Carte au 1/100 000. Thèse 3 ème cycle, Université de Nice, Nice, France, 1975; p. 285. (In French)

50. National Agency of Hydrologic Resources (ANRH). *Map of Annual Rainfall in the North of Algeria*; ANRH: Alger, Algeria, 2003.

51. National Agency of Hydrologic Resources (ANRH). *Daily Data Flow in the Outlet of Tafna Catchment from 2000 to 2011*; ANRH: Alger, Algeria, 2012.

52. Taleb, A.; Belaidi, N.; Gagneur, J. Water Quality before and after AAM building on a heavily polluted river in semi-arid Algeria. *River Res. Appl.* **2004**, *20*, 943–956. [CrossRef]

53. Ministère Algérien de l'agriculture. *Type et Superficie de l'Agriculture de la Wilaya de Tlemcen et Ain Temouchent*; Ministère Algérien de l'agriculture: Alger, Algérie, 2011. (In French)

54. Laurent, F.; Ruelland, D. Assessing impacts of alternative land use and agricultural practices on nitrate pollution at the catchment scale. *J. Hydrol.* **2011**, *409*, 440–450. [CrossRef]

55. Chow, V.; Maidment, D.; Mays, L. *Applied Hydrology*; Chow, V., Maidment, D., Eds.; McGraw Hill: New York, NY, USA, 1988.

56. Williams, J.R. Flood routing with variable travel time or variable storage coefficients. *Trans. ASAE* **1969**, *12*, 100–103. [CrossRef]

57. Cunge, J.A. On the subject of a flood propagation method (Muskingum method). *J. Hydraul. Res.* **1969**, *7*, 205–230. [CrossRef]

58. Williams, J.R. Sediment routing for agricultural watersheds. *J. Am. Water Resour. Assoc.* **1975**, *11*, 965–974. [CrossRef]

59. Neitsch, S.L.; Arnold, J.G.; Kiniry, J.R.; Williams, J.R.; King, K.W. *Soil and Water Assessment Tools: Theoretical Documentation Version*; Grassland, Soil and Water Reasearch Laboratory, ARS: Temple, TX, USA, 2005; p. 494.

60. Bagnold, R.A. Bed load transport by natural rivers. *Water Resour. Res.* **1977**, *13*, 303–312. [CrossRef]

61. Barbut, M.M.; Durand, M.J.-H. *Carte des Sols d'Algérie. Oran. Feuille N.I. 30-N.E*; Service Géographique de l'Armée: Alger, France, 1952.

62. Land-Use Map, the European Space Agency. Available online: http://due.esrin.esa.int/page_globcover.php (accessed on 13 March 2017).

63. Aouissi, J.; Benabdallah, S.; Chabaâne, Z.; Cudennec, C. Evaluation of potential evapotranspiration assessment methods for hydrological modelling with SWAT—Application in data-scarce rural Tunisia. *Agr. Water Manag.* **2016**, *174*, 39–51. [CrossRef]

64. Abbaspour, K.C. *User Manual for SWAT-CUP SWAT Calibration and Uncertainty Analysis Programs*; Swiss Federal Institute of Aquatic Science and Technology: Dübendorf, Switzerland, 2007.

65. Nash, J.E.; Sutcliffe, V. River flow forecasting through conceptual models: Part I. A discussion of principles. *J. Hydrol.* **1970**, *10*, 282–290. [CrossRef]

66. Moriasi, D.N.; Arnold, J.G.; Van Liew, M.W.; Bingner, R.L.; Harmel, R.D.; Veith, T.L. Model evaluation guidelines for systematic quantification of accuracy in watershed simulation. *Am. Soc. Agric. Biol. Eng.* **2007**, *50*, 885–900.

67. Khaldi, A. Impacts de la Sécheresse sur le Régime des Ecoulements Souterrains Dans les Massifs Calcaires de l'Ouest Algérien "Monts de Tlemcen-Saida". Thèse de doctorat d'état, Université d'Oran, Oran, Algérie, 2005; p. 239. (In French)

68. Nerantzaki, S.D.; Giannakis, G.V.; Efstathiou, D.; Nikolaidis, N.P.; Sibetheros, I.A.; Karatzas, G.P.; Zacharia, I. Modeling suspended sediment transport and assessing the impacts of climate change in a karstic Mediterranean watershed. *Sci. Total Environ.* **2015**, *538*, 288–297. [CrossRef] [PubMed]

69. Bouanani, A. Hydrologie, Transport Solide et Modélisation étude de Quelques Sous Bassins de la Tafna (NW–Algérie). Ph.D. Thèse, Université Tlemcen, Tlemcen, Algérie, 2000; p. 250. (In French)

70. Chaâbane Ben Salah, N.; Abida, H. Modélisation des écoulements et de Transport Solide du Bassin d'Oued Hathab en Tunisie Centrale: Couplage d'un SIG avec le Modèle Agro Hydrologique SWAT. In Proceedings of the Actes du Séminaire sur les Systèmes d'Information Géographique pour l'Etude de l'Environnement, Djerba, Tunisia, 21–23 Mai 2012; Revue des Régions Arides—Numéro Spécial—n° 33. Institue des Régions Arides: Médenine, Tunisia, 2014; pp. 87–92. (In French)

71. Akhavan, S.; Abedi-Koupaia, J.; Mousavia, S.; Afyunib, M.; Eslamiana, S.; Abbaspour, K. Application of SWAT model to investigate nitrate leaching in Hamadan–Bahar Watershed, Iran. *Agric. Ecosyst. Environ.* **2010**, *139*, 675–688. [CrossRef]

72. Lin, K.; Lv, F.; Chen, L.; Singh, V.P.; Zhang, Q.; Chen, X. Xinanjiang model combined with Curve Number to simulate the effect of land use change on environmental flow. *J. Hydrol.* **2014**, *519*, 3142–3152. [CrossRef]

73. Niraula, R.; Norman, L.M.; Meixner, T. and Callegary, J.B. Multi-gauge Calibration for modeling the Semi-Arid Santa Cruz Watershed in Arizona-Mexico Border Area Using SWAT. *Air Soil Water Res.* **2012**, *5*, 41–57. [CrossRef]

74. Nie, W.; Yuan, Y.; Kepner, W.; Nash, M.; Jackson, M.; Erickson, C. Assessing impacts of land use and land cover changes on hydrology for the upper San Pedro watershed. *J. Hydrol.* **2011**, *407*, 105–114. [CrossRef]

75. Rostamian, R.; Jaleh, A.; Afyuni, M.; Farhad, M.; Heidarpour, M.; Jalalian, A.; Abbaspour, K.C. Application of a SWAT model for estimating runoff and sediment in two mountainous basins in central Iran. *Hydrol. Sci. J.* **2008**, *53*, 977–988. [CrossRef]

76. Ghorbal, A.; Claude, J. *Mesure de l'Envasement Dans les Retenues de Sept Barrages en Tunisie: Estimation des Transports Solides*; IAHS Publication: Wallingford, UK, 1977; Volume 122, pp. 219–232. (In French)

77. Kassoul, M.; Abdelgader, A.; Belorgey, M. Caractérisation de la sédimentation des barrages en Algérie. *Rev. Sci. Eau* **1997**, *3*, 339–358. [CrossRef]

78. Probst, J.L.; Amiotte-Suchet, P. Fluvial suspended sediment transport and mechanical erosion in the Maghreb (North Africa). *Hydrol. Sci. J.* **1992**, *37*, 621–637. [CrossRef]

Article

Statistical Analysis of the Spatial Distribution of Multi-Elements in an Island Arc Region: Complicating Factors and Transfer by Water Currents

Atsuyuki Ohta *, Noboru Imai, Yoshiko Tachibana and Ken Ikehara

National Institute of Advanced Industrial Science and Technology, Geological Survey of Japan, Central 7, 1-1-1 Higashi, Tsukuba, Ibaraki 305-8567, Japan; noboru.imai@aist.go.jp (N.I.); y.tachibana@aist.go.jp (Y.T.); k-ikehara@aist.go.jp (K.I.)
* Correspondence: a.ohta@aist.go.jp; Tel.: +81-29-861-3848; Fax: +81-29-861-3566

Academic Editor: Sylvain Ouillon
Received: 13 October 2016; Accepted: 5 January 2017; Published: 10 January 2017

Abstract: The compositions and transfer processes affecting coastal sea sediments from the Seto Inland Sea and the Pacific Ocean are examined through the construction of comprehensive terrestrial and marine geochemical maps for western Japan. Two-way analysis of variance (ANOVA) suggests that the elemental concentrations of marine sediments vary with particle size, and that this has a greater effect than the regional provenance of the terrestrial material. Cluster analysis is employed to reveal similarities and differences in the geochemistry of coastal sea and stream sediments. This analysis suggests that the geochemical features of fine sands and silts in the marine environment reflect those of stream sediments in the adjacent terrestrial areas. However, gravels and coarse sands do not show this direct relationship, which is likely a result of mineral segregation by strong tidal currents and the denudation of old basement rocks. Finally, the transport processes for the fine-grained sediments are discussed, using the spatial distribution patterns of outliers for those elements enriched in silt and clay. Silty and clayey sediments are found to be transported and dispersed widely by a periodic current in the inner sea, and are selectively deposited at the boundary of different water masses in the outer sea.

Keywords: geochemical map; particle transfer process; tidal current; analysis of variance (ANOVA); Cluster analysis; Mahalanobis' generalized distances; Seto Inland Sea

1. Introduction

The Geological Survey of Japan, part of the National Institute of Advanced Industrial Science and Technology (AIST), provides nationwide geochemical maps of elements within stream and marine sediments [1]. These geochemical maps have been utilized to explore mineral occurrences and determine the natural abundance of elements, and nationwide and cross-boundary geochemical maps have been developed for such purposes in many countries (e.g., [2–6]). In addition, Japanese geochemical maps have also been created specifically for the purposes of environmental assessment. Ohta and Imai [7] demonstrate an additional use of land and marine geochemical maps, examining particle transfer processes from the land to the sea, or within the marine environment. However, determining provenances with geochemical maps is typically challenging, due to the mixing and homogenization of marine sediments, especially fine sediments such as silt, during transport (e.g., [8]).

In this study, we sought to use geochemical maps of the provenance and transfer analyses of coastal sea sediments in the Chugoku and Shikoku regions, including the Seto Inland Sea and nearby Pacific Ocean as inner and outer seas, respectively (Figure 1a). This region is appropriate for clarifying the influence of terrestrial source materials on the adjacent marine environment, because most rock types found in Japan are distributed in the study area. The Seto Inland Sea is subject to significant tidal variation, while the Pacific Ocean off the Shikoku region is influenced by the Kuroshio Current and the Kuroshio Counter-Current. A number of previous studies of the Seto Inland Sea have analyzed the seafloor topography [9], periodic currents [10], marine organization [11], surface sediments [12], particle transport by marine currents [13,14], and contamination processes [15]. Therefore, this region is well-understood and is considered suitable for investigations into water-current transport processes, especially those between an inner and outer sea. The present study is intended to objectively reveal the complex factors affecting the spatial distributions of elements in coastal sea sediments using various statistical analyses.

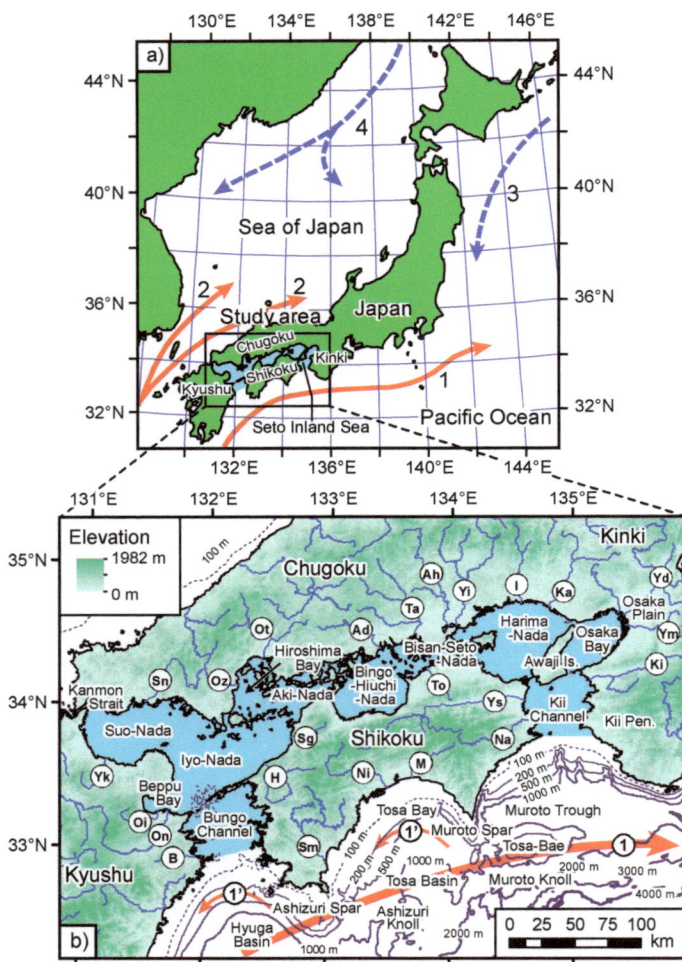

Figure 1. (**a**) Generalized location map of the study area; (**b**) Geographic map of the study area. Solid blue lines in terrestrial areas show major rivers. The abbreviations indicate the river names, as listed in Table A1. The blue colored area is the Seto Inland Sea. 1: Kuroshio Current; 1′: Kuroshio Counter-Current; 2: Tsushima-Current; 3: Oyashio-Current; 4: Liman-Current. Bathymetric depth contours are delineated using a dataset provided by the Japan Oceanographic Data Center.

2. Geological and Marine Settings

2.1. Riverine System

Figure 1b presents a generalized location map of the study area showing the 24 major rivers. The region is mainly mountainous with small tracts of flat land. The bed slope of the streams on Shikoku Island and the Kii Peninsula is very steep, so rainwater is immediately discharged through the rivers to the sea. Table 1 presents the potential sediment yield from each terrestrial region, which is calculated using the data of Akimoto et al. [16] (Appendix A). Several rivers in the Chugoku and Kyushu regions that flow directly into the Sea of Japan, and from which marine sediments were collected, are not discussed in this study. In this regard, the sediment yield data of these rivers are excluded from Table 1. The Yodo, Yoshino, and Shimanto Rivers are the three largest rivers, each with a high sediment yield (Yd, Ys, and Sm, respectively, in Figure 1b), and the Chugoku, Kinki, Shikoku, and Kyushu regions supply 15%, 5%, 5%, and 1% of the total sediment yield to the Seto Inland Sea, respectively. Therefore, it can be seen that most sediments in the Seto Inland Sea originate from the Chugoku region. The Yodo River flows through several large cities, including Kyoto and Osaka, and discharges into Osaka Bay. Together, the Yodo and Yamato Rivers supply 19% of the total sediment yield to Osaka Bay. Rivers on Shikoku Island are the primary source of sediments to the Pacific Ocean side with 22% of the total sediment yield supplied to Tosa Bay. The Kii Channel also receives a high sediment discharge of 23% from Shikoku Island and the Kii Peninsula. In contrast, little sediment is discharged from rivers into the Bungo Channel.

Table 1. Sediment yield of each region and discharge to adjacent marine environment.

Region	Discharged Area			
	Pacific Ocean	Kii or Bungo Channel	Seto Inland Sea	Inner Bay
Chugoku region	0%	0%	15%	4% to Hiroshima Bay
Kinki region	0%	5% to Kii Channel	5%	19% to Osaka Bay
Shikoku region	22%	18% to Kii Channel	5%	0%
Kyushu region	0%	1% to Bungo Channel	1%	5% to Beppu Bay

2.2. Geology and Terrestrial Metalliferous Deposits

Figure 2 presents a geological map of the study area, simplified from the Geological Map of Japan 1:1,000,000 [17]. The geology varies considerably between the Chugoku, Kinki, Shikoku, and Kyusyu regions. Rhyolitic-dacitic volcanic rocks and granitic rocks of Cretaceous and Paleogene age are widely distributed in the Chugoku region. In addition, Permian accretionary complexes associated with large limestone blocks, and Triassic high-pressure Sangun metamorphic rocks outcrop in western and central Chugoku. On Shikoku Island, the zonal arrangement of rock units from north to south is as follows: (1) Cretaceous granite-granodiorite; (2) Cretaceous sedimentary rocks; (3) Jurassic–Cretaceous high-pressure Sambagawa metamorphic rocks, comprising quartz schist and greenschist, associated with Mikabu greenstones consisting of basaltic, pyroclastic and ultramafic rocks; and (4) sedimentary rocks of accretionary complexes dated mainly to the Cretaceous-Paleogene. These rock types also outcrop on southern Kyushu Island and on the Kii Peninsula, while andesitic volcanic rocks and debris-pyroclastic rocks of Neogene-Quaternary age extensively outcrop in the northeast of Kyushu Island. Sedimentary rocks of accretionary complexes dating mostly to the Jurassic-Cretaceous outcrop in the northern Kinki region. Unconsolidated Quaternary sediments are restricted in distribution, occurring mainly in the Kinki region. The Osaka Plain is formed of these Quaternary sediments and is the widest plain in the entire study area, with a population of approximately 10 million people.

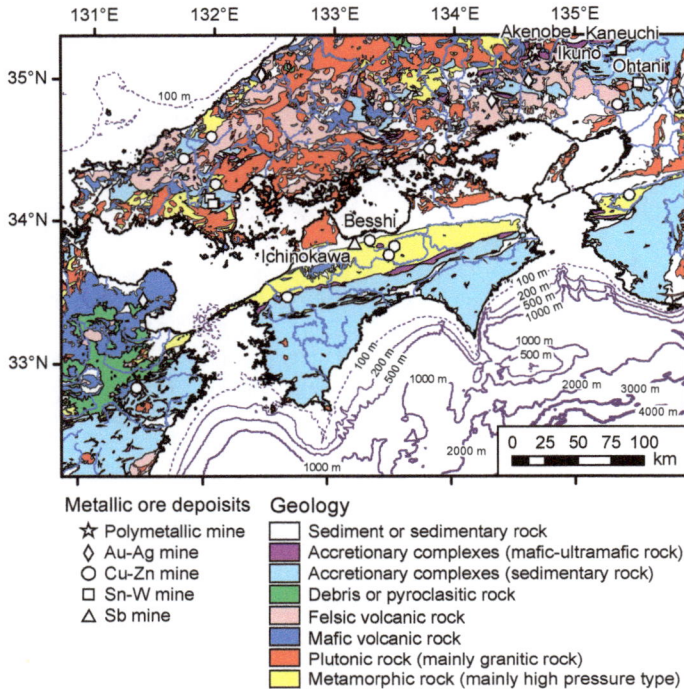

Figure 2. Geological map of the study area simplified from the Geological Map of Japan 1:1,000,000 [17].

Figure 2 also shows the locations of some major economic metalliferous deposits. The Besshi mine is the largest Copper (Cu) mine in Japan, and the Kaneuchi and Ohtani mines are large-scale Tungsten-Tin (W-Sn) mines. The Ikuno and Akenobe mines are the largest polymetallic mineralization mines, and the Ichinokawa mine has the highest levels of Antimony (Sb) in Japan.

2.3. Marine Topography, Hydrographic Condition, and Geology

The Seto Inland Sea is mostly less than 60 m deep, and passes through to the Pacific Ocean through the Bungo and Kii Channels, and to the Sea of Japan through the Kan-mon Strait. The Seto Inland Sea is divided into several regions, termed "Nada", for example, Suo-Nada, Iyo-Nada, Aki-Nada, Bingo-Nada, Hiuchi-Nada, Bisan-Seto-Nada, and Harima-Nada (Figure 1b). Periodic currents flow from the Pacific Ocean through the Bungo and Kii Channels, meeting at Hiuchi-Nada and concentrating fine sediments in the water mass there [10,18]. The marine geology of the Seto Inland Sea comprises mainly glacial to Holocene sediments [12].

The Pacific side of Shikoku Island is characterized by a narrow continental shelf. Topographic highs include Ashizuri Spar, Muroto Spar, Tosa-Bae, Ashizuri Sea Knoll, and Muroto Sea Knoll, and consist of Miocene-Pliocene siltstones, which include benthic foraminifer fossils that are distributed in the deep sub-bottom of the basin [19] In contrast, Tosa Basin, Hyuga Basin, and Muroto Trough are deep-sea basins in which the water depth exceeds 1000 m. Tosa Basin is covered thickly by Quaternary turbidite deposits. The continental shelf and slope of Tosa Bay, the Kii Channel, and the Bungo Channel comprise delta deposits formed during Quaternary regression and transgression cycles [19]. Within the water mass, the Kuroshio Current flows off the coast of Shikoku Island from the southwest to the northeast, while the Kuroshio Counter-Current flows anticlockwise in Tosa Bay and Hyuga Basin [20,21] (Figure 1b).

3. Materials and Methods

3.1. Samples, Sampling Methods, and Processing

Terrestrial and marine sampling locations are presented in Figure 3. A total of 554 stream samples were collected from the study area for a regional geochemical mapping project during 1999–2004 [22]. In addition, for the current study, 22 stream sediments were collected in 2008 from small islands, including Awaji Island in the Seto Inland Sea. Sediment sample were air-dried and sieved using a 180 μm (83-mesh) screen. Magnetic minerals were removed from the sieved samples using a magnet, to minimize the effect of magnetic mineral accumulation [22,23].

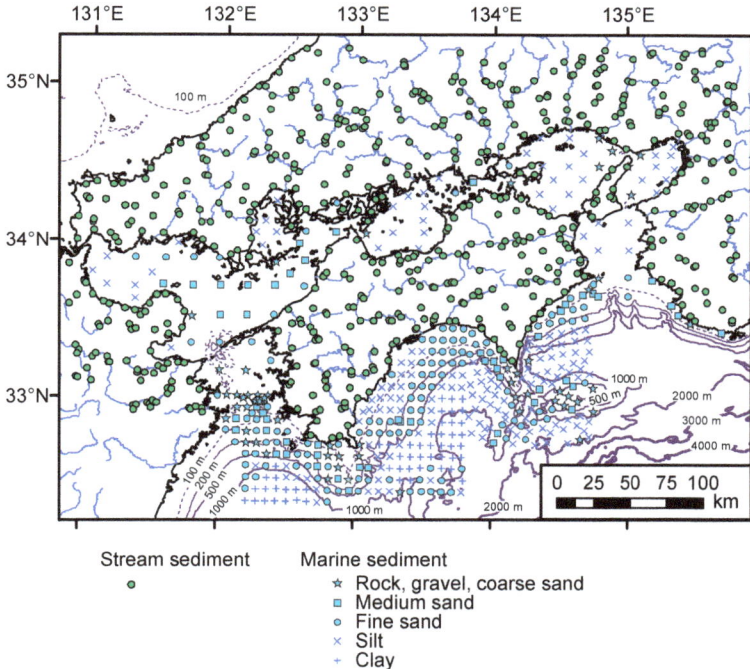

Figure 3. Sample locations for stream and coastal sea sediments.

A total of 366 marine samples were collected from the Pacific Ocean during cruises GH82-1 and GH83-2, in 1982 and 1983, respectively [24–26]. In addition, 97 samples were collected from the Seto Inland Sea and around the Kii Peninsula in 2005. The total 463 samples were collected using a K-grab sampler, and the uppermost 3 cm of each sediment sample was separated, air-dried, ground with an agate mortar and pestle, and retained for chemical analysis. Some samples were composed mainly of rock fragments and gravels, in which case, the sandy infilling materials were collected. The particle sizes of 223 marine sediments were determined, based on the median particle diameter of the surface sediments, and were classified as: coarse sediments, comprising lithic fragments, gravels, and coarse sand; medium sand; fine sand; silt; and clay (Figure 3). The median particle diameter was not measured for the remaining 240 samples, so their classification is based solely on a visual inspection of texture.

Figure 3 shows that silty sediments occur in Suo-Nada, Hiuchi-Nada, Bingo-Nada, and Harima-Nada, as well as basins affected by small periodic current flows or eddies [10]. These localities are also associated with high turbidity [12,27]. Coarse sands were collected from Aki-Nada and Bisan-Seto-Nada, and in northeastern Iyo-Nada. Rocks, gravels, and coarse sands occur in the Bungo Channel. In contrast, the Kii Channel is dominated by silty sediments, despite the presence of high velocity periodic currents similar to those in the Bungo Channel. This is likely due to the greater terrestrial sediment supply to the Kii Channel (18%) compared with the Bungo Channel (1%) (Table 1). On the Pacific side, silty sediments are widely distributed obliquely across the depth contour in the western part of Tosa Bay. Fine sands are found: (1) near to river mouths; (2) at water depths of 100–500 m on the shelf and slope off Tosa Bay; and (3) around Ashizuri Knoll and Muroto Knoll where the water depth is 500–1600 m. Silty and clayey sediments are found in Tosa Basin, Hyuga Basin, and Muroto Trough, where the water depth exceeds 1000 m, and rock, gravels, and coarse sands were sampled from the topographic highs of Ashizuri Spar, Muroto Spar, and Tosa-Bae.

3.2. Geochemical and Spatial Analyses

The geochemical analytical method is detailed by Imai [28]. Each 0.2 g sample was digested using HF, HNO_3, and $HClO_4$ solutions at 120 °C for 2 h. The degraded solution was evaporated to dryness at 200 °C, before the residue was dissolved in 100 mL of 0.35 M HNO_3 solution. Concentrations of Na_2O, MgO, Al_2O_3, P_2O_5, K_2O, CaO, MnO, total (T-) Fe_2O_3, V, Sr, and Ba were determined using ICP atomic emission spectrometry, while contents of Li, Be, Sc, Cr, Co, Ni, Cu, Zn, Ga, Rb, Nb, Y, Mo, Cd, Sn, Sb, Cs, lanthanides (Ln: La–Lu), Ta, Tl, Pb, Bi, Th, and U were measured using ICP mass spectrometry. Analyses of As in all samples and Hg in stream sediment samples were subcontracted to ALS Chemex in Vancouver, B.C. Levels of Hg in marine sediments were determined using an atomic absorption spectrometer that measured the quantity of Hg vapor generated by direct thermal decomposition. Table 2 presents a summary of the analytical results obtained from the marine and stream sediments. The Na_2O component of marine sediments should be used only as a guide, since these marine sediments were not desalinated.

Table 2. Summary for the geochemistry of marine sediments ($n = 463$) and stream sediments ($n = 576$).

Element	Marine Sediment				Stream Sediment			
	Min	Mean	Median	Max	Min	Mean	Median	Max
wt %								
Na_2O	0.91	2.97	2.82	6.35	0.46	2.10	2.04	4.42
MgO	0.49	2.77	2.68	6.71	0.19	2.10	1.65	9.79
Al_2O_3	2.01	9.49	9.60	16.62	3.91	11.39	11.52	17.94
P_2O_5	0.025	0.113	0.108	2.56	0.021	0.125	0.109	1.22
K_2O	0.53	2.03	2.06	3.12	0.66	2.18	2.24	3.79
CaO	0.33	6.51	5.22	34.1	0.12	1.83	1.41	8.40
TiO_2	0.087	0.501	0.492	2.15	0.14	0.71	0.60	3.28
MnO	0.012	0.082	0.066	0.451	0.014	0.128	0.116	1.10
$T\text{-}Fe_2O_3$	0.73	4.91	4.38	23.7	1.41	5.24	4.67	15.8
mg/kg								
Li	7.58	51.9	50.0	136	9.33	38.5	38.3	107
Be	0.32	1.41	1.44	2.25	0.71	1.77	1.67	4.75
Sc	1.77	9.82	9.69	35.1	2.31	11.6	9.57	46.8
V	11.7	84.8	80.1	680	10.9	103	83.4	420
Cr	6.50	58.0	59.0	172	4.47	84.9	52.8	884
Co	1.76	10.9	9.90	41.2	1.79	13.8	11.5	60.7
Ni	1.35	28.2	28.8	138	1.38	34.3	21.7	349
Cu	2.42	20.5	18.4	86.6	5.66	49.7	31.5	2599
Zn	16.1	88.1	80.9	347	17.1	165	118	11,444
Ga	2.77	13.5	14.2	19.0	6.97	17.4	17.2	31.7
As	0.2	6.3	5.0	68	1.0	18.2	8.1	1578
Rb	17.7	73.2	74.9	114	12.3	105	103	265
Sr	70.2	326	259	2525	23.4	133	111	1321
Y	4.95	12.9	12.6	28.8	2.24	19.6	18.7	63.6
Nb	1.59	6.79	6.80	13.9	2.88	9.75	8.86	33.3
Mo	0.14	0.76	0.63	14.8	0.25	1.34	0.99	27.4
Cd	0.018	0.101	0.081	0.480	0.019	0.29	0.16	28.7
Sn	0.37	2.15	1.98	18.9	0.87	5.05	3.47	170
Sb	0.092	0.59	0.55	1.89	0.099	0.98	0.74	17.2
Cs	0.60	4.06	4.25	7.65	0.41	5.63	5.03	33.6
As	0.2	6.3	5.0	68	1.0	18.2	8.1	1578
Ba	67.4	340	340	1649	136	429	431	855
La	5.31	16.0	16.5	24.7	4.73	21.2	19.9	69.2
Ce	12.8	33.8	34.6	58.7	7.20	39.2	36.9	170
Pr	1.28	3.78	3.87	5.75	1.06	4.90	4.63	20.8
Nd	5.22	15.0	15.3	22.2	4.08	19.2	18.2	79.9
Sm	1.02	3.02	3.09	4.59	0.78	3.94	3.71	17.4
Eu	0.29	0.68	0.68	1.15	0.18	0.81	0.78	1.72
Gd	0.92	2.72	2.74	4.44	0.61	3.63	3.39	12.9
Tb	0.16	0.45	0.45	0.78	0.10	0.63	0.59	2.02
Dy	0.80	2.23	2.20	4.07	0.51	3.21	3.03	10.3
Ho	0.16	0.42	0.42	0.80	0.091	0.62	0.59	2.01
Er	0.47	1.23	1.19	2.34	0.26	1.83	1.76	6.29
Tm	0.069	0.19	0.19	0.36	0.036	0.29	0.28	1.05
Yb	0.41	1.18	1.15	2.20	0.21	1.82	1.74	6.88
Lu	0.051	0.17	0.16	0.31	0.032	0.26	0.25	1.03
Ta	0.13	0.61	0.60	1.23	0.28	0.85	0.75	5.23
Hg	0.0005	0.139	0.138	5.19	0.010	0.117	0.060	4.60
Tl	0.068	0.48	0.49	1.00	0.075	0.64	0.61	2.52
Pb	5.00	20.8	18.6	126	7.19	43.8	27.3	4177
Bi	0.036	0.32	0.28	1.43	0.033	0.52	0.30	32.1
Th	1.50	5.81	6.04	26.6	1.52	9.53	7.30	259
U	0.28	1.35	1.26	4.51	0.53	2.13	1.70	31.9

The spatial distributions of elemental concentrations in both the terrestrial and marine environments were plotted using geographic information system software (ArcGIS 10.3; Environmental Systems Research Institute (ESRI) Japan Corporation, Tokyo, Japan) after Ohta et al. [29]. Different classes of elemental concentrations are applied to the terrestrial and marine environments, because the chemical and mineralogical compositions of sediments differ according to particle size see the details in [7,8]. This simply improves geovisualization of the geochemical maps. If the same classification is applied to the geochemical maps of both terrestrial and marine environments, the regional geochemical differences (as depicted by color variation) of most of the elements in the land or the sea will be obscured [7,8]. For example, for CaO and Sr, which show higher concentrations in the marine sediments (Table 2), their geochemical differences in the terrestrial areas would be obscured in resulting maps using the same classification, as shown in the Legend. Percentile ranges are used for the selection of elemental concentration intervals in the geochemical maps: $0 \leq x \leq 5$, $5 < x \leq 10$, $10 < x \leq 25$, $25 < x \leq 50$, $50 < x \leq 75$, $75 < x \leq 90$, $90 < x \leq 95$, and $95 < x \leq 100\%$, where x represents the elemental concentration [30]. This class selection is advantageous in that the same range of percentiles (e.g., 90%–95%) implies the same statistical weight, even at different numerical scales [7,8,30]. Subsequent statistical analysis of geochemical data is performed using EXCEL TOUKEI 7.0 (ESUMI Co. Ltd., Tokyo, Japan).

4. Results

4.1. Spatial Distribution of Terrestrial Elemental Concentrations

Figure 4 shows the terrestrial and marine geochemical maps for 12 elements. Mikoshiba et al. [23] and Ohta et al. [31,32] have examined the factors controlling spatial distribution patterns of terrestrial elemental concentrations. Be, Na_2O, K_2O, Rb, Y, Nb, Ba, Ln, Ta, Th, and U are abundant in the Chugoku region and in the northern part of Shikoku Island where predominantly granitic and felsic volcanic rocks outcrop. Stream sediments derived from granitic rocks are enriched in Li, Be, Na_2O, K_2O, Rb, Y, Nb, Ln, Ta, Th, and U. Felsic volcanic rocks also elevate the K_2O, Rb, and Ba contents of stream sediments. Li and Cs are abundant in muddy sedimentary rocks and the mélange matrix of accretionary complexes in northern Kinki. In contrast, the sandstone-dominated sedimentary rocks of accretionary complexes in the southern part of Shikoku Island and the Kii Peninsula contain fewer elements. Mafic volcanic rocks associated with debris flows and pyroclastic rocks around Aso volcano, and metamorphic rocks, dominantly greenschist, on Shikoku Island and the Kii Peninsula, elevate MgO, Al_2O_3, P_2O_5, CaO, Sc, V, TiO_2, MnO, $T\text{-}Fe_2O_3$, Cr, Co, Ni, and Cu concentrations in the stream sediments. Additionally, extreme enrichments of MgO, Cr, Co, and Ni are caused by the presence of mafic-ultramafic rocks associated with accretionary complexes. Finally, elevated concentrations of P_2O_5, Cu, Zn, Mo, Cd, Sn, Sb, Hg, Pb and Bi are found in Osaka Plain, which is polluted by anthropogenic activity [32].

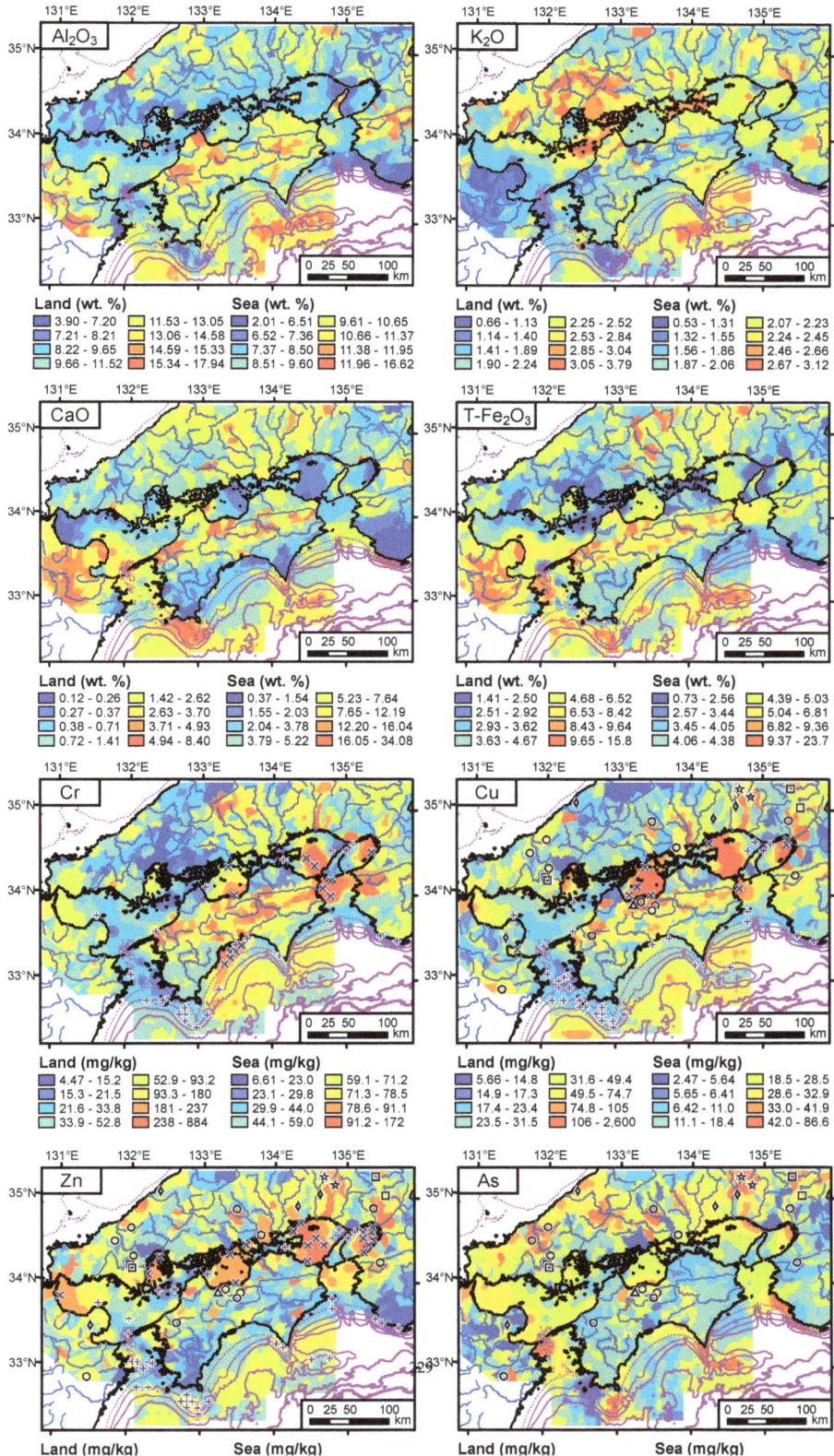

Al₂O₃ / Al_2O_3

Land (wt. %)		Sea (wt. %)	
3.90 - 7.20	11.53 - 13.05	2.01 - 6.51	9.61 - 10.65
7.21 - 8.21	13.06 - 14.58	6.52 - 7.36	10.66 - 11.37
8.22 - 9.65	14.59 - 15.33	7.37 - 8.50	11.38 - 11.95
9.66 - 11.52	15.34 - 17.94	8.51 - 9.60	11.96 - 16.62

K_2O

Land (wt. %)		Sea (wt. %)	
0.66 - 1.13	2.25 - 2.52	0.53 - 1.31	2.07 - 2.23
1.14 - 1.40	2.53 - 2.84	1.32 - 1.55	2.24 - 2.45
1.41 - 1.89	2.85 - 3.04	1.56 - 1.86	2.46 - 2.66
1.90 - 2.24	3.05 - 3.79	1.87 - 2.06	2.67 - 3.12

CaO

Land (wt. %)		Sea (wt. %)	
0.12 - 0.26	1.42 - 2.62	0.37 - 1.54	5.23 - 7.64
0.27 - 0.37	2.63 - 3.70	1.55 - 2.03	7.65 - 12.19
0.38 - 0.71	3.71 - 4.93	2.04 - 3.78	12.20 - 16.04
0.72 - 1.41	4.94 - 8.40	3.79 - 5.22	16.05 - 34.08

$T\text{-}Fe_2O_3$

Land (wt. %)		Sea (wt. %)	
1.41 - 2.50	4.68 - 6.52	0.73 - 2.56	4.39 - 5.03
2.51 - 2.92	6.53 - 8.42	2.57 - 3.44	5.04 - 6.81
2.93 - 3.62	8.43 - 9.64	3.45 - 4.05	6.82 - 9.36
3.63 - 4.67	9.65 - 15.8	4.06 - 4.38	9.37 - 23.7

Cr

Land (mg/kg)		Sea (mg/kg)	
4.47 - 15.2	52.9 - 93.2	6.61 - 23.0	59.1 - 71.2
15.3 - 21.5	93.3 - 180	23.1 - 29.8	71.3 - 78.5
21.6 - 33.8	181 - 237	29.9 - 44.0	78.6 - 91.1
33.9 - 52.8	238 - 884	44.1 - 59.0	91.2 - 172

Cu

Land (mg/kg)		Sea (mg/kg)	
5.66 - 14.8	31.6 - 49.4	2.47 - 5.64	18.5 - 28.5
14.9 - 17.3	49.5 - 74.7	5.65 - 6.41	28.6 - 32.9
17.4 - 23.4	74.8 - 105	6.42 - 11.0	33.0 - 41.9
23.5 - 31.5	106 - 2,600	11.1 - 18.4	42.0 - 86.6

Zn

Land (mg/kg)   Sea (mg/kg)

As

Land (mg/kg)   Sea (mg/kg)

Therefore, lithology is considered to be the main factor affecting the concentrations of elements in stream sediments. For reference, the median elemental concentrations of stream sediments according to their representative rock type, which outcrops over more than half of the drainage basin, are summarized in Appendix B. It should be noted that Cu, Zn, As, Mo, Cd, Sn, Sb, Hg, Pb, and Bi are extremely enriched near to known metalliferous deposits, however, the effects of these mineral deposits on the geochemical maps are restricted to limited areas.

4.2. Spatial Distribution of Elemental Concentrations in Coastal Sea Sediments

Within the Seto Inland Sea, high levels of K_2O, Rb, Ba, and Tl are found in the coarse sands of Aki-Nada and Bisan-Seto-Nada, and in northeastern Iyo-Nada (Figure 4). Coarse sediments with rock fragments, gravels, and coarse sands in western Iyo-Nada and the Bungo Channel are enriched in MgO, P_2O_5, CaO, TiO_2, MnO, Fe_2O_3, Sc, V, Co, Zn, and Sr. In contrast, the sandy sediments in eastern Iyo-Nada and the Bungo Channel are depleted in almost all of the measured elements, except CaO and Sr. Silt and clay in Hiuchi-Nada, Harima-Nada, and the Kii Channel are abundant in Li, Be, MgO, P_2O_5, 3d transition metals, Ga, Y, Nb, Sn, Cs, Ln, Ta, Hg, Pb, Bi, and Th. Suo-Nada is enriched in Li, Zn, Ga, Sn, Mo, Cd, Ta, Pb, Bi, and U. Concentrations of P_2O_5, Cr, Ni, Cu, Zn, Mo, Cd, Sn, Hg, Pb, and Bi are especially elevated in the silt of Osaka Bay. Finally, MnO and Tl are abundant in all sediments within the Seto Inland Sea, irrespective of particle size.

On the Pacific Ocean side, coarse sediments with rock fragments, gravels, and coarse sands distributed around the topographic highs of Ashizuri Spar, Muroto Spar, Tosa-Bae, and the Muroto Sea Knoll are abundant in MgO, CaO, Al_2O_3, P_2O_5, TiO_2, MnO, $T\text{-}Fe_2O_3$, Sc, V, Co, As, Sr, Mo, Sb, and U. Sediments abundant in TiO_2, Cr, Co, Ni, Nb, light REEs (lanthanides from La to Sm), and Ta, occur in Tosa Bay. The silty and clayey sediments of Tosa Basin, Hyuga Basin, and the Muroto trough are enriched in Li, Cu, Nb, Cd, Sb, Cs, Hg, and U. Fine sands collected from the Ashizuri and Muroto Sea Knolls are richer in P_2O_5, CaO, Sc, TiO_2, V, MnO, Fe_2O_3, Co, Cu, As, Sr, Mo, Cd, Sb, Cs, heavy REEs (Y and lanthanides from Gd to Lu), and Bi than the fine sands distributed across the continental shelf.

5. Discussion

5.1. Analysis of Variance (ANOVA) to Reveal the Effects of Particle Size and Regional Difference

Coastal sea sediments typically originate from their adjacent terrestrial materials. Therefore, the geochemical features of terrestrial lithology are fundamentally reflected in those of nearby marine sediments. This phenomenon is referred to herein as "regional difference". Additionally, elemental concentrations in marine sediments are known to vary with particle size, in a phenomenon referred to herein as the "particle size effect". This variation is caused simply by the dilution effects of quartz and biogenic calcareous materials, which are abundant in coarse grains but less abundant in fine grains. Thus, we apply a two-way analysis of variance (ANOVA) to determine the factor that most significantly controls the chemical compositions of marine sediments: regional differences or the particle size effect [8]. The procedure is detailed by [8] (Appendix C).

The marine samples are grouped into those from the Seto Inland Sea region, including the Kii and Bungo Channels, and those from the Pacific Ocean region. They are further classified into: coarse sediments, including rock fragments, gravels, coarse sands, and medium sands; fine sands; and silt-clay. We assume that the coarse sediments, fine sands, and silt-clay are moved by water power over short, middle, and long distances, respectively. Median elemental concentrations of marine sediments are calculated for each region and particle size (Table 3). Table 3 also presents the median concentrations of stream sediments from two regions, for reference. In this regard, 96 stream sediment samples were collected from rivers that flow directly into the Sea of Japan, and so these samples are excluded from Table 3.

Table 3. Median elemental concentrations of marine sediments and stream sediments.

Element	Seto Inland Sea				Pacific Ocean			
	Stream Sed.	Marine Sed.			Stream Sed.	Marine Sed.		
		Coarse Sed.	Fine Sand	Silt and Clay		Coarse Sed.	Fine Sand	Silt and Clay
	n = 321	*n* = 26	*n* = 15	*n* = 47	*n* = 159	*n* = 75	*n* = 107	*n* = 193
wt %								
MgO	1.72	1.45	2.76	3.31	1.80	3.12	2.39	2.68
Al_2O_3	11.43	7.34	8.47	8.48	12.17	9.29	9.23	10.50
P_2O_5	0.115	0.059	0.114	0.129	0.105	0.105	0.095	0.111
K_2O	2.25	2.26	2.02	2.03	2.20	1.80	1.95	2.18
CaO	1.67	4.83	5.21	2.01	0.77	7.41	6.93	5.23
TiO_2	0.64	0.25	0.43	0.54	0.54	0.49	0.42	0.51
MnO	0.119	0.086	0.079	0.109	0.096	0.100	0.060	0.056
$T\text{-}Fe_2O_3$	4.88	2.39	3.54	4.59	4.45	5.13	4.50	4.29
mg/kg								
Li	35	24	55	102	42	29	42	53
Be	1.7	1.2	1.4	1.7	1.6	1.2	1.4	1.5
Sc	10	4.9	8.4	9.3	9.8	10	8.3	10
V	78	37	57	75	94	82	69	90
Cr	51	25	40	74	60	39	47	68
Co	12	6.7	9.3	13	12	12	10	9.5
Ni	19	11	17	32	27	16	22	36
Cu	31	7.9	13	41	36	7.2	13	28
Zn	126	45	83	166	96	71	73	85
Ga	17	11	14	16	17	12	13	15
As	8.0	4.9	5.9	7.0	7.1	8.3	4.3	4.2
Rb	107	83	74	60	95	62	76	80
Sr	113	300	231	121	112	439	319	242
Y	21	9.6	15	15	14	12	11	13
Nb	9.1	4.0	6.3	9.4	8.5	4.4	5.7	7.9
Mo	0.98	0.47	0.74	1.1	0.96	0.46	0.47	0.68
Cd	0.18	0.041	0.079	0.19	0.10	0.054	0.050	0.13
mg/kg								
Sn	3.9	1.3	2.4	4.3	2.7	1.2	1.6	2.2
Sb	0.74	0.30	0.39	0.55	0.71	0.43	0.44	0.76
Cs	5.1	2.0	3.8	5.4	4.6	2.2	3.5	5.1
Ba	433	344	274	247	438	300	323	386
La	20	12	14	17	20	13	16	18
Ce	37	25	32	33	39	27	34	37
Pr	4.8	2.7	3.7	4.2	4.6	3.2	3.7	4.1
Nd	19	11	15	16	18	13	15	16
Sm	3.9	2.1	3.2	3.5	3.6	2.7	2.9	3.2
Eu	0.83	0.60	0.71	0.67	0.77	0.73	0.64	0.70
Gd	3.6	1.9	2.9	3.3	3.2	2.5	2.5	2.8
Tb	0.63	0.31	0.51	0.56	0.51	0.40	0.40	0.45
Dy	3.3	1.6	2.6	2.9	2.6	2.0	1.9	2.2
Ho	0.64	0.31	0.49	0.54	0.48	0.38	0.35	0.42
Er	1.9	0.9	1.4	1.6	1.3	1.1	1.0	1.2
Tm	0.31	0.15	0.22	0.26	0.21	0.17	0.16	0.19
Yb	1.9	0.9	1.4	1.5	1.3	1.0	1.0	1.2
Lu	0.28	0.14	0.20	0.22	0.18	0.15	0.14	0.16
Ta	0.75	0.40	0.60	0.92	0.76	0.38	0.51	0.73
Hg	0.050	0.030	0.079	0.167	0.110	0.026	0.100	0.179
Tl	0.62	0.56	0.60	0.68	0.55	0.34	0.41	0.51
Pb	28	20	24	39	23	17	17	19
Bi	0.31	0.18	0.32	0.77	0.26	0.18	0.19	0.34
Th	7.9	3.7	5.9	7.0	6.6	3.7	5.4	6.6
U	1.9	0.8	1.5	1.8	1.4	0.9	1.1	1.4

Table 4 presents the results of a two-way ANOVA: the variance ratios (F), probabilities (p), and effect size (η^2) due to regional difference (factor A), particle size (factor B), and the interaction effect (factor A × B). When the estimated probability is lower than 0.01, we conclude that the factor makes a significant difference to the chemical compositions, and in fact, each factor was statistically significant ($p < 0.01$) in most cases (Table 4). This is because a statistical test with a large amount of input data, such as $n = 463$ in this study, is highly sensitive to very small differences. Therefore, we calculate η^2 to form a plausible estimation of p value irrespective of sample number [33,34]. The η^2 is an easy-to-understand metric that is calculated as the ratio of the sum of squares (SS) for each factor and the total SS. Magnitudes of $\eta^2 < 0.01$, $0.01 \leq \eta^2 < 0.06$, $0.06 \leq \eta^2 < 0.14$, and $\eta^2 \geq 0.14$ can be considered to represent no effect, a small effect, an intermediate effect, and a large effect, respectively [33,34]. In this study, we conclude that each factor with $\eta^2 \geq 0.06$ has a significant effect on the elemental concentrations of the sediments. Furthermore, this indicator can be used to decide the most dominant factor, on the basis of the highest η^2 score [33].

Results of the ANOVA suggest that particle size is the dominant influencing factor for many elements in the sediments (Table 4). The regional difference effect is more significant than the particle size effect only for Al_2O_3, CaO, MnO, Sc, V, Mo, and Pb. This suggests that consideration of the particle size effect must be made in order to fully understand the influence of terrestrial materials on coastal sea sediments. However, it should be noted that the interaction effect is more influential than either of the two main factors alone for MgO, P_2O_5, Fe_2O_3, Sc, Co, Rb, Ba, and Eu, and is significant for 14 elements, including K_2O and TiO_2. This interaction effect refers to the effect of one factor on the other factor, such that the two main factors synergistically affect the data e.g., [35]. To visualize the interaction effect, variations in the median values of Al_2O_3, MgO, K_2O, Co, Cs, and Ba are presented in Figure 5. When the interaction effect is insignificant, the median values vary approximately in parallel between the Seto Inland Sea and the Pacific Ocean, irrespective of particle size (Al_2O_3 and Cs in Figure 5). In the cases of large η^2 values for the interaction effect, such as for MgO and Co, the elemental concentrations in the Pacific Ocean sediments decrease with decreasing particle size, but increase in the sediments of the Seto Inland Sea with decreasing particle size. Variations in the K_2O and Ba concentrations with particle size are exactly opposite to those of MgO and Co, and still indicate a significant contribution of the interaction effect. These results suggest that regional difference and particle size effects are not independent for many elements. Therefore, we need to evaluate the geochemical similarities and differences among groups subdivided by region and particle size.

Table 4. Variance ratio (F), probability (p), and size effect (η^2) of two-way ANOVA type II.

Element	Two-Way ANOVA									MF
	F (A)	F (B)	F (A × B)	p (A)	p (B)	p (A × B)	η^2 (A)	η^2 (B)	η^2 (A × B)	
MgO	2	13	67	0.20	<0.01	<0.01	<0.01	0.04	**0.22**	A × B
Al$_2$O$_3$	74	34	2	<0.01	<0.01	0.13	**0.12**	**0.11**	<0.01	Both
P$_2$O$_5$	10	19	28	<0.01	<0.01	<0.01	0.02	**0.07**	**0.10**	A × B
K$_2$O	1	46	31	0.36	<0.01	<0.01	<0.01	**0.15**	**0.10**	B
CaO	110	34	6	<0.01	<0.01	<0.01	**0.17**	**0.11**	0.02	A
TiO$_2$	14	26	19	<0.01	<0.01	<0.01	0.02	**0.09**	**0.07**	B
MnO	53	26	17	<0.01	<0.01	<0.01	**0.09**	**0.09**	**0.06**	Both
T-Fe$_2$O$_3$	37	1	38	<0.01	0.60	<0.01	**0.06**	<0.01	**0.13**	A × B
Li	98	366	59	<0.01	<0.01	<0.01	**0.07**	**0.53**	**0.09**	B
Be	44	95	13	<0.01	<0.01	<0.01	**0.06**	**0.27**	0.04	B
Sc	33	13	26	<0.01	<0.01	<0.01	**0.06**	0.05	**0.09**	A × B
V	77	18	23	<0.01	<0.01	<0.01	**0.13**	**0.06**	**0.08**	A
Cr	2	149	21	0.16	<0.01	<0.01	<0.01	**0.37**	0.05	B
Co	0	0	50	0.89	0.83	<0.01	<0.01	<0.01	**0.18**	A × B
Ni	29	280	15	<0.01	<0.01	<0.01	0.03	**0.52**	0.03	B
Cu	30	595	17	<0.01	<0.01	<0.01	0.02	**0.70**	0.02	B
Zn	78	104	115	<0.01	<0.01	<0.01	**0.08**	**0.22**	**0.24**	A × B
Ga	21	191	12	<0.01	<0.01	<0.01	0.02	**0.43**	0.03	B
As	11	28	16	<0.01	<0.01	<0.01	0.02	**0.10**	**0.06**	B
Rb	5	22	26	0.02	<0.01	<0.01	<0.01	**0.08**	**0.09**	A × B
Sr	120	113	5	<0.01	<0.01	<0.01	**0.15**	**0.28**	0.01	B
Y	19	25	20	<0.01	<0.01	<0.01	0.03	**0.09**	**0.07**	B
Nb	22	152	12	<0.01	<0.01	<0.01	0.03	**0.38**	0.03	B
Mo	20	16	8	<0.01	<0.01	<0.01	0.04	**0.06**	0.03	B
Cd	20	124	16	<0.01	<0.01	<0.01	0.03	**0.33**	0.04	B
Sn	180	239	42	<0.01	<0.01	<0.01	**0.15**	**0.41**	**0.07**	B
Sb	44	106	3	<0.01	<0.01	0.03	**0.06**	**0.29**	<0.01	B
Cs	2	467	5	0.14	<0.01	<0.01	<0.01	**0.67**	<0.01	B
Ba	29	34	58	<0.01	<0.01	<0.01	0.04	**0.10**	**0.17**	A × B
La	3	116	5	0.07	<0.01	<0.01	<0.01	**0.33**	0.01	B
Ce	24	89	0	<0.01	<0.01	0.86	0.04	**0.27**	<0.01	B
Pr	2	128	9	0.19	<0.01	<0.01	<0.01	**0.35**	0.02	B
Nd	1	105	12	0.26	<0.01	<0.01	<0.01	**0.30**	0.03	B
Sm	2	80	21	0.13	<0.01	<0.01	<0.01	**0.24**	**0.06**	B
Eu	4	8	6	0.07	<0.01	<0.01	<0.01	0.03	0.03	None
Gd	13	50	27	<0.01	<0.01	<0.01	0.02	**0.16**	**0.09**	B
Tb	38	43	32	<0.01	<0.01	<0.01	**0.06**	**0.13**	**0.10**	B
Dy	25	44	30	<0.01	<0.01	<0.01	0.04	**0.14**	**0.09**	B
Ho	30	39	25	<0.01	<0.01	<0.01	0.05	**0.13**	**0.08**	B
Er	38	36	20	<0.01	<0.01	<0.01	**0.06**	**0.12**	**0.07**	B
Tm	44	37	18	<0.01	<0.01	<0.01	**0.07**	**0.12**	**0.06**	B
Yb	42	36	16	<0.01	<0.01	<0.01	**0.07**	**0.12**	0.05	B
Lu	58	26	18	<0.01	<0.01	<0.01	**0.10**	**0.09**	**0.06**	A
Ta	68	163	19	<0.01	<0.01	<0.01	**0.08**	**0.37**	0.04	B
Hg	0	270	2	0.85	<0.01	0.14	<0.01	**0.54**	<0.01	B
Tl	241	107	1	<0.01	<0.01	0.59	**0.27**	**0.24**	<0.01	Both
Pb	359	58	36	<0.01	<0.01	<0.01	**0.36**	**0.11**	**0.07**	A
Bi	165	214	57	<0.01	<0.01	<0.01	**0.14**	**0.37**	**0.10**	B
Th	5	206	2	0.03	<0.01	0.14	<0.01	**0.47**	<0.01	B
U	9	114	11	<0.01	<0.01	<0.01	0.01	**0.32**	0.03	B

Notes: The factor A, B, and A × B indicate the regional difference effect, particle size effect and interaction effect, respectively. MF means major factor. The boldface in η^2 scores indicates that each factor with $\eta^2 \geq 0.06$ has a significant effect on the elemental concentrations of the sediments.

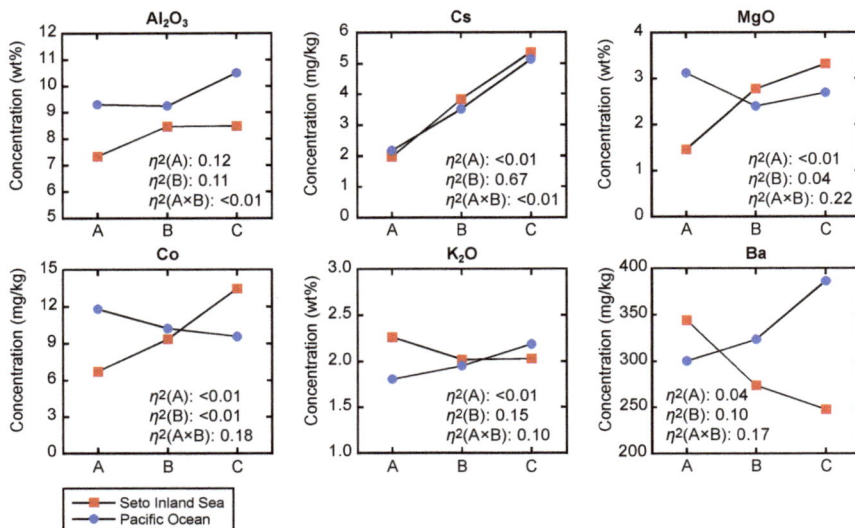

Figure 5. Median concentrations of elements and oxides in sediments from the Seto Inland Sea and the Pacific Ocean. A, B, and C represent coarse sediment, fine sand, and silty-clayey sediments, respectively. η^2 (A), η^2 (B), and η^2 (A × B) are the size effects of regional differences, particle sizes, and the interaction effect, respectively, obtained from the two-way ANOVA (Table 4).

5.2. Comparison of Elemental Abundances in Coastal Sea and Stream Sediments between the Seto Inland Sea and the Pacific Ocean

The chemical and mineralogical compositions of marine sediments differ from those of stream sediments [7,8]. Furthermore, the range in concentration completely differs for each element: major and minor elements in wt % oxide amounts, and trace elements in mg/kg or µg/kg amounts. Therefore, to eliminate the effect of scale and units, the data have to be normalized to elucidate differences amongst the elements or the samples. Thus, in this study, enrichment factors (EF) of the marine and stream sediments compared with the upper continental crust (UCC) [36] are employed to allow direct comparison (Figure 6). The EF is calculated as follows:

$$\text{EF} = ([C]_{\text{sample}}/[Al_2O_3]_{\text{sample}})/([C]_{\text{UCC}}/[Al_2O_3]_{\text{UCC}}), \tag{1}$$

where [C] and [Al_2O_3] are the concentrations of a given element and Al_2O_3, respectively. Al_2O_3 is a major element group, and a dominant constituent of many minerals such as plagioclase. Therefore by using it to normalize the EFs, we can effectively remove the effects of quartz, calcareous materials, and organic matter [8]. The systematically high EFs of Li, As, Sb, Hg, and Bi, and low EFs of Nb, Sn, and Ta for all samples, as well as CaO and Sr in stream sediments, are fundamental geochemical signatures of upper crust materials in an island arc setting [37] (Figure 6a,b). The high abundance ratios of CaO and Sr in marine sediments indicate the presence of biogenic calcareous materials, such as shell fragments and foraminifera tests. In the Seto Inland Sea, the UCC normalized patterns of fine sands are similar to those of stream sediments (Figure 6a). Abundance ratios of most elements in marine sediments decrease with increasing grain size. CaO, K_2O, Rb, Sr, and Ba are especially dominant in coarse sediments, whereas silt and clay have higher abundance ratios for Li, Cr, Co, Ni, Zn, Mo, Cd, Cs, Hg, Pb, and Bi. In the Pacific Ocean, the UCC normalized patterns of fine sands and silt-clay samples are most similar to those of stream sediments (Figure 6b). The variations in EF of most elements with particle size of the marine sediments in the Pacific Ocean are likely smaller than those for samples in the Seto Inland Sea, except for MnO, As, Ce, Eu, and Tl. Coarse sediments have relatively high abundance ratios for MnO, Fe_2O_3, Co, and As, and low abundance ratios for Li, Cu, Nb, Sn, Cs, light REEs, Ta, Hg, Tl, Th, and U. The silty and clayey sediments are particularly enriched in Cr, Ni, Cu, Cd, Sb, Hg, and Bi.

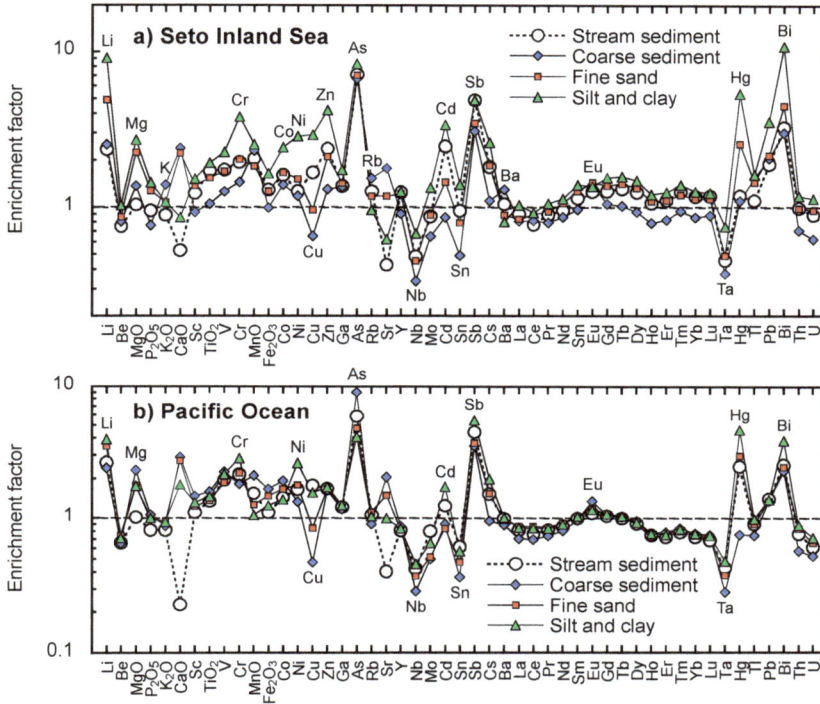

Figure 6. Enrichment factors of 49 elements and oxides in stream and marine sediments of (**a**) Seto Inland Sea and (**b**) Pacific Ocean, compared to the upper continental crust [36] (details in the text).

5.3. Discrimination of Coastal Sea Sediments Using Cluster Analysis

Coarse sediments have UCC normalized patterns that notably differ from those of terrestrial materials (Figure 6a,b), which implies that they have different origins to the modern terrestrial materials or are supplied by a particular lithology. Thus, we refine the identification of probable source materials for marine sediments using cluster analysis. The classification of marine sediment samples is as presented in Table 3. Stream sediments grouped according to the seven rock types summarized in Table A2 are used for comparison. The distance between data points was calculated as a Euclidian distance, requiring the geochemical data from marine and stream sediments to be standardized or transformed appropriately. This is because the compositional data constitute "closed" data that sum to a constant [38,39]. To mitigate biased analysis arising from data closure, there is a requirement for data transformation to yield data with normal distributions suitable for parametric statistical analysis. Additive log-ratio, the centered log-ratio, and the isometric log-ratio transformation of compositional data have been proposed [38,39]. We used a log transformation of the EF data to generate logarithmic

EFs for each sample, which were used as the input data for cluster analysis. The logarithmic EF corresponds to the additive log-ratio transformation of the compositional data [38,39]:

$$\log EF = alr(C)_{sample} - alr(C)_{UCC},$$ (2)

where $alr(C)$ stands for additive log-ratio transformation ($\log \{[C]/[Al_2O_3]\}$). We have confirmed that the EF is effective in eliminating the effect of scale and units, and in removing the dilution effect, although Al_2O_3 concentration is empirically used for denominator in Equation (1). The logarithmic EF also incorporates data normality and is thus amenable to multivariate parametric tests such as cluster analysis.

Figure 7 presents dendrograms that show the distances between two points. The distances between clusters were obtained by Ward's method [40]. We calculated two kinds of dendrograms using different combinations of elements. Figure 7a uses a dataset that excludes sea salt (Na_2O), biogenic carbonate materials (CaO and Sr), and heavy metals related to mining and anthropogenic activities (As, Mo, Cd, Sn, Sb, Hg, Pb, and Bi), which do not reflect the geochemistry of the parent lithology. Figure 7b uses the EFs of immobile elements (Sc, TiO_2, Cr, Nb, Ta, Y, Ln, and Th), which are not strongly influenced by weathering processes. The obtained dendrograms are mutually similar (Figure 7a,b). Coarse sediments from the Pacific Ocean plot in the same group as stream sediments derived from mafic volcanic and metamorphic rocks in Figure 7a. However, Figure 7b shows that all marine sediments in the Pacific Ocean cluster together with coarse sediments from Seto Inland Sea and accretionary complexes. Fine sand and silt-clay in the Pacific Ocean and coarse sands in the Seto Inland Sea are always grouped together with stream sediments derived from sedimentary rocks of accretionary complexes. These results for fine sands and silt-clay are reasonable, because such sedimentary rocks outcrop extensively across the southern part of Shikoku Island (Figure 2), which supplies the highest sediment yield to the Pacific Ocean (Table 1). In contrast, the fine sands and silt-clay in the Seto Inland Sea plot with those stream sediments derived from granitic and felsic volcanic rocks. This result is also expected, due to the extensive outcropping of these rock types across the Chugoku region, which supplies the highest sediment yield to the Seto Inland Sea (Table 1).

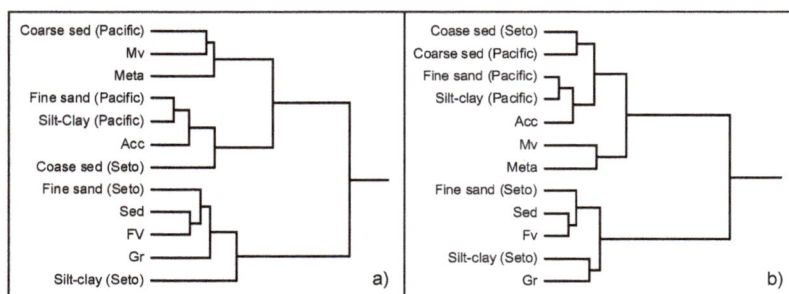

Figure 7. (**a**) Cluster dendrogram obtained using a dataset of 39 elements (excepting CaO, Sr, and heavy metals); (**b**) Cluster dendrogram obtained using a dataset of 21 elements (immobile elements). Sed: sediments and sedimentary rocks; Acc: sedimentary rocks of accretionary complexes; Mv: mafic volcanic rocks; Fv: felsic volcanic rocks; Gr: granitic rocks; Meta: metamorphic rocks.

5.4. Discrepancies in the Geochemistry of Coarse Marine Sediments and Stream Sediments

5.4.1. Fractionation of Mineralogical Compositions in Coarse Sediments by a Strong Tidal Current

Cluster analysis suggests that the coarse sediments in the Seto Inland Sea are related to stream sediments sourced from sedimentary rocks of accretionary complexes. However, granitic and felsic volcanic rocks are geographically more suitable sources for the coarse sediments, in analogy with the fine sands. To reveal their geochemical features in more detail, Figure 8a shows the abundance patterns of elements in coarse sediments collected from the narrow sea channels (Aki-Nada, Bisan-Seto-Nada and around Awaji Island), Iyo-Nada, and Bungo Channel, normalized to the stream sediments in the Seto Inland Sea region.

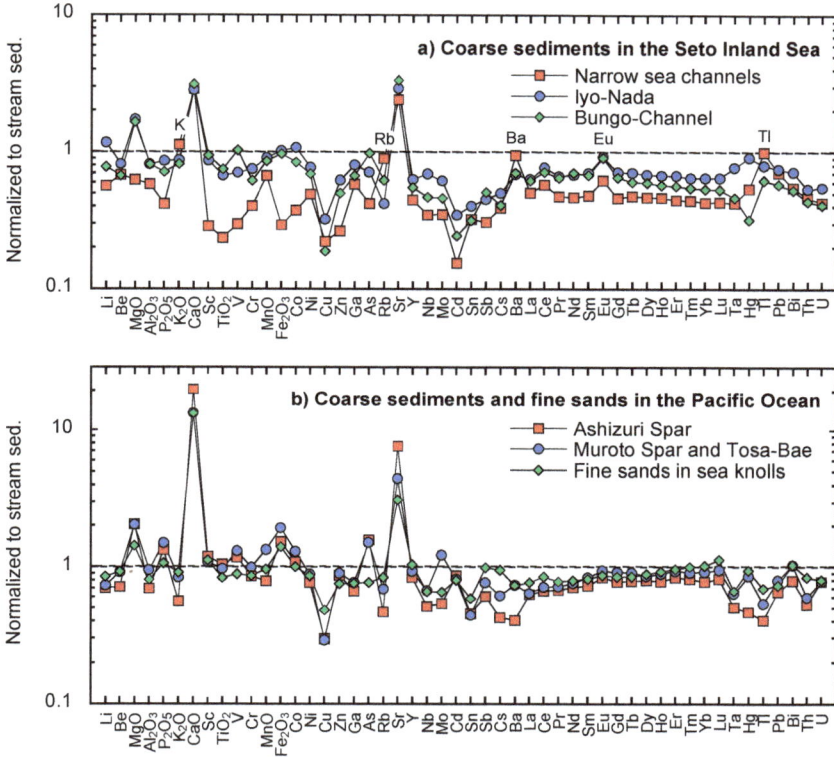

Figure 8. (**a**) Profiles of chemical compositions of coarse sediments in the narrow sea channels (around Aki-Nada, Bisan-Seto-Nada, and Awaji Island), Iyo-Nada, and the Bungo Channel, normalized to stream sediments in the Seto Inland Sea region; (**b**) Profiles of chemical compositions of coarse sediments in Ashizuri Spar, Muroto Spar and Tosa Bae, and fine sands in sea knolls normalized to stream sediments in the Pacific Ocean region.

Coarse sands in the narrow sea channels are particularly rich in K_2O, CaO, Rb, Sr, Ba, and Tl, and have a distinctive positive Eu anomaly, but are extremely depleted in many elements, such as Sc and TiO_2 (Figure 8a). These features are indicative of the selective accumulation of biogenic calcareous materials, quartz, and alkali-feldspars. Quartz and alkali-feldspar are abundant in granitic and felsic volcanic rocks, and are highly resistant to weathering processes, so are preserved as coarse particles e.g., [41]. In contrast, the coarse sediments in Iyo-Nada and the Bungo Channel are relatively abundant in MgO, P_2O_5, CaO, Sc, TiO_2, V, Fe_2O_3, Co, and Zn in addition to the elements mentioned above (Figure 8a). These additional elements are found abundantly in mafic minerals such as biotite, hornblende, and pyroxene, as well as accessary minerals such as magnetite and ilmenite. We therefore assume that the fine particles of mafic and accessary minerals are swept away from the narrow sea channels by a strong tidal current, but this does not occur in Iyo-Nada and the Bungo Channel. In contrast, the coarse grains in all areas are rarely moved over any long distances by water power.

5.4.2. Denudation of the Basement Rocks of the Topographic Highs

Cluster analysis provides different results for coarse sediments from the Pacific Ocean (compare Figure 7a,b). This is because only three elements (Sc, TiO$_2$, and Cr) abundant in sediments derived from mafic volcanic rocks and metamorphic rocks (dominantly greenschist) are used in the cluster analysis of the immobile element dataset (Figure 7b). Thus, cluster analysis using immobile elements may be poorly sensitive to such rock types. Consequently, we conclude that Figure 7a would provide a more plausible result: mafic volcanic and metamorphic rocks are the probable sources of coarse sediments in the Pacific Ocean. However, these rock types are located relatively distantly from the coarse sediment deposition sites (Figure 2). The high EFs of MnO, Fe$_2$O$_3$, Co, and As in coarse sediments from the Pacific Ocean (Figure 6b) can be explained by Fe hydroxide and Mn dioxide coatings on coarse particles and gravel (for example, T-Fe$_2$O$_3$ and As; Figure 4). In this regard, relict quartz particles coated with Fe hydroxides are found in the Bungo Channel [26]. However, this explanation does not fully account for the enrichment of MgO, P$_2$O$_5$, Sc, TiO$_2$, and V. According to Okamura [19], the topographic highs in this region are formed from basement rocks of Miocene–Pliocene siltstone. These highs typically have a low sedimentation rate and are located in a region of erosion, due to the strong bottom flow associated with the Kuroshio Current [24]. The spar and knoll sediments are therefore assumed to have formed by denudation of the basement rocks (siltstone), which have a different lithology to the southern part of Shikoku Island. Therefore, we assume that basement rocks of siltstone might be influenced by mafic volcanic activity at Miocene-Pliocene age. Figure 8b shows the abundance of elements in coarse or gravelly spar sediments and knoll sediments (mainly fine sands), normalized to the stream sediments in the Pacific Ocean region. Despite the fact that Ashizuri Spar is more than 100 km away from Muroto Spar and Tosa-Bae, the abundance patterns of the coarse spar sediments are very similar. Furthermore, the fine sands collected around Ashizuri Knoll and Muroto Knoll have similar abundance patterns to those of the coarse spar sediments. As such, the unexpected cluster analysis result for coarse sediments in this region can be explained by mineral segregation under a strong tidal current unique to the Seto Inland Sea, and by a denudation of old basement rocks (Miocene–Pliocene siltstone) in the Pacific Ocean.

5.5. The Transfer of Silty and Clayey Sediments Influenced by Periodic Currents, Tidal Currents, and a Water Mass Boundary

5.5.1. Mahalanobis' Generalized Distances for Finding Outliers

Silty and clayey sediments supplied from rivers take a long time to reach the sea floor. Therefore, their spatial distribution is a result of conveyance and dispersion by coastal sea currents. Ohta and Imai [7] and Ohta et al. [8] simply focused on the spatial distribution patterns of high Cr and Ni concentrations in silty sediments, which are indicative of transport of fine particles from land to sea or marine environments. Their spatial distribution can be easily discriminated because they are extremely enriched in silty marine sediments derived from ultramafic rocks in the adjacent terrestrial area. Ohta et al. [42] tried to visualize the particle transfer process using the spatial distribution patterns of the other elements. However, many elements are enriched in finer sediments due to the particle size effect, and so in order to highlight the particle transfer process, it is useful to detect outliers using the relationships between elemental concentrations and median diameter or mud content [42].

Unfortunately, in this study, data of median diameter or mud content are available for only half of the samples. ANOVA analyses suggest that Cs, Hg and Th concentrations in marine sediments are determined only by particle size and are not sensitive at all to either regional differences or interaction effects ($\eta^2 < 0.01$) (Table 4). The particle size effect of Cs has the largest η^2 score of the three elements, suggesting that the particle size effect has the maximum effect on Cs concentrations. Although concentrations of 13 elements, including Cr, Cu, and La, in marine sediments are also controlled dominantly by particle size effects, they are further influenced weakly by regional difference effects and interaction effects ($0.01 \leq \eta^2 < 0.06$) (Table 4). In addition, Hg, Cu, and Cd concentrations in marine sediments might be influenced by metalliferous deposits and anthropogenic activity in the adjacent terrestrial area. For these reasons, we conveniently use Cs concentration as a proxy for the median diameter or mud content.

Figure 9 shows the relationships between Cr, Cu, Zn, Cd, Sb, and La concentrations and that of Cs. The concentrations of all six elements correlate positively and linearly with Cs, and it can be assumed that those samples plotting outside the linear trends are subject to influencing factors in addition to the particle size effect. Ohta et al. [42] detected outliers from a scatter diagram according to their best judgment. In this study, the outliers appearing in the scatter diagram are objectively obtained using Mahalanobis' generalized distance (D) [43]. This method presupposes that the data follow a normal distribution similar to ANOVA. Thus, the data transformation is given in Table A3. The obtained D values gradually increase from the inner part of the data cluster to the outer part, giving an onion-like structure (Figure 9). D values of $D \leq 1.117$, $D \leq 1.665$, and $D \leq 2.146$ indicate that 50%, 75%, and 90% of the total data points are included, respectively, and data points are defined as an outlier herein if $D > 2.146$, which includes the upper and lower 5% of multivariate outliers. The outliers were determined from all datasets, so silt and clay classified as outliers are expressed as cross symbols; while coarse sediments and fine sands classified as outliers are expressed as plus symbols in Figure 4.

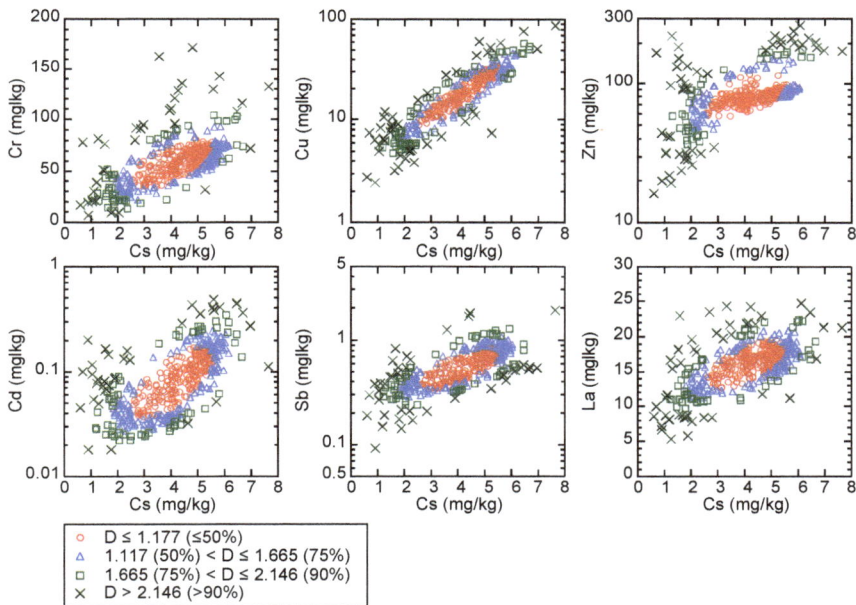

Figure 9. Relationship between elemental concentrations and Cs concentration. Mahalanobis' generalized distances (D) increase from the central part to the outer part of the plot in an onion-like structure. Values of $D \leq 1.117$, $D \leq 1.665$, and $D \leq 2.146$ indicate that 50%, 75%, and 90% of the total data points are included, respectively.

5.5.2. Transport of Elements in Silt Derived from the Parent Lithology

Figure 4 shows that the locations of the outliers calculated for Cr (cross symbols) correspond with a NNW-SSE distribution of silty-clayey sediments with high Cr concentrations in the Kii Channel. This distribution of Cr outliers indicates that silty particles that originated from ultramafic rocks are supplied through the Yoshino, Naka, and Kino Rivers to the Kii Channel, and are then widely dispersed by a periodic current, as Ohta and Imai [7] noted. Hiuchi-Nada also contains abundant silty sediments, however the distribution of outliers (corresponding to high concentration areas) for MgO, Cr, Co, and Ni is restricted to the vicinity of Shikoku Island (shown by Cr in Figure 4). This can be explained by the fact that the terrestrial sediment discharge yield to Hiuchi-Nada is minimal compared to the output of the Yoshino River, Naka River and Kino River systems (Table 1).

Y, Nb, Ln, Ta and Th are also enriched in the silty sediments of Hiuchi-Nada, especially in the eastern part (La in Figure 4). The differing distributions of La and Cr outliers suggest that Y, Nb, Ln, Ta and Th have a different origin from MgO, Cr, Co, and Ni. Minakawa et al. [44] reported that rare earth minerals, including allanite and monazite, occur within pegmatite veins of the Ryoke granitic rocks, which outcrop in northwest Shikoku Island, on the small islands in Aki-Nada and Bisan-Seto-Nada, and are partly exposed on the seafloor [12]. Yanagi [10] suggested that fine particles transported by periodic currents accumulate in eastern Hiuchi-Nada where the amplitude of the tidal current is small and a counterclockwise circulation occurs. Consequently, we assume that silt enriched in rare earth minerals has been swept toward the eastern side of Hiuchi-Nada by periodic currents and accumulated there over a long period of time.

In Tosa Bay, tongue-shaped distribution patterns of high MgO, Cr, Co, Ni, Nb, Y, Ln, Ta, and Th concentrations and their outliers are apparent (shown by Cr and La in Figure 4). MgO, Cr, Co, and Ni originated from ultramafic rocks, while the other elements are derived from greenstones wedged between the Sambagawa metamorphic rocks and accretionary complexes e.g., [45]. Ikehara [25] and Hoshino [46] reported that silty and clayey sediments are selectively deposited at the boundary between coastal waters with low salinity (due to River input) and the outer sea with higher salinity. This is supported by the fact that the water discharges from the Niyodo and Shimanto Rivers into Tosa Bay are among the largest in the region (Table A1). Thus, on the basis of this evidence, sedimentation at the boundary between water masses can explain this distinctive distribution [7]. In addition, the countercurrent of the Kuroshio Current, which flows in a counterclockwise direction on the continental shelf (Figure 1b), may also contribute to the transport of fine particles enriched in the elements mentioned above [21].

5.5.3. Transfer of Materials Related to Mineral Deposits and Anthropogenic Activity

A number of silty sediment samples in Hiuchi-Nada are abundant in Cu (40.6–86.6 mg/kg) and are classified as outliers (Figure 4); these reflect the mining activity at the nearby Besshi Cu mines over the past 240 years [47]. Similarly, the outlier observed within the Kii Channel sediments is related to activity at the Cu mine in the watershed of the Yoshino River. In addition, one outlier for Sb (1.89 mg/kg) is found near to the shore in southern Hiuchi-Nada. This enrichment of Sb suggests contamination from the Ichinose Sb mine, which is located close to the Besshi mine.

Osaka Bay is adjacent to Osaka City, which is the second largest city in Japan and has a long history of industrial development. Ohta et al. [32] found that Cu, Zn, Cd, Sn, Sb, Hg, and Pb are significantly enriched in stream sediments collected from the urban areas of Osaka Plain. Similarly, the silty sediments in Osaka Bay are highly enriched in Cr, Cu, Zn, As, Mo, Cd, Sn, Hg, Pb, and Bi (as shown by Cr, Cu, Zn, and As in Figure 4). Samples with the top 1% of concentrations for these elements, which are mostly classified as outliers, are found in Osaka Bay, and the area of high concentrations near the mouth of the Yodo and Yamato Rivers is clearly due to the discharge of contaminated sediments. Hoshika and Shiozawa [27], Hoshika et al. [48], and Manabe [49] reported the same results, showing large accumulations of Cr, Cu, Zn, Cd, Pb, and Hg in Osaka Bay.

A number of outliers of Cu, Zn, Cd, Sn, Pb, and Bi are also found in Harima-Nada, Hiroshima Bay and Suo-Nada (shown by Cu, Zn, and Cd in Figure 4). Their enrichment may be a result of the nearby metalliferous deposits and contamination by the coastal industrial zones [49,50]. In particular, Ikuno mine is considered a possible source for the elements abundant in Harima-Nada because it is one of the largest hydrothermal-type polymetallic mines, bearing Cu, Zn, As, Cd, Sb, Sn, Pb, and Bi [31] (Figure 2). However, the spatial distributions of the outliers are different for each enriched element. A likely explanation for this is that the rivers flowing through the mineralized zones are small and have much smaller sediment discharges. Alternatively, the elements may have been released in the water during the oxidation of sulfide ores, and have subsequently absorbed into the surface sediment of the coastal seas [51].

5.5.4. Geochemical Features of Silt and Clay in the Deep Sea Basins of the Pacific Ocean

Silt and clay deposited in deep Pacific basins are typically enriched in MnO, V, Ni, Co, Mo, Sb, Pb, and Bi, as a result of early diagenetic processes in an oxic environment [7,42]. In contrast, Cu, Sb, Cd, Hg, and U are enriched in deep-sea sediments during early diagenesis under anoxic conditions, as they are immobile in reducing waters [52]. In addition, Cu and Hg are taken up by living organisms, transported to sediments, and bound within residual organic matter in bottom sediments [53–55]. Results of this study reveal that Cu, Sb, Cd, Hg, and U are abundant in the silt and clay of the Tosa Basin, Hyuga Basin, and the Muroto Trough (shown by Cu, Sb and Cd in Figure 4). This suggests that these deep-sea basins are under anoxic conditions. However, outliers for these elements are scarce in the Pacific Ocean, in contrast to the Seto Inland Sea (Figure 4). Ikehara [25] reported that these deep-sea basins are hemi-pelagic and are partly associated with turbidites. Therefore, we assume that the enrichment of heavy metals is lost when gradual sedimentation is interrupted by turbidity flows, and that early diagenetic processes and the deposition of organic matter begin again after the turbidity flow event [56]. As a result, no significant enrichment of heavy metals is visible in the Pacific Ocean.

6. Conclusions

The composition and transfer processes of coastal sea sediments are analyzed using a comprehensive geochemical database of many elements from the Seto Inland Sea and the Pacific Ocean in western Japan. The geochemical features of fine sands and silty-clayey sediments reflect those of stream sediments in the adjacent terrestrial areas. This pattern is comparable to the potential sediment yield from terrestrial areas and its supply to the adjacent marine environment. The spatial distribution of anomalous elemental concentrations in silty and clayey sediments suggests horizontal dispersion by a periodic current, as well as precipitation processes occurring at the boundary between water masses. However, the geochemical features of coarse sands and gravels differ significantly from those of stream sediments in the adjacent terrestrial areas, and fractionation of the mineralogical composition due to strong tidal currents, in addition to the denudation of Miocene–Pliocene basement rocks of the underwater topographic highs are possible reason for this inconsistency in geochemistry. As such, comprehensive geochemical maps of both the land and the sea are demonstrated as being useful for determining the source of fine marine sediments, as well as for tracking the transfer and deposition of silty-clayey sediments.

Acknowledgments: The authors extend special thanks to Takashi Okai, Yutaka Kanai, Masumi Mikoshiba, and Ran Kubota (Geological Survey of Japan, AIST) for their useful suggestions, which have helped to improve the manuscript. We are grateful to the Japan Oceanographic Data Center (JODC) for providing data files.

Author Contributions: Noboru Imai conceived the study and organized the sampling on land; Ken Ikehara organized the sampling in the coastal seas; Yoshiko Tachibana measured elemental concentrations; and Atsuyuki Ohta analyzed the data and wrote the paper.

Conflicts of Interest: The authors declare no conflict of interest.

Appendix A

The sediment yield, drainage basin area, and river water discharge of 24 major rivers in the study area are summarized in Table A1. The abbreviation of each river name is shown in Figure 1b. The sediment yield of each river system is calculated using the drainage basin area and the average rate of erosion in each river basin after Akimoto et al. [16].

Table A1. Sediment yield and river water discharge data in each river system of the study area.

River System	Sediment Yield ($\times 10^3$ m^3/Year)	River Water Discharge [1] ($\times 10^6$ m^3/Year)	Discharged Area	Relative Rate to Total Sediment Yield
Shikoku region				
Yoshino (Ys)	870	3202	Kii Channel	11%
Naka (Na)	510	1628	Kii Channel	7%
Toki (To)	83	(34) [2]	Ibi-Seto-Nada	1%
Shigenobu (Sg)	97	46	Iyo-Nada	1%
Hiji (H)	215	1111	Iyo-Nada	3%
Mononobe (M)	458	777	Tosa Bay	6%
Niyodo (Ni)	534	3101	Tosa Bay	7%
Shimanto (Sm)	699	4712	Tosa Bay	9%
Chugoku region				
Yoshii (Yi)	290	(1125) [2]	Ibi-Seto-Nada	4%
Asahi (Ah)	238	1492	Ibi-Seto-Nada	3%
Takahari (Ta)	358	1605	Ibi-Seto-Nada	5%
Ashido (Ad)	142	187	Bingo-Nada	2%
Ohta (Ot)	(238) [3]	-	Hiroshima Bay	3%
Oze (Oz)	48	255	Hiroshima Bay	1%
Sanami (Sn)	95	261	Suo-Nada	1%
Kinki region				
Ibo (I)	174	676	Harima-Nada	2%
Kako (Ka)	247	918	Harima-Nada	3%
Yodo (Yd)	1310	5376	Osaka Bay	17%
Yamato (Ym)	123	(426) [2]	Osaka Bay	2%
Kino (Ki)	377	1184	Kii Channel	5%
Kyushu region				
Yamakuni (Yk)	99	418	Suo-Nada	1%
Ohita (Oi)	150	498	Beppu Bay	2%
Ohno (On)	240	1616	Beppu Bay	3%
Banjo (B)	89	392	Bungo Channel	1%

Notes: [1] River water discharge data in 2000, which are taken from [57]; [2] River water discharge data in 1994 [57]; [3] Average rate of sediment yield of Ohta River is missing [16], so the value is assumed to be the same as that of Oze River.

Appendix B

Ohta et al. [29] and Ohta et al. [32] assumed that when a specific rock type outcrops over more than half of the drainage basin area, it is representative of the surface rock types in the watershed and is the dominant control of elemental abundances in stream sediments. In the study area, there are seven rock types exposed in more than half of the river basin areas: sediments and sedimentary rocks (Sed); sedimentary rocks of accretionary complexes (Acc); granitic rocks (Gr); felsic volcanic rocks (Fv); mafic volcanic rocks (Mv); debris and pyroclastic rocks (Phy); and high-pressure type metamorphic rocks (Meta). We further separated those stream sediments collected in urban areas (Sed_u), from sediments and sedimentary rocks (Sed), using a land use map, because anthropogenic contamination significantly enhances certain elements in the sediments, including Cu, Zn and Cd [32]. Table A2 presents the median elemental concentrations of stream sediments classified by the above rock types. In this, "other" represents those cases in which no specific rock type occupies more than half of the catchment area, and is not used in the cluster analysis. The Sed_u and Phy categories are also excluded from cluster analysis because contaminated sediments reflect only local conditions, and debris and pyroclastic rocks represent a mixture of mafic and felsic eruptive products.

Table A2. Median elemental concentrations in stream sediments classified according to the dominant rock type in the drainage basin.

Element	Sed	Sed_u	Acc	Gr	Mv	Fv	Phy	Meta	Other
	$n = 70$	$n = 11$	$n = 157$	$n = 71$	$n = 41$	$n = 61$	$n = 13$	$n = 47$	$n = 105$
wt %									
Na_2O	2.06	2.21	1.74	2.89	2.44	1.82	2.34	2.26	2.08
MgO	1.34	1.63	1.54	1.03	3.17	1.12	3.80	3.57	1.98
Al_2O_3	11.54	9.89	11.47	11.43	12.14	10.35	13.43	13.35	11.05
P_2O_5	0.109	0.157	0.093	0.098	0.156	0.085	0.181	0.148	0.103
K_2O	2.21	2.22	2.28	2.57	1.13	2.65	1.34	1.74	2.18
CaO	1.21	1.41	0.52	1.58	3.60	1.07	4.33	2.87	1.80
TiO_2	0.56	0.65	0.49	0.50	1.10	0.48	1.17	0.73	0.68
MnO	0.088	0.083	0.091	0.116	0.152	0.117	0.187	0.168	0.126
$T\text{-}Fe_2O_3$	3.94	4.59	4.13	3.82	7.00	4.10	9.42	7.21	5.19
mg/kg									
Li	35.2	32.5	42.7	30.4	25.5	38.4	24.2	37.3	37.6
Be	1.54	1.70	1.66	2.34	1.20	1.95	1.32	1.38	1.64
Sc	8.33	8.23	8.55	7.41	14.1	7.80	20.3	18.3	11.0
V	69	59	82	52	155	62	237	155	97
Cr	47.1	110	55.6	24.1	53.7	32.8	38.6	169	61.1
Co	9.05	11.9	10.4	7.1	18.5	8.6	22.7	23.8	13.1
Ni	18.7	38.4	26.2	9.3	17.2	13.0	12.6	59.5	25.3
Cu	29.5	67.7	35.9	25.2	27.0	27.7	31.9	62.3	35.2
Zn	108	194	101	114	130	137	131	122	130
Ga	16.3	16.8	16.6	19.2	17.9	17.8	18.8	17.9	17.1
As	6	7	8	7	4	22	8	8	13
Rb	102	109	102	121	39	153	49.8	73	104
Sr	110	97	90	100	254	91	301	144	122
Y	18.0	20.3	11.9	24.7	17.2	20.5	23.2	24.8	19.5
Nb	8.77	10.05	8.12	10.8	11.4	8.82	8.82	8.29	8.63
Mo	0.77	1.79	1.00	0.91	1.03	0.98	1.46	0.94	1.07
Cd	0.13	0.20	0.11	0.17	0.15	0.26	0.16	0.18	0.20
Sn	3.62	7.72	2.95	4.60	2.36	4.30	2.17	2.81	4.04
Sb	0.80	1.21	0.75	0.38	0.52	0.92	0.48	0.85	0.92
Cs	4.84	3.89	4.97	4.44	2.39	7.65	3.24	5.06	5.59
Ba	469	491	448	401	352	456	402	351	414
La	20.7	30.9	19.4	24.1	16.4	20.9	18.2	20.0	19.4
Ce	38.9	57.1	37.3	41.2	29.2	35.3	37.7	39.4	34.0
Pr	4.63	6.74	4.42	5.42	3.87	4.66	4.94	4.80	4.55
Nd	17.8	26.2	16.9	22.0	15.9	17.8	20.6	19.6	17.9
Sm	3.72	5.44	3.21	4.59	3.27	3.62	4.46	4.27	3.73
Eu	0.78	0.88	0.65	0.73	0.94	0.71	1.24	1.07	0.79
Gd	3.40	4.37	2.78	4.31	3.09	3.33	4.40	4.28	3.48
Tb	0.59	0.72	0.45	0.75	0.54	0.59	0.75	0.77	0.59
Dy	3.10	3.61	2.14	4.00	2.70	3.05	3.87	4.05	3.10
Ho	0.60	0.67	0.41	0.79	0.52	0.61	0.75	0.81	0.60
Er	1.77	1.89	1.14	2.34	1.56	1.81	2.24	2.29	1.80
Tm	0.28	0.29	0.18	0.40	0.25	0.30	0.36	0.36	0.29
Yb	1.75	1.74	1.11	2.58	1.53	1.89	2.23	1.97	1.82
Lu	0.26	0.27	0.16	0.39	0.24	0.28	0.33	0.27	0.26
Ta	0.77	0.87	0.73	1.11	0.80	0.70	0.72	0.71	0.68
Hg	0.06	0.12	0.12	0.03	0.06	0.04	0.06	0.07	0.04
Tl	0.63	0.62	0.58	0.69	0.33	0.85	0.47	0.44	0.64
Pb	26.0	37.1	24.2	32.4	22.6	44.1	19.0	20.6	29.6
Bi	0.23	0.32	0.28	0.40	0.17	0.52	0.25	0.25	0.39
Th	8.21	12.8	6.62	14.4	4.37	8.46	5.78	6.74	7.33
U	1.83	2.27	1.51	3.31	1.03	2.41	1.61	1.21	1.79

Appendix C

Statistical analysis assumes that the data follow a normal distribution. However, it is known that the elemental concentrations of minor elements follow a lognormal distribution, such that their logarithmic data follow a normal distribution. The Shapiro–Wilk test is applied to examine the data distribution for each data set at a confidence level of $p = 0.01$ [58,59]. When the estimated probability is greater than 0.01, we conclude that the data follow a normal distribution. However, in this study, the estimated probabilities were all smaller than 0.01, and the geochemical data are concluded to follow neither a normal nor a lognormal distribution. This is because statistical tests with a large amount of data, such as the 463 sets used in this study, are highly sensitive to very small differences. Thus, we determine a probable distribution using the W-value of the Shapiro–Wilk statistic; the closer this value is to 1.0, the more closely it follows a normal distribution. As an alternative, Ohta et al. [8] and Ohta et al. [32] proposed that the data with skewness close to zero, which indicates a symmetrical distribution, is used for ANOVA. Both the Shapiro–Wilk statistics and skewness are summarized in Table A3. For example, K_2O data are unchanged for ANOVA because the Shapiro–Wilk statistics and the skewness of unaltered K_2O data are closer to 1.0 and zero, respectively, than those of log-transformed K_2O data (Table A3). On the basis of this analysis, data for 24 elements within the marine sediments were log-transformed, with the remaining data left unaltered for ANOVA (Table A3).

Table A4 shows the two-way layout of ANOVA in this study. The table shows that the sample numbers are not equal for each condition, this is known as an "unbalanced sample size". Some methods have been proposed to calculate the sum of squares for an unbalanced two-way ANOVA [59,60]. Ohta et al. [8] suggested that a two-way ANOVA using the Type II method [60,61] is preferable for these data because our interest lies in understanding the effects of the two main factors.

Table A3. The Shapiro–Wilk test applied to the unchanged and log-transformed data and their skewness.

Element	W-Value of Shapiro–Wilk Statistics		Skewness		Data Transformation
	Unchanged	Log-Transformed	Unchanged	Log-Transformed	
MgO	0.884	0.872	1.3	−1.3	Unchanged
Al_2O_3	0.971	0.869	−0.48	−2.0	Unchanged
P_2O_5	0.168	0.870	19	0.96	Log-transformed
K_2O	0.949	0.820	−0.78	−2.3	Unchanged
CaO	0.777	0.980	2.5	−0.34	Log-transformed
TiO_2	0.722	0.873	3.3	−0.80	Log-transformed
MnO	0.731	0.946	2.9	0.72	Log-transformed
$T-Fe_2O_3$	0.667	0.839	3.4	−0.32	Log-transformed
Li	0.924	0.975	1.1	−0.46	Log-transformed
Be	0.957	0.854	−0.60	−2.0	Unchanged
Sc	0.769	0.874	2.6	−0.84	Log-transformed
V	0.580	0.914	6.1	−0.19	Log-transformed
Cr	0.952	0.923	0.78	−1.2	Unchanged
Co	0.804	0.905	2.5	−0.55	Log-transformed
Ni	0.883	0.930	1.9	−1.0	Log-transformed
Cu	0.922	0.965	1.1	−0.44	Log-transformed
Zn	0.734	0.896	2.7	−0.02	Log-transformed
Ga	0.919	0.781	−1.2	−2.6	Unchanged
As	0.615	0.899	4.7	−0.98	Log-transformed
Rb	0.985	0.897	−0.44	−1.5	Unchanged
Sr	0.647	0.964	3.7	0.69	Log-transformed
Y	0.977	0.991	0.66	−0.29	Log-transformed
Nb	0.990	0.945	0.01	−0.93	Unchanged
Mo	0.378	0.961	11.22	0.85	Log-transformed
Cd	0.835	0.978	1.8	0.14	Log-transformed
Sn	0.519	0.936	6.8	0.53	Log-transformed
Sb	0.931	0.980	1.1	−0.27	Log-transformed
Cs	0.961	0.885	−0.42	−1.3	Unchanged
Ba	0.771	0.886	4.4	−1.2	Log-transformed
La	0.981	0.917	−0.45	−1.3	Unchanged
Ce	0.981	0.924	−0.38	−1.2	Unchanged
Pr	0.973	0.897	−0.61	−1.5	Unchanged
Nd	0.969	0.888	−0.64	−1.6	Unchanged
Sm	0.975	0.905	−0.44	−1.4	Unchanged
Eu	0.987	0.971	0.20	−0.64	Unchanged
Gd	0.983	0.942	0.01	−1.0	Unchanged
Tb	0.982	0.971	0.36	−0.62	Unchanged
Dy	0.981	0.982	0.48	−0.46	Log-transformed
Ho	0.980	0.989	0.56	−0.34	Log-transformed
Er	0.982	0.988	0.53	−0.36	Log-transformed
Tm	0.983	0.989	0.50	−0.36	Log-transformed
Yb	0.985	0.986	0.45	−0.42	Log-transformed
Lu	0.987	0.983	0.40	−0.50	Unchanged
Ta	0.986	0.948	0.10	−0.87	Unchanged
Hg	0.187	0.819	19	−1.6	Log-transformed
Tl	0.964	0.876	0.20	−1.6	Unchanged
Pb	0.637	0.893	4.3	1.2	Log-transformed
Bi	0.753	0.975	2.5	0.31	Log-transformed
Th	0.804	0.890	3.2	−1.0	Log-transformed
U	0.886	0.988	1.8	0.10	Log-transformed

Table A4. The two-way layout for ANOVA.

	Coarse Sediment	Fine Sand	Silt and Clay
Seto Inland Sea	$n = 26$	$n = 15$	$n = 47$
Pacific Ocean	$n = 75$	$n = 107$	$n = 193$

References

1. Imai, N.; Terashima, S.; Ohta, A.; Mikoshiba, M.; Okai, T.; Tachibana, Y.; Togashi, S.; Matsuhisa, Y.; Kanai, Y.; Kamioka, H. *Geochemical Map of Sea and Land of Japan*; Geological Survey of Japan, AIST: Tsukuba, Japan, 2010.
2. Webb, J.S.; Thornton, I.; Thompson, M.; Howarth, R.J.; Lowenstein, P.L. *The Wolfson Geochemical Atlas of England and Wales*; Clarendon Press: Oxford, UK, 1978.
3. Salminen, R.; Batista, M.J.; Bidovec, M.; Demetriades, A.; De Vivo, B.; De Vos, W.; Duris, M.; Gilucis, A.; Gregorauskiene, V.; Halamic, J.; et al. *Geochemical Atlas of Europe. Part 1—Background Information, Methodology and Maps*; Geological Survey of Finland: Espoo, Finland, 2005.
4. Caritat, P.D.; Cooper, M. *National Geochemical Survey of Australia: Data Quality Assessment*; Geoscience Australia Record 2011/21; p. 478. Available online: http://www.ga.gov.au/corporate_data/71971/Rec2011_021_Vol1.pdf (accessed on 17 November 2014).
5. Smith, D.B.; Cannon, W.F.; Woodruff, L.G.; Solano, F.; Ellefsen, K.J. *Geochemical and Mineralogical Maps for Soils of the Conterminous United States*; U.S. Geological Survey Open-File Report 2014-1082. Available online: http://pubs.usgs.gov/of/2014/1082/pdf/ofr2014--1082.pdf (accessed on 19 May 2014).
6. Xie, X.J.; Chen, H.X. Global geochemical mapping and its implementation in the Asia-Pacific region. *Appl. Geochem.* **2001**, *16*, 1309–1321.
7. Ohta, A.; Imai, N. Comprehensive survey of multi-elements in coastal sea and stream sediments in the island arc region of Japan: Mass transfer from terrestrial to marine environments. In *Advanced Topics in Mass Transfer*; El-Amin, M., Ed.; InTech: Rijeka, Croatia, 2011; pp. 373–398.
8. Ohta, A.; Imai, N.; Terashima, S.; Tachibana, Y.; Ikehara, K.; Okai, T.; Ujiie-Mikoshiba, M.; Kubota, R. Elemental distribution of coastal sea and stream sediments in the island-arc region of Japan and mass transfer processes from terrestrial to marine environments. *Appl. Geochem.* **2007**, *22*, 2872–2891. [CrossRef]
9. Kuwashiro, I. Submarine topography of Japanese Inlandsea Setonaikai. *Geogr. Rev. Jpn.* **1959**, *32*, 24–35. [CrossRef]
10. Yanagi, T. Currents and sediment transport in the Seto Inland Sea, Japan. In *Residual Currents and Long-Term Transport*; Cheng, R.T., Ed.; Springer: Beilin, Germany, 1990; Volume 38, pp. 348–355.
11. Sano, S.; Inouchi, Y.; Kanai, Y.; Maruoka, N. Geochemical properties of surface sediments in Seto Inland Sea part 1: Aki-nada surface sediments. *Mem. Fac. Educ. Ehime Univ. Ser. III Nat. Sci.* **2000**, *20*, 1–9.
12. Inouchi, Y. Distribution of bottom sediments in the Seto Inland Sea—The influence of tidal currents on the distribution of bottom sediments. *J. Geol. Soc. Jpn.* **1982**, *88*, 665–681. [CrossRef]
13. Yanagi, T.; Hagita, T.; Saino, T. Episodic outflow of suspended sediments from the Kii channel to the Pacific Ocean in winter. *J. Oceanogr.* **1994**, *50*, 99–108. [CrossRef]
14. Kawana, K.; Tanimoto, T. Turbid bottom water layer and bottom sediment in the Seto Inland Sea. *J. Oceanogr. Soc. Jpn.* **1984**, *40*, 175–183. [CrossRef]
15. Hoshika, A.; Shiozawa, T.; Kawana, K.; Tanimoto, T. Heavy metal pollution in sediment from the Seto Inland Sea, Japan. *Mar. Pollut. Bull.* **1991**, *23*, 101–105. [CrossRef]
16. Akimoto, T.; Kawagoe, S.; Kazama, S. Estimation of sediment yield in Japan by using climate projection model. *Proc. Hydraul. Eng.* **2009**, *53*, 655–660.
17. *Geological Map of Japan, 1:1,000,000*, 3rd ed.; Geological Survey of Japan, AIST: Tsukuba, Japan, 1992.
18. Kato, M. Recent foramininifera in the surface sediments in the inland sea of Japan. *Ann. Sci. Coll. Lib. Arts Kanazawa Univ.* **1995**, *19*, 63–73.
19. Okamura, Y. Geologic structure of the upper continental slope off Shikoku and Quaternary tectonic movement of the outer zone of southwest Japan. *J. Geol. Soc. Jpn.* **1990**, *96*, 223–237. [CrossRef]
20. Fujimoto, M. On the flow types and current stability in Tosa bay and adjacent seas. *Sea Sky* **1987**, *62*, 127–140.

21. Miyata, K.; Sakamoto, H.; Momota, M. Oceanographycal structure in the Tosa bay-1. On the tidal current. *Bull. Nansei Natl. Fish. Res. Inst.* **1980**, *12*, 115–124.

22. Imai, N.; Terashima, S.; Ohta, A.; Mikoshiba, M.; Okai, T.; Tachibana, Y.; Togashi, S.; Matsuhisa, Y.; Kanai, Y.; Kamioka, H.; et al. *Geochemical Map of Japan*, 1st ed.; Geological Survey of Japan, AIST: Tsukuba, Japan, 2004.

23. Mikoshiba, M.U.; Imai, N.; Tachibana, Y. Geochemical mapping in Shikoku, southwest Japan. *Appl. Geochem.* **2011**, *26*, 1549–1568. [CrossRef]

24. Arita, M.; Kinoshita, Y. Sedimentological map of cape Muroto. In *1:200,000 Marine Geology Map Series 37*; Geological Survey of Japan: Tsukuba, Japan, 1990.

25. Ikehara, K. Sedimentological map of Tosa Wan. In *1:200,000 Marine Geology Map Series 34*; Geological Survey of Japan: Tsukuba, Japan, 1988.

26. Ikehara, K. Sedimentological map south of Bungo channel. In *1:200,000 Marine Geology Map Series 51*; Geological Survey of Japan: Tsukuba, Japan, 1999.

27. Hoshika, A.; Shiozawa, T. Heavy metals and accumulation rates of sediments in Osaka bay, the Seto Inland Sea, Japan. *J. Oceanogr. Soc. Jpn.* **1986**, *42*, 39–52. [CrossRef]

28. Imai, N. Multielement analysis of rocks with the use of geological certified reference material by inductively coupled plasma mass spectrometry. *Anal. Sci.* **1990**, *6*, 389–395. [CrossRef]

29. Ohta, A.; Imai, N.; Terashima, S.; Tachibana, Y.; Ikehara, K.; Nakajima, T. Geochemical mapping in Hokuriku, Japan: Influence of surface geology, mineral occurrences and mass movement from terrestrial to marine environments. *Appl. Geochem.* **2004**, *19*, 1453–1469. [CrossRef]

30. Reimann, C. Geochemical mapping: Technique or art? *Geochem. Explor. Environ. Anal.* **2005**, *5*, 359–370. [CrossRef]

31. Ohta, A.; Imai, N.; Terashima, S.; Tachibana, Y. Investigation of elemental behaviors in Chugoku region of Japan based on geochemical map utilizing stream sediments. *Chikyukagaku (Geochemistry)* **2004**, *38*, 203–222.

32. Ohta, A.; Imai, N.; Terashima, S.; Tachibana, Y. Application of multi-element statistical analysis for regional geochemical mapping in central Japan. *Appl. Geochem.* **2005**, *20*, 1017–1037. [CrossRef]

33. Fritz, C.O.; Morris, P.E.; Richler, J.J. Effect size estimates: Current use, calculations, and interpretation. *J. Exp. Psychol. Gen.* **2012**, *141*, 2–18. [CrossRef] [PubMed]

34. Richardson, J.T. Eta squared and partial eta squared as measures of effect size in educational research. *Educ. Res. Rev.* **2011**, *6*, 135–147. [CrossRef]

35. Miller, J.C.; Miller, J.N. *Statistics for Analytical Chemistry*, 6th ed.; Pearson Education Canada: Don Mills, ON, USA, 2010.

36. Taylor, S.R.; McLennan, S.M. The geochemical evolution of the continental crust. *Rev. Geophys.* **1995**, *33*, 241–265. [CrossRef]

37. Togashi, S.; Imai, N.; Okuyama-Kusunose, Y.; Tanaka, T.; Okai, T.; Koma, T.; Murata, Y. Young upper crustal chemical composition of the orogenic Japan arc. *Geochem. Geophys. Geosyst.* **2000**, *1*, 2000GC000083. [CrossRef]

38. Aitchison, J. The statistical analysis of compositional data. *J. R. Stat. Soc. Ser. B* **1982**, *44*, 139–177.

39. Pawlowsky-Glahn, V.; Egozcue, J.J. Compositional data and their analysis: An introduction. In *Compositional Data Analysis in the Geosciences: From Theory to Practice*; Buccianti, A., Mateu-Figueras, G.H., Pawlowsky-Glahn, V., Eds.; Geological Society, Publishing House: Bath, UK, 2006; Volume 264, pp. 1–10.

40. Ward, J.H., Jr. Hierarchical grouping to optimize an objective function. *J. Am. Stat. Assoc.* **1963**, *58*, 236–244. [CrossRef]

41. Goldich, S.S. A study in rock-weathering. *J. Geol.* **1938**, *46*, 17–58. [CrossRef]

42. Ohta, A.; Imai, N.; Terashima, S.; Tachibana, Y.; Ikehara, K.; Katayama, H. Elemental distribution of surface sediments around Oki trough including adjacent terrestrial area: Strong impact of Japan Sea Proper Water on silty and clayey sediments. *Bull. Geol. Surv. Jpn.* **2015**, *66*, 81–101. [CrossRef]

43. Mahalanobis, P.C. On the generalized distance in statistics. *Proc. Natl. Inst. Sci.* **1936**, *2*, 49–55.

44. Minakawa, T.; Funakoshi, N.; Morioka, H. Chemical properties of allanite from the Ryoke and Hiroshima granite pegmatites in Shikoku, Japan. *Men. Fac. Sci. Ehime Univ.* **2001**, *7*, 1–13.

45. Nozaki, T.; Nakamura, K.; Osawa, H.; Fujinaga, K.; Kato, Y. Geochemical features and tectonic setting of greenstones from Kunimiyama, northern Chichibu belt, central Shikoku, Japan. *Resour. Geol.* **2005**, *55*, 301–310. [CrossRef]

46. Hoshino, M. On the muddy sediments of the continental shelf adjacent to Japan. *J. Geol. Soc. Jpn.* **1952**, *58*, 41–53.

47. Hoshika, A.; Shiozawa, T. Sedimentation rates and heavy metal pollution of sediments in the Seto Inland Sea. Part 3. Hiuchi-nada. *J. Oceanogr. Soc. Jpn.* **1984**, *40*, 334–342. [CrossRef]

48. Hoshika, A.; Tanimoto, T.; Mishima, Y. Sedimentation processes of particulate matter in the Osaka bay. *Oceanogr. Jpn.* **1994**, *3*, 419–425. [CrossRef]

49. Manabe, T. Distribution of sediments contamination in the eastern Seto Inland Sea. *Sea Sky* **1991**, *67*, 1–9.

50. Hoshika, A.; Shiozawa, T. Sedimentation rates and heavy metal pollution of sediments in the Seto Inland Sea. Part 4. Suo-nada. *J. Oceanogr. Soc. Jpn.* **1985**, *41*, 283–290. [CrossRef]

51. Hudson-Edwards, K.A.; Macklin, M.G.; Curtis, C.D.; Vaughan, D.J. Processes of formation and distribution of Pb-, Zn-, Cd-, and Cu-bearing minerals in the Tyne basin, northeast England: Implications for metal-contaminated river systems. *Environ. Sci. Technol.* **1996**, *30*, 72–80. [CrossRef]

52. Rosenthal, Y.; Lam, P.; Boyle, E.A.; Thomson, J. Authigenic cadmium enrichments in suboxic sediments: Precipitation and postdepositional mobility. *Earth Planet. Sci. Lett.* **1995**, *132*, 99–111. [CrossRef]

53. Mason, R.P.; Fitzgerald, W.F.; Morel, F.M.M. The biogeochemical cycling of elemental mercury: Anthropogenic influences. *Geochim. Cosmochim. Acta* **1994**, *58*, 3191–3198. [CrossRef]

54. Shaw, T.J.; Gieskes, J.M.; Jahnke, R.A. Early diagenesis in differing depositional environments: The response of transition metals in pore water. *Geochim. Cosmochim. Acta* **1990**, *54*, 1233–1246. [CrossRef]

55. Klinkhammer, G.P. Early diagenesis in sediments from the eastern equatorial pacific. II. Pore water metal results. *Earth Planet. Sci. Lett.* **1980**, *49*, 81–101. [CrossRef]

56. Wilson, T.R.S.; Thomson, J.; Colley, S.; Hydes, D.J.; Higgs, N.C.; Sorensen, J. Early organic diagenesis: The significance of progressive subsurface oxidation fronts in pelagic sediments. *Geochim. Cosmochim. Acta* **1985**, *49*, 811–822. [CrossRef]

57. Ministry of Land, Infrastructure, Transport and Tourism. Available online: http://www.mlit.go.jp/river/toukei_chousa/ (accessed on 19 November 2015).

58. Shapiro, S.S.; Wilk, M.B. An analysis of variance test for normality (complete samples). *Biometrika* **1965**, *52*, 591–611. [CrossRef]

59. Royston, P. A remark on algorithm as181: The w-test for normality. *Appl. Stat. J. R. Stat. Soc.* **1995**, *44*, 547–551.

60. Shaw, R.G.; Mitchell-Olds, T. ANOVA for unbalanced data: An overview. *Ecology* **1993**, *74*, 1638–1645. [CrossRef]

61. Littell, R.C.; Stroup, W.W.; Freund, R. *SAS for Linear Models*; SAS Institute: Cary, NC, USA, 2002.

water

MDPI

Article

Dynamics of Suspended Sediments during a Dry Season and Their Consequences on Metal Transportation in a Coral Reef Lagoon Impacted by Mining Activities, New Caledonia

Jean-Michel Fernandez [1,*], Jean-Dominique Meunier [2], Sylvain Ouillon [3], Benjamin Moreton [1], Pascal Douillet [4] and Olivier Grauby [5]

[1] Analytical and Environmetal Laboratory (AEL),
    Institut de Recherche pour le Développement (IRD)-Nouméa, BP A5,
    98800 Nouméa, Nouvelle-Calédonie; bmoreton@ael-environnement.nc
[2] Aix Marseille University, Centre National de la Recherche Scientifique (CNRS), IRD,
    Unité Mixte de Recherche CEREGE, 13545 Aix en Provence CEDEX 05, France; meunier@cerege.fr
[3] Unité Mixte de Recherche LEGOS, Université de Toulouse, IRD, Centre National d'Etudes Spatiales (CNES),
    CNRS, Université Paul Sabatier (UPS), 14 avenue Edouard Belin, 31400 Toulouse, France;
    sylvain.ouillon@legos.obs-mip.fr
[4] Mediterranean Institute of Oceanography (MIO), Unité Mixte de Recherche 110, IRD, CNRS/Institut
    National des Sciences de l'Univers (INSU), Aix Marseille Université, Université de Toulon,
    13284 Marseille, France; pascal.douillet@ird.fr
[5] CINaM-CNRS-Aix-Marseille Université, Campus de Luminy Case 913, 13288 Marseille CEDEX 9, France;
    grauby@cinam.univ-mrs.fr
*   Correspondence: jmfernandez@ael-environnement.nc; Tel.: +687-76-84-30

Academic Editor: Roger A. Falconer
Received: 30 March 2017; Accepted: 8 May 2017; Published: 10 May 2017

Abstract: Coral reef lagoons of New Caledonia form the second longest barrier reef in the world. The island of New Caledonia is also one of the main producers of nickel (Ni) worldwide. Therefore, understanding the fate of metals in its lagoon waters generated from mining production is essential to improving the management of the mining activities and to preserve the ecosystems. In this paper, the vertical fluxes of suspended particulate matter (SPM) and metals were quantified in three bays during a dry season. The vertical particulate flux (on average 37.70 ± 14.60 g·m^2·d^{-1}) showed fractions rich in fine particles. In Boulari Bay (moderately impacted by the mining activities), fluxes were mostly influenced by winds and SPM loads. In the highly impacted bay of St Vincent and in the weakly impacted bay of Dumbéa, tide cycles clearly constrained the SPM and metal dynamics. Metals were associated with clay and iron minerals transported by rivers and lagoonal minerals, such as carbonates, and possibly neoformed clay as suggested by an unusually Ni-rich serpentine. Particle aggregation phenomena led to a reduction in the metal concentrations in the SPM, as identified by the decline in the metal distribution constants (K_d).

Keywords: suspended sediment; sediment transport; lagoon; geochemistry; Ni mining; sediment trap; hydrodynamics; New Caledonia; dry season

1. Introduction

The mining industry in New Caledonia is one of the most important environmental concerns for the tropical island lagoonal ecosystem [1–9].

With about 85% endemism among terrestrial plants, 24 different species of mangroves among the 70 listed throughout the world, about 2800 species of molluscs and the second longest barrier reef in the world [10–15], New Caledonia's ecosystems and biodiversity are highly sensitive to anthropogenic activities (e.g., [16–23] for its lagoons). Since the beginning of mining in New Caledonia, more than 160×10^6 tonnes of ore have been extracted. This has led to the mobilization and transport of approximately 300 million m^3 of soil material (laterites). Opencast Ni mines have enhanced soil erosion and transportation of sediments and metals into the lagoon [19,24–27] with several consequences on the lagoonal ecosystems, including increased sedimentation rates; decreased light penetration and dissolved oxygen levels; and an increased metal contamination in the food web which may affect humans [23,28,29]. The Ni mining industry has flourished for over 25 years, and New Caledonia will remain one of the major worldwide Ni producers for the foreseeable future, with global Ni reserves estimated at around 20–25% [30]. As a consequence, environmental studies are required in order to mitigate the effects of 400–500 km^2 of deforestation specifically related to the mining industry in New Caledonia.

Numerous studies of the south-western lagoon of New Caledonia have been conducted investigating hydrodynamics, sediment transport, sedimentation dynamics, metal fluxes, accumulation zones, and particle sources [23,27,31–37]. In complement to these works, this paper aims at characterizing the suspended sediment mineralogy and geochemistry (including metals) in three bays; analysing the relationships between their composition and the mining activities; determining how hydrodynamics forced by wind regimes affect the transportation of particulate metals bounded to the lateritic Suspended Particulate Matter (SPM) into the lagoon. Three contrasting bays in the south west lagoon, where hydrodynamics modelling has been carried out [38–40] were selected: Boulari Bay, Dumbéa Bay and Saint Vincent Bay. Samples were collected during a dry season in order to limit the influence of riverine inputs which could affect the understanding of hydrodynamic regimes, during distinct wind regimes (trade wind and west-breezes) and two neap/spring tide cycles.

2. Study Area

New Caledonia is located at the southern end of the Melanesian Arc, near the Tropic of Capricorn. In New Caledonia, mining activities are almost exclusively conducted on the main island (16,642 km^2, [41]). In its south-western part, host rocks are composed of peridotites and harzburgites incorporating metals like Ni, Co, Cr, Fe and Mn [42,43] in Mg and Fe-minerals. Elements like Pb and Zn are only present in significant quantities in rocks from the northernmost part of the island [44–49]. The weathering of peridotites results in the accumulation of transition metals in the saprolite (also called "garnierite") and the yellow lateritic layers which are subjected to mining extraction. On the top of the series, the red lateritic layer corresponds to a more advanced weathering state of the peridotites where the structure of the bedrock is no longer visible [50]. Mg and Si are very low and the main constituents are ferric hydroxides more or less widely crystallized in goethite. In the upper part of the profile, the ultimate term of the weathering process is represented by a ferricrete composed mostly of goethite and, in lower proportions, hematite.

The climate of New-Caledonia is dry-tropical [51] with alternating dry and wet seasons. South-East trade winds blow from October to May with a mean speed of 8 m·s^{-1} and from April to September a variable northern wind blows. The temperatures vary moderately between dry and wet seasons.

In the south-west lagoon of New Caledonia, the tide is mixed and mainly semi diurnal [38]. Due to the interaction between the different components, spring tide and neap tide periods alternate during a lunar month. The maximum tidal amplitude is 1.5 m during a spring tide.

Similar to most of the New Caledonian Rivers, the Coulée, Dumbéa and Tontouta Rivers have steep upper courses and much flatter lower courses where deposits of weathered bedrock products accumulate (Figure 1). Due to the tropical climate conditions in New Caledonia, the hydrological regime is of torrential type. During the dry season, sediment loads carried by the rivers are low because of the low energy for erosion and the weak transport capacity [52,53]. Rain events reaching 700 mm and more over a 24-h period lead to intense weathering of the slopes and flushing of large quantities of suspended matter to the lagoon. Baltzer and Trescases [52] reported that during cyclone Brenda in 1968, over 20,000 t of particles were discharged in a single day through the Dumbéa River estuary. The present study focuses on the three above-mentioned estuaries located on the south-west coast of New Caledonia influenced by their respective watersheds (Figure 1).

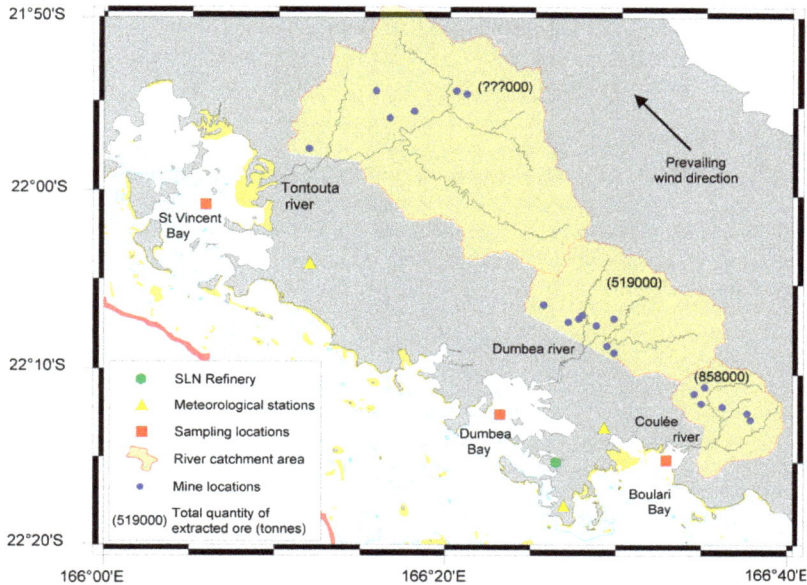

Figure 1. Map location of the study area in the west coast of New-Caledonia: Boulari Bay, influenced by a medium-scale mine activities until 1981; Dumbéa Bay, halted mining activity in order to maintain the water supply of Nouméa (the peninsula between Dumbéa Bay and Boulari Bay); St Vincent Bay, affected by intense opencast mining activities.

2.1. Boulari Bay

The Coulée River catchment (92 km^2) is located almost entirely in the ultrabasic Grand Massif of the South New-Caledonia. An intermediate-scale mining operation was active in the area until 1981, but erosion from the initial prospecting and extraction sites has continued. The present terrigenous inputs delivered to Boulari Bay by the Coulée River result from combined natural and anthropogenic influences [19,25]. The river is extending its delta into the southern part of Boulari Bay where tidal mudflats are being formed.

2.2. Dumbéa Bay

The catchment area of Dumbéa River covers about 233 km^2 and only a few small-scale localized garnierite extractions have occurred in the headwater regions. Similar to other drainage basins near the main city, Nouméa, any mining activities in the area have been forbidden since 1927 to maintain a quality water supply for the city. The sediment load yielded at the river mouth—where a mangrove extends—and delivered to Dumbéa Bay consists of clay, silts and sand, and the effects of mining activities have been limited [25,26].

2.3. St Vincent Bay

The Tontouta River and its tributaries form the largest of the three river catchments (476 km^2) and drain a peridotic hinterland where opencast mining is still intense today. These activities extend to the mountain crests, and on hillslopes, only a few kilometres from the coast. The Tontouta River carries substantial amounts of fine terrigenous material that has resulted in a shallowing of the Saint Vincent Bay, particularly nearshore. The impact of mining activities appears to be stronger than in the Coulée catchment because of the lack of conservation work along the river and tributaries between 1960s and 1980s. This has led to a drastic increase in the sediment load at the river mouth.

3. Materials and Methods

During the dry season between 21 November and 14 December, 2005, SPM was sampled in the three bays every two days, and currents were measured continuously (Figure 1, Table 1). Rainfall rates, and wind direction and velocity were obtained from Météo-France's meteorological stations at Magenta airport, Mont Coffin and Tontouta airport close to Boulari Bay, Dumbéa Bay and Saint Vincent Bay, respectively.

Table 1. Sampling sites, depth and localisation.

Site	Longitude	Latitude	Depth (m)
Boulari Bay	E 166°32.126	S 22°15.355	13.2
Dumbéa Bay	E 166°23.243	S 22°12.291	13.0
St Vincent Bay	E 166°06.635	S 22°00.561	12.8

3.1. SPM Sampling

Three sequential sediment traps (model PPS 4/3; section of 0.05 m^2, Technicap, La Turbie, France) were used for suspended particulate matter (SPM) sampling. They were moored at sites of ~13 m depth downstream of the mouth of the Coulée, Dumbéa and Tontouta Rivers (Figure 1, Table 1). Samples were collected at a frequency of 48 h, 3 m above the seabed. The sediment traps were equipped with twelve 250 mL polypropylene vials filled with 5% formaldehyde-filtered seawater solution before mooring in order to preserve the particles from microbiological activity [54]. After sampling, the samples were placed in a refrigerator at 2–4 °C before analysis. Particles fluxes were calculated using the formula:

$$\text{Flux } (g \cdot m^{-2} \cdot d^{-1}) = \text{sample load (g)}/(\text{Section area } (m^2) * \text{Collecting time (d) per flask)} \tag{1}$$

3.2. Current Measurement

Currents were measured using an Acoustic Doppler Current Profilers (RDI Workhorse Monitor ADCP, Teledyne RD Instruments, Poway, USA 300 kHz, 12 cells, 1-m resolution) placed on the seabed in Dumbéa and St Vincent Bays (Figure 1). In Boulari Bay, local currents were measured using an Acoustic Doppler Velocimeter (Sontek) located 3 m above the seabed. Moored in the vicinity of the 3 sediment traps, the three current meters simultaneously recorded measurements during the SPM sampling period (one month). Unfortunately, due to technical problems, measurements are not available for Dumbéa Bay during the last ten days of the field campaign.

3.3. In Situ Laser Grain Size and CTD Profiling

Turbidity was measured regularly at each station by the use of a Seapoint Optical Backscattering Sensor (Seapoint Sensors, Inc., Brentwood, NH, USA) (λ = 880 nm) connected to a Seabird SBE19 CTD profiler. The Seapoint sensor was factory-adjusted for a consistent response to Formazin Turbidity Standard measured in Formazin Turbidity Units (FTU). A former calibration showed that, in the south-west lagoon of New Caledonia, turbidity is related to the mass concentration (C) of SPM following [55]:

$$\text{Turbidity (FTU)} = 1.85 \text{ C (mg·L}^{-1})\tag{2}$$

An in situ Laser Scattering and Transmissometry device (LISST-100X; Sequoia Scientific Inc., Bellevue, WA, USA) was used in situ to quantify the SPM and the Particle Size Distribution (PSD). The LISST-100X provides the distribution of particle volume concentrations in 32 size classes logarithmically spaced within the range 1.25–250 μm (e.g., [56]). Jouon et al. [55] gave an extended presentation of its first application in the lagoon of New Caledonia.

Synthetic parameters were defined to characterize the particle distribution: (1) the median diameter (D_{50}) as the diameter of a particle for which the cumulative volumetric distribution reaches 50% of the SPM volume concentration; (2) the Junge parameter (s) characterizing the slope of the particle size distribution (PSD) (e.g., [57,58]): high values correspond to SPM dominated by fine particles or aggregates, while low values correspond to macro-flocs; (3) the percentage of particles with diameter > 60 μm that was shown to be an indicator of the state of aggregation [55].

3.4. Geochemistry

All apparatus was acid soaked (10% nitric acid) for a minimum of five days and rinsed with ultrapure water (Milli-Q), and then stored in acid cleaned plastic bags until needed. While analytical acid grades were used for all cleaning steps, high purity reagents were used for all parts of the analytical procedure.

Seawater samples: Seawater was collected from the three bays using 5L teflon lined Go-Flo™ water samplers (General Oceanics Inc., Miami, FL, USA). The Go-Flo™ water samplers were primed to be open at the site and lowered into the water, rinsed thoroughly and closed using a teflon-coated messenger. Once at the surface, samples were transferred in situ into acid cleaned HDPE bottles and sealed in clean plastic bags. After an on-line filtration at 0.45 μm (Millipore acetate filters, Merck Millipore, Billerica, MA, USA), samples were then preconcentrated and analysed using ICP-OES following the procedure described by Moreton et al. [35]. Only the results for Fe, Mn and Ni, which represent the main elements used to trace watershed lixiviation, are presented in this article.

The accuracy and precision of the analytical results was controlled by assaying a SLEW-3 certified water sample (National Research Council, Canada), to check the preconcentration method. The stability of the ICP-OES was controlled inserting independent standards in the sample series: in our case, one at the beginning and one at the end. The quantification limits (LQ) of the method for the 3 metals, obtained after deduction of blanks, are given in Table 2.

Table 2. Results of the analysis of the reference material SLEW-3 and LQ of the method.

Metal	Reference Material SLEW-3 (μg·L⁻¹)		LQ (μg·L⁻¹)
	Analysed ($n = 1$)	Certified	
Fe	0.32	0.57 ± 0.06	0.068
Mn	1.92	1.61 ± 0.22	0.028
Ni	1.17	1.23 ± 0.07	0.022

Particulate samples: Swimmers were removed from SPM collected at each site with sediment traps by sieving at 40 μm. The formaldehyde solution and salt were removed by rinsing several times and centrifuging. Organic matter and faecal pellets were destroyed using a solution of 30% hydrogen peroxide. The purified sediments were then oven dried at 60 °C for a period of 72 h.

Particulate metals were then dissolved by an alkaline fusion digestion performed using 0.5 g of lithium tetraborate mixed with 100 mg of SPM and heated in a muffle furnace (1100 °C) for 15 min. The resulting amalgam was dissolved into 0.5M HCl, and the metals analysed.

Analysis of 9 elements (Al, Ca, Co, Cr, Fe, Mg, Mn, Ni and Si) in SPM was performed using an inductively coupled plasma optical emission spectrometer (Vista, Varian, Inc., Palo Alto, CA, USA).

The validity of the analysis was verified by assaying a MESS-3 certified sediment sample (National Research Council, Canada). The quantification limits (LQ) of the method for the 9 metals, obtained after deduction of blanks, are given in Table 3.

Table 3. Results of the analysis of the reference material MESS-3. The Quantification Limits of the method were not estimated because of the high levels of concentrations measured in SPM.

Reference Material MESS-3 (mg·kg^{-1}·dw)		
Metal	Analyzed	Certified
Al	90,053	85,900 ± 2300
Ca	13,746	14,700 ± 600
Co	15.2	14.4 ± 2.0
Cr	97	105 ± 4
Fe	37,815	32,400 ± 1200
Mg	16,905	16
Mn	308	324 ± 12
Ni	40.6	46.9 ± 2.2
Si	232,765	270,000 *

Note: * Information value only.

3.5. Kd Calculation

Trace metal mobility in the lagoon water column was quantified through its distribution coefficient (K_d, in mL·g^{-1}), given by the following general formula:

$$K_d = \frac{C_p}{C_w} \tag{3}$$

with C_p = metal concentration in SPM, C_w = dissolved metal concentration in sea water.

3.6. Mineralogy

The mineralogical composition of the suspended sediments was determined using X-ray diffractometry (XRD), and Transmission Electron Microscopy (TEM). XRD analyses were done on slightly ground samples using Philips (PW1050/81) equipment (Philips, Eindhoven, The Netherlands) with a Cu anticathode. TEM observations were carried out on a JEOL-2000 FX microscope (JEOL USA, Inc., Peabody, MA, USA), operating with a beam intensity of 126 mA and an accelerating voltage of 200 kV. Microanalyses were acquired with a Si(Li) detector filled with a UTW and a Brucker Esprit EDS System. Quantitative data were obtained by the method developed by Cliff and Lorimer [59] after calibration of the $k_{x,Si}$ factors (x = Al, Mg . . .) against natural and synthetic layer silicates of known and homogeneous composition.

4. Results

4.1. Rainfall

During the study period (21 November to 14 December 2005), rainfall was low, scarce and irregular over all 3 sites. Only one day of significant rainfall (12 December) was recorded at the meteorological stations at Magenta airport (18 mm) and Mont Coffin (10 mm). At the Tontouta airport station, the maximum rainfall was 6 mm on 9 December. Besides this, only 2 mm were recorded at the 3 stations on the 20 and 21 November and on 26 November. Generally, rainfall at the Tontouta station was systematically lower than at the two other stations.

4.2. Wind

The meteorological stations at Magenta airport (near Boulari Bay) and Mont Coffin (between Dumbéa Bay and Boulari Bay, but representative of Dumbéa Bay conditions) recorded mainly two distinct regimes (Figure 2):

- A typical dominant trade wind regime during the study period, with the direction changing from NE during the night to SE in the day, increasing in strength until the beginning of the afternoon and reaching a maximum of 10 m·s⁻¹ (periods B, D);
- A regime characterized by variable and weaker winds (below 5 m·s⁻¹) (periods A, C and E).

The meteorological station at Tontouta airport (St Vincent Bay) recorded winds that were systematically weaker than those recorded at Magenta airport (Boulari Bay) and Mont Coffin (Dumbéa Bay). In the Saint Vincent Bay, wind speeds were lower than 6 m·s⁻¹ and wind direction was irregular. In this area, trade winds are weakened by relief and coastal thermal breezes.

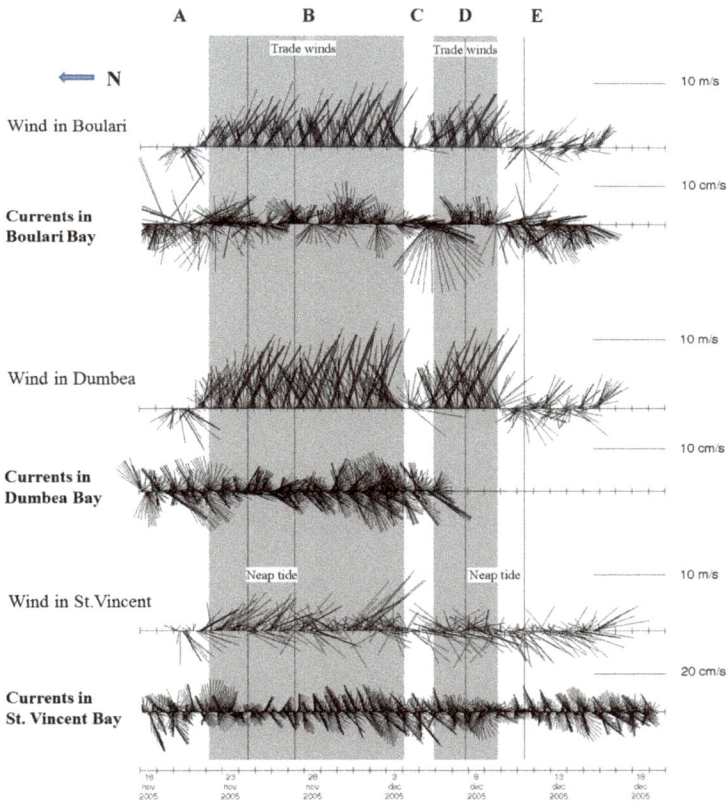

Figure 2. Wind and current speeds and directions for Boulari, Dumbéa and St Vincent bays during the study period (21 November to 14 December 2005).

4.3. Hydrodynamics

During the study period, the amplitudes of the semi-diurnal tides changed from 0.6 to 1.2 m. Neap tide periods are identified in Figures 2 and 3.

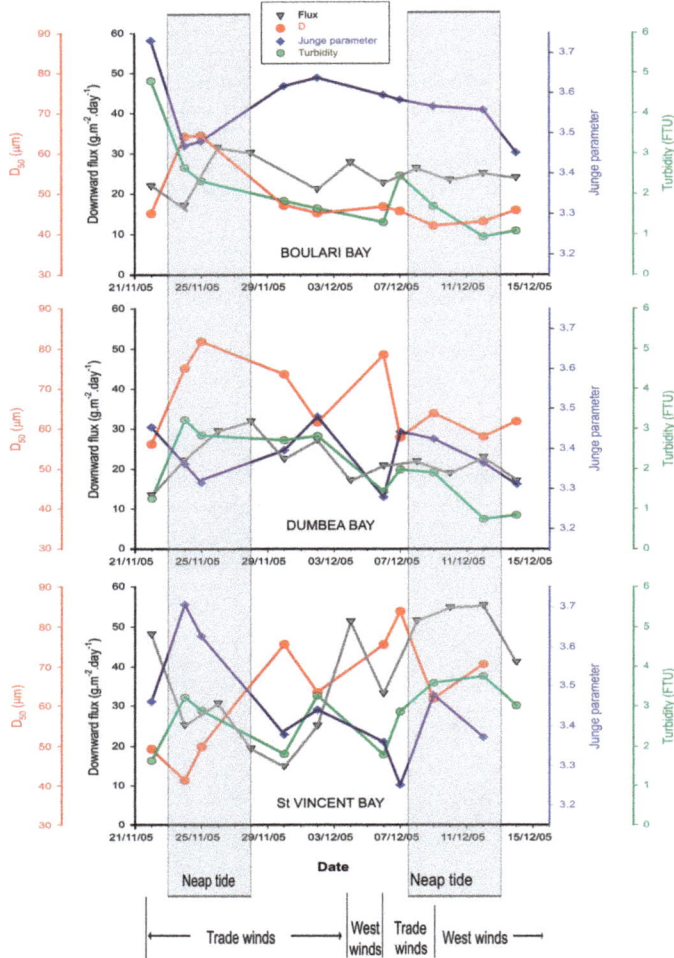

Figure 3. Median diameter (D50), Junge parameter (*s*), flux and turbidity for Boulari, Dumbéa and St Vincent bays over the study period (21 November to 14 December 2005), 3 m above the seabed.

In Boulari Bay, the neap/spring tide cycles had a non-significant influence on the currents measured 3 m above the seabed (Figure 2). In the absence of trade winds (periods C and E), a strong westward flow was observed during several days and may indicate the development of a cyclonic gyre circulation along the isobaths from Mont-Dore (SW of Boulari Bay) towards Nouméa (as described by Fernandez et al. [27]). This gyre results from the conjunction of the propagation of the tide along the coastline of the bay. During trade winds (periods B and D), which blew from an E-NE direction in the Coulée River valley in the morning and from the SE in the afternoon, an anticyclonic gyre generated flows toward the East (as described by Douillet et al. [39]). However, during short periods (28 and 30 November and 2 December), flows in the opposite direction were observed. The present data suggest the strong impact of winds on currents in Boulari Bay, and the formation of a drive out phenomena of waters which were accumulated at the bottom of the bay only when trade winds blow.

In Dumbéa Bay, the strong tidal influence and the weak effects of the wind on the direction and the strength of the currents 3 m above seabed were noticed: flows were the highest during spring tides and lowest during neap tides (Figure 2). Furthermore, currents were similar at the same periods of the tide but with different wind forcings, for example, during a neap tide, with low wind (20 and 21 November) and with a trade wind (3 and 4 December). This suggests that the wind has little influence on the water circulation in an area that is partly protected from the trade winds by the topography.

In St Vincent Bay, the strength of the currents 3 m above the seabed strongly depended on tide cycles (Figure 2); currents rotated 180 degrees during a neap-spring tide cycle, the currents being stronger during spring tides and lower during neap tides. The weakest flows were measured between 24 to 27 November and 24 to 27 December during neap tides. The strongest currents were recorded during spring tides around 3 December. Tides are thus the major factor influencing hydrodynamics in St Vincent Bay. A residual drift of the current to the S-W was observed; however, its value was low.

4.4. SPM Collection

In Boulari Bay, the SPM load collected over 48-h periods ranged between 1.72 and 3.16 g, corresponding to downward fluxes in the range 17.23 to 31.63 $g \cdot m^{-2} \cdot d^{-1}$ (Table 4), with an average value of 24.14 $g \cdot m^{-2} \cdot d^{-1}$ (σ = 4.50 $g \cdot m^{-2} \cdot d^{-1}$). The maximum fluxes were recorded over the period of 4 days from 25 to 28 November (F > 30 $g \cdot m^{-2} \cdot d^{-1}$) and the minimum on 23 to 24 November and 29 to 30 November (F $\approx$ 17 $g\ m^{-2} \cdot d^{-1}$).

Table 4. Mass (g) of suspended particulate matter collected over 48 h in sediment traps in the three sampling bays during the study period (21 November to 14 December 2005).

Date	Boulari Bay	Dumbéa Bay	St Vincent Bay
21–22 November	2.21	1.29	4.83
23–24 November	1.72	2.16	2.53
25–26 November	3.16	2.84	3.07
27–28 November	3.03	3.19	1.96
29–30 November	1.74	2.26	1.51
1–2 December	2.12	2.78	2.52
3–4 December	2.79	1.95	5.15
5–6 December	2.27	2.09	3.32
7–8 December	2.64	2.20	5.17
9–10 December	2.35	1.90	5.50
11–12 December	2.51	2.30	5.55
13–14 December	2.40	1.71	4.12

In Dumbéa Bay, the SPM load was similar to that of Boulari Bay with fluxes between 12.92 and 31.93 $g \cdot m^{-2} \cdot d^{-1}$, and a mean value of 22.24 $g \cdot m^{-2} \cdot d^{-1}$ (σ = 5.20 $g \cdot m^{-2} \cdot d^{-1}$). The maximum fluxes were recorded over the period of 4 days from 25 to 28 November (F $\approx$ 32 $g \cdot m^{-2} \cdot d^{-1}$) and the minimum on 22 November (F $\approx$ 13 $g \cdot m^{-2} \cdot d^{-1}$).

The values in St Vincent Bay were clearly different with a higher average value of 37.70 $g \cdot m^{-2} \cdot d^{-1}$ (σ = 14.60 $g \cdot m^{-2} \cdot d^{-1}$). Variations around the average value were large with frequent loads higher than 50 $g \cdot m^{-2} \cdot d^{-1}$. Except on 21 and 22 November, the first half of the sampling period was characterised by low fluxes (15 < F < 31 $g \cdot m^{-2} \cdot d^{-1}$) and from 4 December onwards, the values were much higher (33 < F < 56 $g \cdot m^{-2} \cdot d^{-1}$).

For each bay, variations in the fluxes, turbidity, D_{50} (mean diameter of SPM from measurements in the range 1.25–250 μm) and Junge parameter (s) during the study period are presented in Figure 3.

4.5. Turbidity, Water Column Structure, and Particle Dynamics

Turbidity was systematically higher in the bay of St Vincent with an average value of 2.8 FTU, compared with average values around 2.0 FTU in the other two bays (Table 5).

In Boulari Bay between 21 and 24 November, the median diameter (D_{50}) increased from 45 to 64 μm while the Junge parameter (s) decreased from 3.75 to 3.48 (Figure 3). While fine particles dominated initially, coarser and medium sizes suddenly increased (24 and 25 November) two days after the beginning of the trade-winds. After 25 November, the decrease of D_{50} was fairly constant up to the end of the study period, when the value reached 42 μm. Conversely, the Junge parameter increased until December 2 ($s = 3.63$), then decreased gradually up to 12 December and then increased sharply until December 14. The downward flux of particles increased just after the peak of coarser particles (26 November) and slightly decreased afterwards (Figure 3). Although generally ranging between 1 and 3 FTU, turbidity showed values around 7.5 FTU in the bottom first four metres above seabed, at the beginning of the study period. Another nepheloid layer, of weaker intensity (4.5 FTU), was also observed around 7 December. The particle grain size distribution was fairly homogeneous throughout the water column during the study period except on the 12 and 14 December when the concentration of fine particles (<7.75 μm) increasing towards the seabed was observed.

Table 5. Main characteristics (mean temperature, salinity and turbidity) recorded in the three sites during the study period (21 November to 14 December, 2005).

Site	Statistics	Temperature (°C)	Salinity (‰)	Turbidity (FTU)
Boulari Bay	Mean ± Std Dev.	26.0 ± 0.6	35.9 ± 0.3	2.1 ± 1.1
	Min.–Max.	25.3–27.5	35.4–36.9	0.5–7.6
Dumbéa Bay	Mean ± Std Dev.	26.5 ± 0.5	36.1 ± 0.1	2.0 ± 0.9
	Min.–Max.	24.8–27.6	35.3–36.6	0.5–5.5
St Vincent Bay	Mean ± Std Dev.	26.9 ± 1.0	36.1 ± 0.1	2.8 ± 0.8
	Min.–Max.	25.3–28.8	35.2–36.5	1.4–7.8

In Dumbéa Bay, the evolution of the median diameter and the Junge parameter were almost inversely related (Figure 3): for example, the two maximum values of D_{50} measured on 25 November (82 μm) and 6 December (79 μm) corresponded with the minimal values of s (3.32 and 3.28), respectively. Turbidity stayed fairly homogeneous throughout the water column, but decreased with time from 3.2 FTU to 0.8 FTU. Only two profiles (11 November and 12 December) showed a clear increase in turbidity towards the bottom. A general decrease in the volumetric concentration, detected between 24 November (10 μL·L^{-1}) and 14 December (5 μL·L^{-1}), combined with a decrease in turbidity, was caused by a reduction in the largest particle-size ranges ($\varnothing > 40.6$ μm). After 9 December, the reduction in the volumetric concentration was due to a decrease in both the smallest ($\varnothing < 7.75$ μm) and largest particle-size ($\varnothing > 40.6$ μm) populations. At the very end of the measurement period, an increase in the amount of fine particles ($\varnothing < 7.75$ μm) was observed at depth, with large particles ($\varnothing > 40.6$ μm) towards the surface.

In St Vincent Bay, the minimum median diameter (42 μm) was measured at the beginning of the study period (24 November) (Figure 3). D_{50} increased gradually until 7 December (85 μm) with an intermediate maximum value observed on 30 November (76 μm). From the 9 to the 12 December, the median diameter increased from 62 to 70 μm. The Junge parameter followed an exact opposite evolution. The maximum value was 3.70 on 24 November and the minimal value was 3.25 on 7 December. From the 9 to the 12 December, the parameter s decreased from 3.46 to 3.38. Turbidity ranged between 1.5 and 3.0 FTU in the first few metres below the surface. From 2 December until the end of the study period, turbidity systematically increased towards the seabed with a consistently higher total volumetric concentration. A significant population of particles above 40 μm and high downward fluxes of particles were observed throughout the study period (Figure 3). High downward fluxes varied similar to turbidity after a short delay.

The values of the median diameter and the Junge parameter strongly differed from one bay to another (Figure 3). D_{50} values ranged between 42 and 65 μm in Boulari Bay (median D_{50} = 48.9 μm), between 55 and 82 in Dumbéa Bay (median D_{50} = 66.8 μm) and between 42 and 84 μm in St Vincent

Bay (median D_{50} = 63.4 µm). The Junge parameter *s* values ranged between 3.45 and 3.73 in Boulari Bay (median *s* = 3.58), 3.28 and 3.48 in Dumbéa Bay (median *s* = 3.38) and 3.25 and 3.70 in St Vincent Bay (median *s* = 3.45). The *s* parameter variation was minimal in Dumbéa Bay and maximal in St Vincent Bay, the variation of D_{50} was minimal in Boulari Bay and maximal in St Vincent Bay.

4.6. Geochemistry

The chemical composition of the seawaters in the 3 bays (Table 6) differed in their dissolved Fe, Mn and Ni concentration. The respective highest and the lowest concentrations were measured in St Vincent Bay and in Boulari Bay. Fe and especially Mn were found at much lower concentration than Ni.

These values are high but reflect the influence of the geology of New Caledonia on the concentrations in dissolved metals. The values show a typical "coast-to-offshore" gradient, with maximum concentrations in bays influenced by rivers and minimum near the coral reef-barrier (Table 7). This evolution is similar to that of the lateritic metals analysed in the sedimentary cover [27].

During the study, Mn concentrations were similar for the 3 bays, with limited variation between 0.33 to 1.24 µg·L^{-1}. The respective Fe and Ni ranges were larger, i.e., 0.23–2.65 µg·L^{-1} and 0.95 to 7.10 µg·L^{-1} in the 3 bays. Indeed, in Boulari Bay, the maximum Fe values were measured on 24 and 30 November, and were slightly higher on 8 December, and seemed to coincide with those of Ni. For the other two bays, the concentrations of theses metals changed differently over time. For example, for Ni, maximum concentrations were measured on 24 November, 2 December and 22 November, for Boulari, Dumbéa and St Vincent bays, respectively.

Table 6. Concentration of the dissolved Fe, Mn and Ni in seawater during the study period from 22 November to 14 December 2005 in the 3 bays.

Sampling Date	Boulari Bay			Dumbea Bay			StVincent Bay		
	Fe (µg·L^{-1})	Mn (µg·L^{-1})	Ni (µg·L^{-1})	Fe (µg·L^{-1})	Mn (µg·L^{-1})	Ni (µg·L^{-1})	Fe (µg·L^{-1})	Mn (µg·L^{-1})	Ni (µg·L^{-1})
22 November 2005	0.58	0.66	2.00	0.73	0.56	2.87	1.09	0.73	7.10
24 November 2005	1.19	0.40	3.92	0.74	0.59	1.30	1.33	0.80	5.49
26 November 2005	0.70	0.48	2.11	1.41	0.46	2.87	1.43	1.07	4.86
28 November 2005	0.24	0.45	1.17	0.64	0.49	3.84	1.59	1.24	5.11
30 November 2005	1.38	0.35	1.60	1.07	0.57	4.52	2.51	0.96	1.95
2 December 2005	0.69	0.33	1.06	1.07	0.55	4.04	1.87	0.96	2.45
6 December 2005	0.32	0.45	1.31	0.77	0.55	3.65	1.43	0.67	3.23
8 December 2005	0.55	0.60	1.70	1.30	0.58	3.35	2.65	1.06	4.92
10 December 2005	0.40	0.53	1.18	0.91	0.57	2.77	1.99	0.88	4.71
12 December 2005	0.30	0.46	0.95	0.23	0.73	1.99	0.71	0.40	6.50
14 December 2005	0.28	0.69	1.24	0.30	0.57	1.78	0.95	0.61	4.87
Min	0.24	0.33	0.95	0.23	0.46	1.30	0.71	0.40	1.95
Max	1.38	0.69	3.92	1.41	0.73	4.52	2.65	1.24	7.10

The analytical results showed that the SPM collected in St Vincent Bay had a distinctly different chemical composition to that of the other two bays (Table 8 and Figure 4). Indeed, in St Vincent Bay's SPM, 7 of the 9 analysed metals (Co, Cr, Fe, Mg, Mn, Ni, Si) were highly enriched, up to one order of magnitude (e.g., Co, Ni or Mn) compared with the two others sites. Only the behaviour of Ca differed, being slightly more concentrated in the SPM collected in Dumbéa Bay, particularly during the second half of the sampling period (Figure 4).

In terms of intra–site variability, the metal and Ca concentrations remained relatively constant in St Vincent Bay's SPM, while they evolved in Dumbéa Bay and particularly in Boulari Bay (Figure 4). This time-variation started with high metal concentrations at the beginning of the study period (21 and 22 November) followed by a strong decrease over a 4-day period (23 to 27 November) before increasing to the highest values at the end of the sampling period. This increasing trend was irregular in Dumbéa Bay, where the highest concentrations were observed from 4 December, while in Boulari

Bay, the increase was slight but continuous to reach the maximum values for all the metals and Ca on 14 December. The mean concentrations increased about 2, 3 and 5 times, in St Vincent, Dumbéa and Boulari bays, respectively.

Table 7. Concentrations of dissolved Fe, Mn and Ni in bays and coral reef barrier (n = 965). Analysis carried out between November 2013 and August 2016 in the frame of marine environmental monitoring along the west coast of New Caledonia (unpublished environmental monitoring data of the surrounding area of the KNS plant). Observed especially in the shallow bays, the extreme Std Deviations demonstrate the high variability of the metal concentration levels generated by the lixiviation of the exploited basins.

Location	Fe ($\mu g \cdot L^{-1}$)	Mn ($\mu g \cdot L^{-1}$)	Ni ($\mu g \cdot L^{-1}$)
Bays (n = 288)	0.241 ± 0.444	4.565 ± 9.802	2.904 ± 4.700
Intermediate (n = 315)	0.123 ± 0.095	0.422 ± 0.659	0.322 ± 0.423
Reef (n = 362)	0.058 ± 0.061	0.103 ± 0.095	0.115 ± 0.100

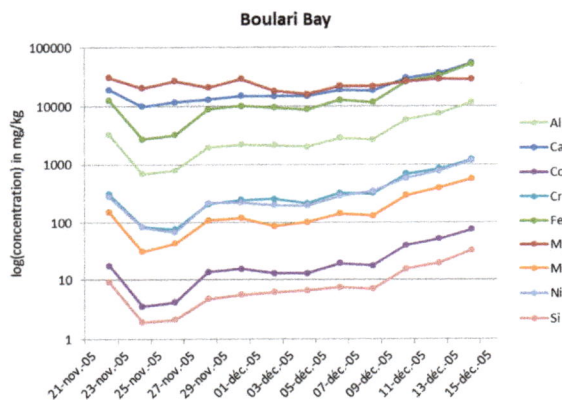

(a)

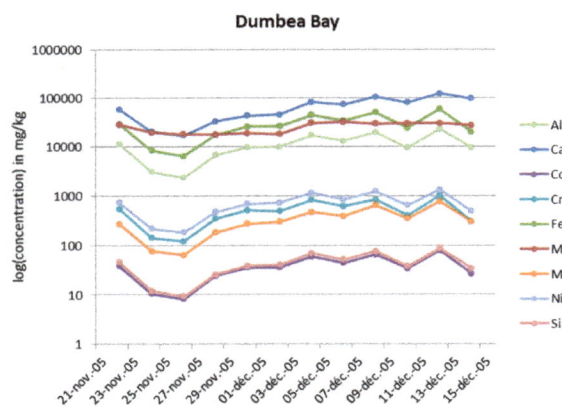

(b)

Figure 4. *Cont.*

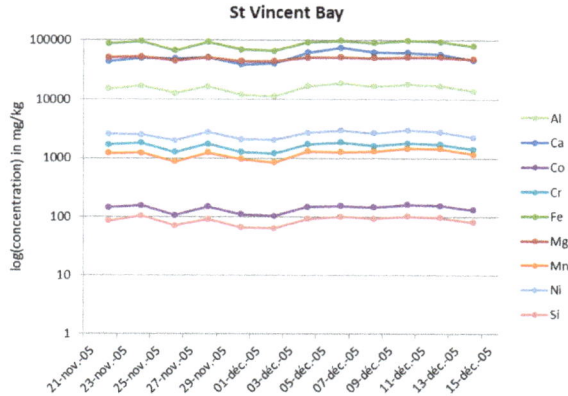

(c)

Figure 4. Time variation of the Ca and the 8 metals analysed in SPM trapped during study period from 21 November to 14 December, 2005 in each sampling site: (**a**) Boulari Bay; (**b**) Dumbéa Bay and (**c**) St Vincent Bay.

Table 8. Minimum and maximum concentrations for the analysed elements in SPM trapped during study period from 21 November to 14 December 2005 in each sampling site.

Concentration (mg kg^{-1})		Al	Ca	Co	Cr	Fe	Mg	Mn	Ni	Si
Boulari bay	Min	1930	12,990	13	208	690	15,874	87	195	4750
	Max	11,520	53,710	76	1209	51,820	29,169	568	1157	31,850
Dumbéa bay	Min	6740	33,060	24	306	17,420	17,672	183	473	25,830
	Max	22,740	119,600	78	1025	58,190	32,064	765	1332	86,590
StVincent bay	Min	11,220	38,360	105	1204	64,670	43,684	844	2033	63,780
	Max	18,440	73,950	164	1856	97,900	51,051	1459	3012	102,940

For each bay, variations in elements concentrations were remarkably correlated ($R^2 > 0.850$) except for (Table 9): (i) Mg, where concentrations showed poor correlations with other SPM metals in Dumbéa Bay (mean $R^2 \approx 0.480$) and no correlation in Boulari Bay (mean $R^2 \approx 0.223$); (ii) Ca and all the other metals in St Vincent Bay (mean $R^2 \approx 0.534$), and, to a lesser extent, in Dumbéa Bay (mean $R^2 \approx 0.710$), and with Mg in Boulari Bay ($R^2 = 0.236$).

Table 9. Correlation coefficients (R^2) for Ca, Mg, Fe, Co, Cr, Mn, Ni, Al and Si concentrations in the suspended matter trapped from 21 November to 14 December in each sampling site: (a) Boulari Bay, (b) Dumbéa Bay and (c) St Vincent Bay.

	Ca	Mg	Fe	Co	Cr	Mn	Ni	Al	Si
	Boulari Bay								
Ca	1	0.236	0.994	0.989	0.987	0.988	0.988	0.995	0.993
Mg		1	0.221	0.223	0.217	0.248	0.209	0.233	0.212
Fe			1	0.998	0.994	0.995	0.993	0.999	0.993
Co				1	0.996	0.998	0.994	0.995	0.986
Cr					1	0.992	0.992	0.993	0.983
Mn						1	0.991	0.994	0.984
Ni							1	0.991	0.982
Al								1	0.996
Si									1

(a)

Table 9. *Cont.*

				Dumbéa Bay					
	Ca	Mg	Fe	Co	Cr	Mn	Ni	Al	Si
Ca	1	0.688	0.702	0.701	0.604	0.826	0.655	0.772	0.734
Mg		1	0.484	0.468	0.419	0.497	0.461	0.514	0.516
Fe			1	0.998	0.985	0.946	0.988	0.992	0.997
Co				1	0.986	0.948	0.990	0.989	0.994
Cr					1	0.888	0.986	0.962	0.977
Mn						1	0.912	0.962	0.945
Ni							1	0.971	0.988
Al								1	0.995
Si									1

(b)

				St Vincent Bay					
	Ca	Mg	Fe	Co	Cr	Mn	Ni	Al	Si
Ca	1	0.438	0.542	0.448	0.483	0.450	0.634	0.725	0.552
Mg		1	0.953	0.958	0.961	0.808	0.813	0.878	0.920
Fe			1	0.986	0.963	0.869	0.905	0.954	0.963
Co				1	0.941	0.903	0.866	0.904	0.949
Cr					1	0.754	0.853	0.922	0.932
Mn						1	0.833	0.796	0.806
Ni							1	0.908	0.803
Al								1	0.948
Si									1

(c)

The geochemical compositions of SPM (Figure 4, Table 8, which differed substantially between the 3 bays, contrast strongly with the average composition of red laterites (Table 10). Comparatively, red laterite showed much lower levels of Ca, Mg and Si (Table 10), being composed principally of Fe with a high proportion of Cr. The concentrations of the other elements in the red laterites were the same order of magnitude as those observed in the SPMs collected in the bays.

Table 10. Mean concentrations ($n = 22$) of the main elements analysed in the red laterite of the south and west coastal ore sites of New Caledonia.

Concentration (mg·kg^{-1})	Al	Ca	Co	Cr	Fe	Mg	Mn	Ni	Si
Mean	26,566	117	437	19,677	586,760	2560	3887	5760	6920
StDev	3170	69	21	1309	46,928	381	221	939	1803

4.7. Mineralogy

The minerals detected in both fractions $\varnothing < 40$ µm and $\varnothing > 40$ µm were not significantly different between each sampled site; the main difference being that clay minerals were enhanced in the finer fraction. The main minerals detected in the suspended sediments of the 3 bays analysed were: carbonates (calcite, Mg-calcite and aragonite), goethite, talc, serpentine and quartz (Figure 5). Smectite was detected in St Vincent and Boulari Bays, but was not significant in Dumbéa Bay. The peaks of talc and serpentine were less intense in Dumbéa Bay than in the two other bays. In the 3 sites, other detected, but less abundant, minerals were: kaolinite, feldspar, pyroxene, and olivine.

TEM observations were mainly focused on the Ni-bearing minerals found in the bays. Carbonates, quartz, feldspar, biogenic silica (diatom tests) detected by TEM did not contain Ni, according to EDS spectra. Ni was detected in goethite and clay minerals (Figure 6 and Table 11).

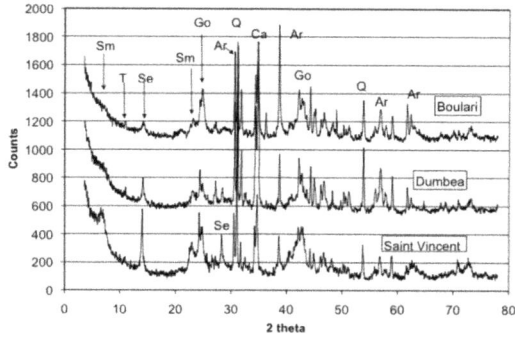

Figure 5. X-ray diffractograms of suspended particulate matter showing the main minerals found in the three study sites (Sm = smectite; T = talc; Se = serpentine; Go = goethite; Ar = aragonite; Q = quartz; Ca = Calcite).

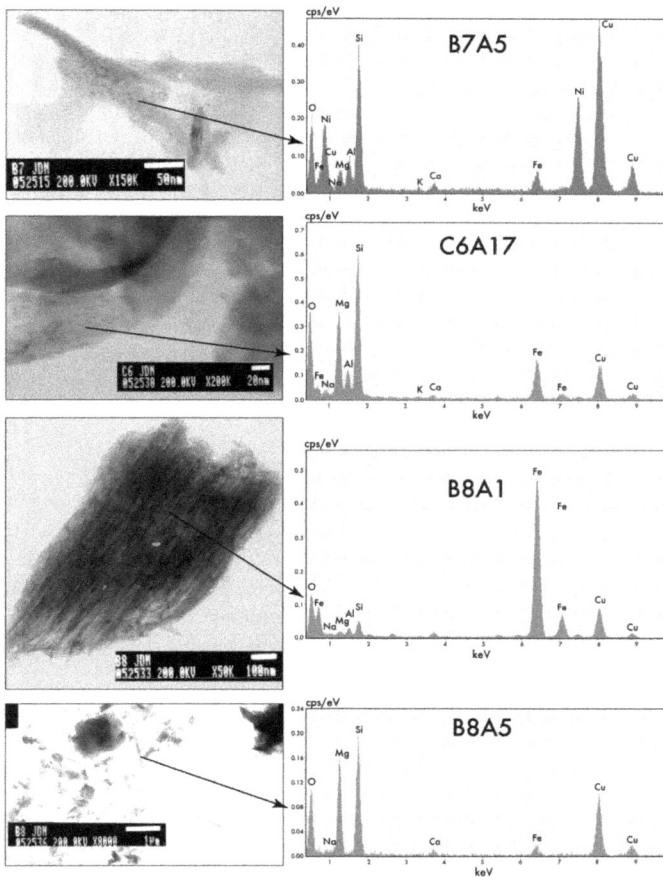

Figure 6. Images and composition determined by transmitted electron microscopy of some particles collected during the study. The chemical formulae are given in Table 11.

Table 11. Chemical formulas of particles (from Figure 6) collected with sequential sediment traps compared to minerals reported in the literature: B7(*A5*) from St Vincent Bay collected on 12 December 2005; B8(*A1*) and B8(*A5*) collected from St Vincent Bay on 6 December, 2005 and C6(*A17*) collected from Dumbéa Bay on 2 December 2005.

Sample	SiO_2 (%)	Al_2O_3	Fe_2O_3	MgO	Cr_2O_3	TiO_2	CaO	Na_2O	K_2O	NiO
B7(*A5*)	42.22	7.11	6.93	4.82	0.21	0.00	1.77	0.17	0.44	36.33
B8(*A1*)	5.77	1.95	86.84	0.58	0.98	0.00	1.75	0.00	0.00	2.14
B8(*A5*)	51.58	0.12	4.95	39.31	0.29	0.00	2.33	0.42	0.25	0.7
C6(*A17*)	48.57	7.4	16.11	25.49	0.49	0.00	0.84	0.13	0.26	0.72
Goethite *	4.86	3.62	88.53	1.69	0.47	0.00	0.00	0.00	0.00	0.83
Lizardite **	42.20	0.15	2.57	35.00	0.00	0.00	0.00	0.00	0.00	4.50
Antigorite *	49.84	0.26	2.04	46.65	0.64	0.00	0.13	0.00	0.00	0.45
Talc *	66.39	0.00	0.00	32.98	0.00	0.00	0.63	0.00	0.00	0.00
CryptoNont *	51.58	8.42	24.21	12.63	0.00	0.00	0.00	0.00	0.00	3.16
CryptoSapo *	50.53	10.53	13.68	25.26	0.00	0.00	0.00	0.00	0.00	0.00
Nontronite *	55.67	4.26	33.02	3.65	0.23	0.00	0.26	0.00	0.00	2.91
Smectite *	55.59	3.87	33.96	6.58	0.00	0.00	0.00	0.00	0.00	0.00

Notes: * Trescases [50]; ** Manceau et Calas [60]; 0.00 = below detection limit or undetermined.

5. Discussion

5.1. Impact of Mining Activities on the Suspended Sediment Composition

The high proportions in Mg and Si content measured in SPM correspond to the geochemical signature of the exploited saprolitic layers, with Mg and Si concentrations being strongly correlated ($R^2 = 0.920$). These enrichments result from weathering phenomena occuring in the upper layers, which lead to the formation of laterites [49,50]. Moreover, the Mg concentrations measured in SPM cannot have a predominantly marine origin (aragonite) since the correlation coefficients between Ca and Mg are not significant, except for Dumbéa Bay where biological activity seems more important than in the other two bays. Studies of sedimentary records [19,26] demonstrate the effects of the weathering mechanisms on the marine environment in terms of SPM composition.

The highest Mg and Si concentrations were measured in St Vincent Bay, which is supplied with SPM from the active mining of the La Tontouta basin. There are few differences between Dumbéa and Boulari bays.

5.1.1. Boulari Bay

The strong correlation obtained between all the major and metal elements ($R^2 > 0.982$), except for Mg, is probably the consequence of the erosion of former mining sites, which have been abandoned for more than 30 years. Indeed, all the correlated elements are present in both the metals-bearing garnierite and the exploitable laterites as the result of the weathering of the ultra-mafic rocks. As for Mg (with Si and Ca), this element is subject to a preferential leaching [50,61], and consequently, the concentration of Mg decreases in the upper lateritic non exploitable layers that are washed away by surface runoff into the lagoon.

Concerning Mg, XRD analysis showed that Mg-bearing minerals may be carbonates or clay minerals. The lack of Ca/Mg correlation and the relatively similar concentrations of these two minerals suggested that Mg is mainly bound to an Mg clay mineral devoid of metals such as talc (Table 9). The good correlation between Ca and metal may be surprising because these elements are not the main metal-bearing minerals transported from the soils (Table 6). Two reasons may explain this correlation: (i) co-precipitation of dissolved metals with coral reef $CaCO_3$ [26,62], suggesting these could have been formed from inputs of SPM from former mines from the beginning of the 20th century until the late 1970s; (ii) Ca is also present in the metal-bearing iron hydroxides and clay minerals (Table 11 and Figure 6).

The high content of metals in SPM collected at the beginning of the sampling period (21 to 22 November) can be correlated with the presence of a large amount of fine particles as suggested by both the Junge parameter ($s = 3.7$) and the high turbidity (4.7 FTU) (Figure 3). In the days that followed, the increase in the mean diameter (D_{50}) and the decrease in the Junge parameter s demonstrated that a fast physical and chemical aggregation occurred from 24 to 26 November. This aggregation, probably with organic matter, was accompanied by a significant solid dilution in the terrigenous metal concentrations in the SPM (Figure 4). This reduction in the metal concentration was highlighted by the sharp decrease in the distribution coefficients (K_d) of the lateritic metal nickel (Figure 7). Later, aggregates became finer (Figure 3) with a higher specific surface area, and relatively stable concentrations in metals (28 November to 7 December, Figure 7). From 8 December onwards, the flux of trapped SPM was fairly constant; however, a drastic increase in metal concentrations was observed. These results suggest that sedimentation resulted mostly from settling of small particles ($\varnothing < 10$ μm).

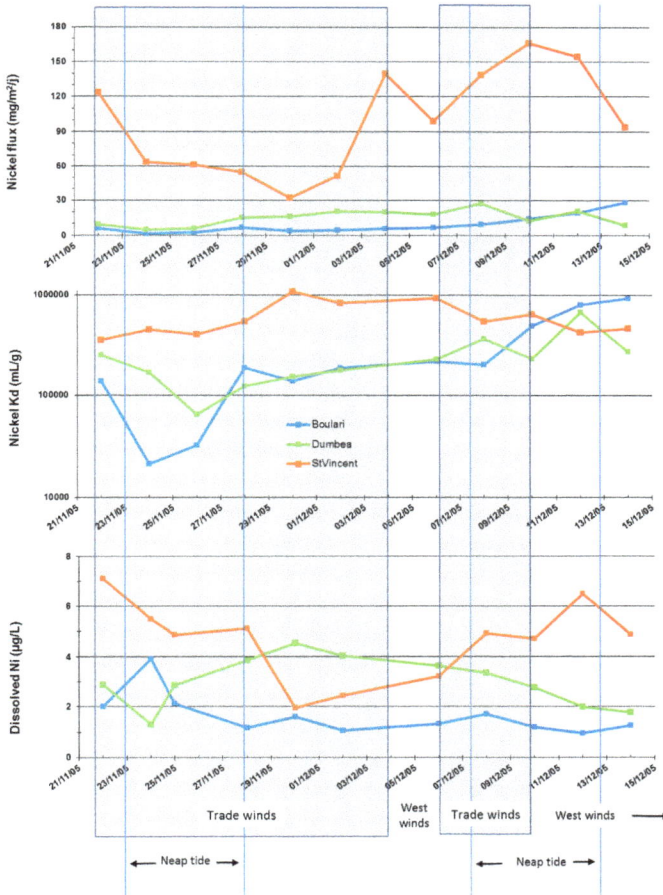

Figure 7. Particulate Ni flux, distribution constant (K_d) of Ni and dissolved concentration of Ni for Boulari, Dumbéa and St Vincent bays over the study period (21 November to 14 December 2005), 3 m above the seabed.

5.1.2. Dumbéa Bay

A strong correlation was observed between the metals and the major elements Si and Fe. For Mg, no correlation was found with the other analysed elements (Table 9). Mg concentrations were similar to those measured in Boulari Bay's SPM but two times lower than the ones observed in St Vincent Bay. This can be interpreted as a low contribution of smectite as shown by the XRD determinations (Figure 5). Regarding Ca, its concentrations in Dumbéa Bay were much higher than in Boulari Bay and is likely to be generated by strong resuspension of carbonated debris from numerous coral reef colonies, by trade winds upstream of the sampling area (Table 8). Indeed, Dumbéa Bay shelters fringing reefs and corals both, alive and dead, on its sea bottom which constitute an important source of carbonates compared to the Coulée River mouth [36]. The metals were only slightly correlated to Ca and this probably reflects the low residence time of seawaters in this bay [63].

The concentrations of metals determined in the SPM were averaged at the beginning of the study period (22 November), and correlated well with the presence of fine particles ($s = 3.45$); in spite of this, the turbidity remained low (1.3 FTU) (Figures 3 and 4). From the 24 of November onwards, the strong increase in median diameter ($D_{50} > 75$ μm) and reduction in the Junge parameter ($s < 3.36$) preceded a strong sedimentation (Figure 3); this increase in particle size led to a reduction in the metal concentrations in the SPM, a phenomenon identified by the decline in the metal distribution constants (K_d), for example Nickel (Figure 7).

The concentrations of particulate metals progressively increased with the reduction in turbidity and median diameter of SPM until the end of the study period. This phenomenon was probably due to the increase in the specific surface area of the particles. The turbidity and especially the SPM flux was correlated with the strength and direction of the wind while the bottom currents were quite low. Hence, the increase in the mass of SPM probably corresponded to the resuspension of carbonate particles originating from the fringing coral patches of shallow depth found south-east of the bay and subjected to trade winds, as shown by the significant increase in Ca concentrations observed (Figure 4). Until 11 December, the trade winds may have been the cause of occasional deposition of aeolian nickel dust generating the high metal concentration increase observed in the SPM; this dust is generated by the nickel processing SLN plant (Figure 1) located on the south-east coast of Dumbéa Bay. After that date, the westerly winds that blew until the end of the study period were probably responsible for the decrease in the concentration of metals (Figures 2 and 4).

5.1.3. St Vincent Bay

Except for Ca, a strong correlation was also observed between the metals and the major elements (Si, Fe and Mg) but the values of R^2 were slightly lower than in Boulari Bay (Table 9). The correlation is well explained by the present-day mining extraction of less weathered lateritic layers enriched with metals. The main difference with Boulari Bay is that a high correlation was observed between Mg and the metals and a lower correlation between Ca and the metals. XRD from St Vincent Bay samples showed the presence of smectite (Figure 5) not observed in Boulari Bay's SPM, which might explain the difference. SPM in St Vincent Bay was also enriched in Fe and Si (and Al, not presented in this paper) compared to the other bays, which might be explained by the higher proportion of clays. The high concentrations of Ca suggested a sizeable contribution of resuspended carbonates as a result of the regular effect of the winds in this shallow bay. Besides this, a significant proportion of former resuspended coral reef debris could explain the lower correlation of Ca with the metals in St Vincent Bay.

Over the study period, some variations were observed in the metal concentrations present in the SPM with no major trends evident and of smaller amplitude than in the 2 other bays. However, a clear correlation between K_d values and the tide was detected, with smaller K_d at neap tides than at spring tides (Figure 7). During spring tides, the resuspension of fine particles and subsequent adsorption of metals (K_d values) increased. Nevertheless, resuspension was not only caused by tides, but also by the wind regimes. Indeed, on the 24 November, an increase in turbidity and in the Junge parameter and a decrease in D_{50} highlighted a resuspension event at the end of a spring tide period, due to the re-establishment of trade winds (Figures 2, 3 and 7). Aggregation of suspended particles and subsequent deposition followed, but was interrupted by a short resuspension event on the 2 December, likely due to the combined effect of spring tides and waves generated by the wind. This resuspension likely induced the high amount of SPM collected on 4 December. Immediately after a short period of trade winds, aggregation and deposition were observed from 8 December during weak westerly and variable winds, hence promoting the deposition of aggregates.

5.2. Origin of the Minerals

All of the detrital minerals detected were previously described [50] from the weathering profiles of the plateaus. The predominance of clay minerals in St Vincent Bay and Boulari Bay may be attributed to the presence of, respectively, actual and former open cast mines in their watershed which erode the deeper lateritic horizons where clay minerals are for the most part located.

In all the lagoon sediments, carbonates minerals (calcite, aragonite, Mg carbonate), absent in the riverine sediments, provide evidence of sediment resuspension [26]. Other authors [64] also showed that in the different typical bottoms of the lagoon, more than 80% of total sedimentation was linked to deposition of resuspended benthic material. Suspended sediments present in St Vincent Bay contain the same main terrestrial minerals as those detected in the Tontouta River: quartz, goethite, talc, serpentine and smectite. However, smectite and serpentinite may also result from neoformation in the delta area or in the bay itself [26,50,65]; the presence of smectite was higher in St Vincent Bay than in the connected Tontouta River. In tBoulari Bay under trade wind conditions, SPM contained the same minerals as those found in St Vincent Bay except that goethite was more present than clays. During a west wind regime, no clay minerals were detected, and goethite and quartz were the only terrestrial minerals found. During the same period, SPM collected in Dumbéa Bay were characterized by the same minerals as in St Vincent Bay, but clays and goethite contents were lower and samples were dominated by lagoonal material (calcites and aragonite). These results therefore show that the mineralogical composition of suspended sediments in the 3 bays was not strictly related to the composition of sediments transported by their connected rivers. These findings may be due to the presence of authigenic minerals in the bay [26,36] besides detrital particles.

A chemical analysis of a goethite particle referenced as B8A1 (Table 11 and Figure 6) yields similar results to the one given by [50]. Chrysotile (a mineral from the serpentine family) particles forming long acicular tubes were also detected (referenced as C6A17 in Table 11 and Figure 6. This mineral is formed in fractures of the ultrabasites and results from an episode of serpentinisation which concentrates Ni [50]. Compared to other serpentinites analysed previously [50,60], the analysed chrysotile particle had a comparable chemical composition (Table 11). A particle with a different composition (less Al and Mg and more Fe) was also detected (referenced as B8A5 in Table 6 and Figure 6. The composition of this particle is comparable to a poorly crystallized smectite named crypto nontronite [50], which is found in sediments of the deltaic plain and probably originates from diagenesis.

The composition of the particle referenced B7A5 (Table 11 and Figure 6) is more intriguing because of its high Ni content. It could be comparable to a clay mineral like that of a serpentine phase with a high degree of Ni substitution, but such a composition has not yet been reported in New Caledonia. Neoformed serpentine has been identified [26] in the lagoonal sediments of Dumbéa Bay but was not quantified. In addition, the serpentinite was of the Fe (III) type and comparable to the authigenic green phyllosilicates described by Odin et al. [66] in the lagoonal sediments of New Caledonia. Authigenic clay minerals in the Amazon delta have also been described by Michalopoulos and Aller [67], who demonstrated that clay minerals may form rapidly in the sediment pores after liberation of Si from the diatoms and Al and Fe from the oxy-hydroxides derived from the drainage basin. The amount of diatoms or other biogenic silica sources is not known in New Caledonia but their presence has been detected here by TEM. Besides this, goethite is abundant. We therefore support the idea that the high amount of Ni in the clay particle B7A5 resulted from Ni incorporation in the structure of the clay during diagenesis. This statement implies that part of the dissolved Ni in the lagoon may be fixed by minerals, which limit its dissemination.

6. Conclusions

Our approach combining mineralogy, geochemistry and hydrodynamics allowed us to determine how driving factors are affecting the dynamics of particulate matter in lagoonal ecosystems influenced by the mining industry (Table 12). During the dry season, the concentrations of metal present in the water of the 3 bays were principally governed by the alternating south-easterly (trade winds) and westerly winds. The spring and neap tides do not appear to play a major role in the conditions observed during the study period, except in St Vincent Bay. The driving forces behind the resuspension of particles were similar in Boulari and Dumbéa bays, but clearly differed in St Vincent Bay. This difference can be attributed to the shallow depths present in the bay, the intense mixing and the resulting aggregation mechanisms. This resuspension phenomenon was responsible for the distribution of dissolved and particulate metals in the water column (K_d).

In St Vincent Bay, during periods of intense resuspension, the adsorption of Ni onto many particles was promoted and reversely, the concentration of dissolved nickel increased during the sedimentation phase as a result of calm meteorological conditions. In Boulari Bay, the sedimentation stages and constant Ni concentrations coincided with west weak wind periods allowing the coastal waters blocked along the coast-line by the long trade winds periods to flow off-shore. This phenomenon was reversed in Dumbéa Bay where the redissolution of Ni seemed to be higher during the period of resuspension of the particles richer in carbonates. Table 12 summarizes the effects of the different wind regimes in these 3 bays during the dry season.

Table 12. Effects of wind regimes on the dynamics of the particulate matter in lagoonal ecosystems influenced by the mining industry in New-Caledonia.

Bay	Trade Wind Regime (5–10 Knots)	Light West Wind Regime (<5 Knots)	Coastal Breeze Regime (<5 Knots)
Boulari Bay	Off-shore water inputs, resuspension of SPM transported eastwards and blocked, settling in-shore	Drainage of blocked coastal waters toward off-shore (westward), SPM aggregation and sedimentation	Settling of a benthic turbid layer and westwards transport of SPM: increase in metal fluxes (dissolved and particulate)
Dumbéa Bay	Resuspension of SPM rich in carbonates debris, followed by sedimentation	Off-shore water inputs low both in SPM and metal content	Sedimentation of SPM and reduction in metal fluxes (dissolved and particulate)
St Vincent Bay	Intense resuspension of SPM by the tide and winds over shallow water, then settling and high particulate metal flux	Sedimentation of a small fraction of SPM, reduction in metal fluxes	Important persistence of resuspension of SPM: high metal fluxes (dissolved and particulate)

In terms of environmental impact, the amounts of lateritic particles that have accumulated over time can modify the geochemical equilibriums in the water column, particularly in shallow and sheltered bays. Reducing the concentration of SPM injected into the lagoon seems essential to limit the effects of the bio-accumulation in exposed marine organisms, for example, dissolved Ni, up to 7 μg·L^{-1} in St Vincent Bay (vs. 2 μg·L^{-1} in Boulari Bay) correspond to the higher particulate Ni fluxes of 170 mg·m^{-2}·d^{-1} observed in the bay.

Acknowledgments: This work was supported by the Institut de Recherche pour le Développement. The authors are grateful to the diving IRD team (J.L. Menou, E. Folcher and C. Jeoffroy), the R/V Captains (M. Clarque, S. Tereua and N. Colombani), Alain Belhandouz and Jean Pierre Lamoureux for their assistance in the field trips and sample collections.

Author Contributions: Jean Michel Fernandez, Jean Dominique Meunier and Sylvain Ouillon conceived and designed the experiments; Jean Michel Fernandez, Benjamin Moreton performed the experiments;

Jean Michel Fernandez, Benjamin Moreton, Pascal Douillet, Olivier Grauby, Jean Dominique Meunier and Sylvain Ouillon analysed the data; Jean Michel Fernandez, Jean Dominique Meunier and Sylvain Ouillon wrote the paper.

Conflicts of Interest: The authors declare no conflict of interest.

References

1. Labrosse, P.; Fichez, R.; Farman, R.; Adams, T. New Caledonia. In *Seas at the Millenium, an Environmental Evaluation*; Sheppard, C., Ed.; Elsevier: Amsterdam, The Netherlands, 2000; Volume 2, pp. 723–736.
2. Pandolfi, J.M.; Bradbury, R.H.; Sala, E.; Hughes, T.P.; Bjorndal, K.A.; Cooke, R.G.; McArdle, D.; McClenachan, L.; Newman, M.J.H.; Paredes, G.; et al. Global trajectories of the long-term decline of coral reef ecosystems. *Science* **2003**, *301*, 955–958. [CrossRef] [PubMed]
3. Doney, S.C. The growing human footprint on coastal and open-ocean biogeochemistry. *Science* **2010**, *328*, 1512–1516. [CrossRef] [PubMed]
4. Burke, L.; Reytar, K.; Spalding, M.; Perry, A. *Reefs at Risk Revisited*; World Resources Institute: Washington, DC, USA, 2011; Available online: http://pdf.wri.org/reefs_at_risk_revisited.pdf (accessed on 3 January 2017).
5. Maina, J.; McClanahan, T.R.; Venus, V.; Ateweberhan, M.; Madin, J. Global gradients of coral exposure to environmental stresses and implications for local management. *PLoS ONE* **2011**, *6*, e23064. [CrossRef] [PubMed]
6. Brodie, J.E.; Kroon, F.J.; Schaffelke, B.; Wolanski, E.C.; Lewis, S.E.; Devlin, M.J.; Bohnet, I.C.; Bainbridge, Z.T.; Waterhouse, J.; Davis, A.M. Terrestrial pollutant runoff to the Great Barrier Reef: An update of issues, priorities and management responses. *Mar. Poll. Bull.* **2012**, *65*, 81–100. [CrossRef] [PubMed]
7. De'ath, G.; Fabricius, K.E.; Sweatman, H.; Puotinen, M. The 27-year decline of coral cover on the Great Barrier Reef and its causes. *PNAS* **2012**, *109*, 17995–17999. [CrossRef] [PubMed]
8. Erftemeijer, P.L.A.; Riegl, B.; Hoeksema, B.W.; Todd, P.A. Environmental impacts of dredging and other sediment disturbances on corals: A review. *Mar Poll. Bull.* **2012**, *64*, 1737–1765. [CrossRef] [PubMed]
9. Morrison, R.J.; Denton, G.R.W.; Bale Tamata, U.; Grignon, J. Anthropogenic biogeochemical impacts on coral reefs in the Pacific Islands-An overview. *Deep-Sea Res. II* **2013**, *96*, 5–12. [CrossRef]
10. Myers, N.; Mittermeier, R.A.; Mittermeier, C.G.; da Fonseca, G.A.B.; Kent, J. Biodiversity hotspots for conservation priorities. *Nature* **2000**, *403*, 853–858. [CrossRef] [PubMed]
11. Alongi, D.M. Present state and future of the world's mangrove forests. *Environ. Conserv.* **2002**, *29*, 331–349. [CrossRef]
12. Bouchet, P.; Lozouet, P.; Maestrati, P.; Heros, V. Assessing the magnitude of species richness in tropical marine environments: Exceptionally high numbers of molluscs at a New Caledonia site. *Biol. J. Linn. Soc.* **2002**, *75*, 421–436. [CrossRef]
13. Nagelkerken, I.; Blaber, S.J.M.; Bouillon, S.; Green, P.; Haywood, M.; Kirton, L.G.; Meynecke, J.O.; Pawlik, J.; Penrose, H.M.; Sasekumar, A.; et al. The habitat function of mangroves for terrestrial and marine fauna: A review. *Aquat. Bot.* **2008**, *89*, 155–185. [CrossRef]
14. Adjeroud, M.; Fernandez, J.M.; Caroll, A.G.; Harrison, P.L.; Penin, L. Spatial patterns and recruitment processes of coral assemblages among contrasting environmental conditions in the southwestern lagoon of New Caledonia. *Mar. Poll. Bull.* **2010**, *61*, 375–386. [CrossRef] [PubMed]
15. Losfeld, G.; L'Huillier, L.; Fogliani, B.; Jaffré, T.; Grison, C. Mining in New Caledonia: Environmental stakes and restoration opportunities. *Environ. Sci. Pollut. Res.* **2015**, *22*, 5592–5607. [CrossRef] [PubMed]
16. Hédouin, L.; Bustamante, P.; Fichez, R.; Warnau, M. The tropical brown alga Lobophora variegate as a bioindicator of mining contamination in the New Caledonia lagoon: A field transplantation study. *Mar. Environ. Res.* **2008**, *66*, 438–444. [CrossRef] [PubMed]
17. Metian, M.; Bustamante, P.; Hédouin, L.; Warnau, M. Accumulation of nine metals and one metalloid in the tropical scallop Comptopallium radula from coral reefs in New Caledonia. *Environ. Pollut. (Oxford, UK)* **2008**, *152*, 543–552. [CrossRef] [PubMed]
18. Hédouin, L.; Bustamante, P.; Churlaud, C.; Pringault, O.; Fichez, R.; Warnau, M. Trends in concentrations of selected metalloid and metals in two bivalves from the SW lagoon of New Caledonia. *Ecotoxicol. Environ. Saf.* **2009**, *72*, 372–381. [CrossRef] [PubMed]

19. Debenay, J.P.; Fernandez, J.M. Benthic foraminifera records of complex anthropogenic environmental changes combined with geochemical data in a tropical bay of New Caledonia. *Mar. Poll. Bull.* **2009**, *59*, 311–322. [CrossRef] [PubMed]

20. Bonnet, X.; Briand, M.; Brischoux, F.; Letourneur, Y.; Fauvel, T.; Bustamante, P. Anguilliform fish reveal large scale contamination by mine trace elements in the coral reefs of New Caledonia. *Sci. Total Environ.* **2014**, *470–471*, 876–882. [CrossRef] [PubMed]

21. Cuif, M.; Kaplan, D.M.; Lefèvre, J.; Faure, V.M.; Caillaud, M.; Verley, P.; Vigliola, L.; Lett, C. Wind-induced variability in larval retention in a coral reef system: A biophysical modelling study in the South-West Lagoon of New Caledonia. *Prog. Oceanogr.* **2014**, *122*, 105–115. [CrossRef]

22. Gilbert, A.; Heintz, T.; Hoeksema, B.W.; Benzoni, F.; Fernandez, J.M.; Fauvelot, C.; Andrefouet, S. Endangered New Caledonian endemic mushroom coral Cantharellus noumeae in turbid, metal-rich, natural and artificial environments. *Mar. Poll. Bull.* **2015**, *100*, 359–369. [CrossRef] [PubMed]

23. Heintz, T.; Haapkylä, J.; Gilbert, A. Coral health on reefs near mining sites in New Caledonia. *Dis. Aquat. Org.* **2015**, *115*, 165–173. [CrossRef] [PubMed]

24. Dugas, F. La sedimentation en baie de St Vincent (Côte ouest de la Nouvelle-Caledonie). In *Cah. ORSTOM, ser. Géol.*; Office de la recherche scientifique et technique outre-mer (ORSTOM): Paris, France, 1974; Volume VI, pp. 41–62.

25. Bird, E.C.F.; Dubois, J.P.; Iltis, J.A. *The Impact of Opencast Mining on the Rivers and Coasts of New Caledonia*; The United Nation University: Shibuya, Japan, 1984; Available online: http://archive.unu.edu/unupress/unupbooks/80505e/80505E00.htm (accessed on 9 May 2017).

26. Ambatsian, P.; Fernex, F.; Bernant, M.; Parron, C.; Lecolle, J. High metal inputs to close seas: The New-Caledonia Lagoon. *J. Geochem. Explor.* **1997**, *59*, 59–74. [CrossRef]

27. Fernandez, J.-M.; Ouillon, S.; Chevillon, C.; Douillet, P.; Fichez, R.; Le Gendre, R. A combined modelling and geochemical study of the fate of terrigenous inputs from mixed natural and mining sources in a coral reef lagoon (New Caledonia). *Mar. Poll. Bull.* **2006**, *52*, 320–331. [CrossRef] [PubMed]

28. Fichez, R.; Adjeroud, M.; Bozec, Y.M.; Breau, L.; Chancerelle, Y.; Chevillon, C.; Douillet, P.; Fernandez, J.M.; Frouin, P.; Kulbicki, M.; et al. A review of selected indicators of particle, nutrient and metal input in coral reef lagoon systems. *Aquat. Living Res.* **2005**, *18*, 125–147. [CrossRef]

29. Grenz, C.; Le Borgne, R.; Fichez, R.; Torreton, J.P. Tropical lagoon multidisciplinary investigations: An overview of the PNEC New Caledonia pilot site. *Mar. Poll. Bull.* **2010**, *61*, 267–268. [CrossRef] [PubMed]

30. A collective of 75 authors. *Atlas de la Nouvelle Calédonie*, IRD ed.; Institut de Recherche pour le Développement (IRD): Marseille, France, 2013; ISBN: 978-2-7099-1740-7.

31. Breau, L. Extractions Séquentielles et Analyses de Métaux dans une Carotte de Sédiments Lagonaires Datée: Mise en Evidence de L'évolution des Apports Terrigènes Liée aux Activités Humaines au Cours des 150 Dernières Années. Master's Thesis, University Aix-Marseille II, Marseille, France, 1998.

32. Magand, O. Contribution à la Modélisation du Transport de la Matière Particulaire dans le Lagon Sud-Ouest de Nouvelle-Calédonie: Etude des flux, Détermination des Signatures Minéralogiques et Evolution Spatiale. Master's Thesis, University Perpignan, Perpignan, France, 1998.

33. Magand, O. Recherche et Définition des Signatures Géochimiques (Métaux Lourds et Lanthanides) des Sources Terrigènes du Lagon Sud-Ouest de Nouvelle-Calédonie. Master's Thesis, University Aix-Marseille II, Marseille, France, 1999.

34. Fernandez, J.-M.; Moreton, B.; Fichez, R.; Breau, L.; Magand, O.; Badie, C. Advantages of combining ^{210}Pb and geochemical signature determinations in sediment record studies, application to coral reef lagoon environments. In *Environmental Changes and Radioactive Tracers*, IRD ed.; Fernandez, J.-M., Fichez, R., Eds.; Institut de Recherche pour le Développement (IRD): Paris, France, 2002; pp. 187–200.

35. Moreton, B.M.; Fernandez, J.-M.; Dolbecq, M.B.D. Development of a field preconcentration/elution unit for routine determination of dissolved metal concentrations by ICP-OES in marine waters: Application for monitoring of the New Caledonia lagoon. *Geostand. Geoanal. Res.* **2009**, *33*, 205–218. [CrossRef]

36. Ouillon, S.; Douillet, P.; Lefebvre, J.P.; Le Gendre, R.; Jouon, A.; Bonneton, P.; Fernandez, J.M.; Chevillon, C.; Magand, O.; Lefèvre, J.; et al. Circulation and suspended sediment transport in a coral reef lagoon: The southwest lagoon of New Caledonia. *Mar. Poll. Bull.* **2010**, *61*, 269–296. [CrossRef] [PubMed]

37. Fichez, R.; Chifflet, S.; Douillet, P.; Gérard, P.; Gutierrez, F.; Jouon, A.; Ouillon, S.; Grenz, C. Biogeochemical typology and temporal variability of lagoon waters in a coral reef ecosystem subject to terrigeneous and anthropogenic inputs (New Caledonia). *Mar. Poll. Bull.* **2010**, *61*, 309–322. [CrossRef] [PubMed]

38. Douillet, P. Tidal dynamics of the south-west lagoon of New Caledonia: Observations and 2D numerical modelling. *Oceanol. Acta* **1998**, *21*, 69–79. [CrossRef]

39. Douillet, P.; Ouillon, S.; Cordier, E. A numerical model for fine suspended sediment transport in the south-west lagoon of New-Caledonia. *Coral Reefs* **2001**, *20*, 361–372. [CrossRef]

40. Jouon, A.; Lefebvre, J.P.; Douillet, P.; Ouillon, S.; Schmied, L. Wind wave measurements and modelling in a fetch-limited semi-enclosed lagoon. *Coast. Eng.* **2009**, *56*, 599–608. [CrossRef]

41. Andréfouët, S.; Cabioch, G.; Flamand, B.; Pelletier, B. A reappraisal of the diversity of geomorphological and genetic processes of New Caledonian coral reefs: A synthesis from optical remote sensing, coring and acoustic multibeam observations. *Coral Reefs* **2009**, *28*, 691–707. [CrossRef]

42. Lillie, A.R.; Brothers, R.N. The geology of New-Caledonia. *N. Zeal. J. Geol. Geophys.* **1970**, *13*, 159–167. [CrossRef]

43. Paris, J.P. *Les Ressources Minérales de Nouvelle-Calédonie*; Mémoire 113, Bureau de Recherches Géologiques et Minières (BRGM): Orléans, France, 1981.

44. Antheaume, B. Chronique de l'Atlas de la Nouvelle-Calédonie, un bilan méthodologique et critique. Cahiers ORSTOM. *Série Schum* **1981**, *18*, 389–398.

45. Paris, J.P. Gîtes minéraux et substances utiles. In *Atlas de Nouvelle-Calédonie*; Sautter, G., Ed.; ORSTOM: Paris, France, 1981.

46. Perrier, N.; Ambrosi, J.P.; Colin, F.; Gilkes, R.J. Biogeochemistry of a regolith: The new Caledonian Koniambo ultramafic massif. *J. Geochem. Explor.* **2006**, *88*, 54–58. [CrossRef]

47. Dublet, G.; Fandeur, D.; Juillot, F.; Morin, G.; Ambrosi, J.P.; Fritsch, E.; Brown, G.E., Jr. Ni speciation in a New Caledonian lateritic regolith: A quantitative X-ray absorption spectroscopy investigation. *Geochim. Cosmochim. Acta* **2012**, *95*, 119–133. [CrossRef]

48. Dublet, G.; Juillot, F.; Morin, G.; Fritsch, E.; Noel, V.; Brest, J.; Brown, G.E., Jr. XAS evidence for Ni sequestration by siderite in a latéritic Ni-deposit from New Caledonia. *Am. Miner.* **2014**, *99*, 225–234. [CrossRef]

49. Dublet, G.; Juillot, F.; Morin, G.; Fritsch, E.; Fandeur, D.; Brown, G.E., Jr. Goethite aging explains Ni depletion in upper units of ultramafic lateritic ores. *Geoch. Cosmoch. Acta* **2015**, *160*, 1–15. [CrossRef]

50. Trescases, J.J. L'évolution Geochimique Supergène des Roches Ultrabasiques en Zone Tropicale. Formation des Gisements Nickélifères de Nouvelle-Calédonie. Ph.D. Thesis, University Louis Pasteur, Strasbourg, France, 1973.

51. Pesin, E.; Blaize, S.; Lacoste, D. *Atlas Climatique de la Nouvelle Calédonie*; Météo France: Nouméa, New Caledonia, 1995.

52. Baltzer, F.; Trescases, J.J. Erosion, transport et sédimentation liés aux cyclones tropicaux dans les massifs d'ultrabasites de Nouvelle-Calédonie. *Cahiers ORSTOM Série Géologie III* **1973**, *2*, 221–244.

53. Dugas, F.; Debenay, J.P. *Carte Sédimentologique et carte Annexe du Lagon de Nouvelle Calédonie au 1/50000—Mont Dore, Tontouta, Prony, Nouméa*; Notices explicatives n° 76, 86, 91 and 95; ORSTOM: Paris, France, 1982.

54. Heussner, S.; Ratti, C.; Carbonne, J. The PPS3 times series sediment trap and the trap sample processing techniques used during the ECOMARGE experiment. *Cont. Shelf Res.* **1990**, *10*, 943–958. [CrossRef]

55. Jouon, A.; Ouillon, S.; Douillet, P.; Lefebvre, J.P.; Fernandez, J.-M.; Mari, X.; Froidefond, J.M. Spatio-temporal variability in suspended particulate matter concentration and the role of aggregation on size distribution in a coral reef lagoon. *Mar. Geol.* **2008**, *256*, 36–48. [CrossRef]

56. Lefebvre, J.-P.; Ouillon, S.; Vinh, V.D.; Arfi, R.; Panche, J.-Y.; Mari, X.; Thuoc, C.V.; Torreton, J.P. Seasonal variability of cohesive sediment aggregation in the Bach Dang-Cam Estuary, Haiphong (Vietnam). *Geo-Mar. Lett.* **2012**, *32*, 103–121. [CrossRef]

57. Mobley, C.D. *Light and Water Radiative Transfer in Natural Water*; Academic Press: San Diego, CA, USA, 1994.

58. Pinet, S.; Martinez, J.-M.; Ouillon, S.; Lartiges, B.; Villar, R.E. Variability of apparent and inherent optical properties of sediment-laden waters in large river basins—Lessons from in situ measurements and bio-optical modeling. *Opt. Express* **2017**, *25*, A283–A310. [CrossRef] [PubMed]

59. Cliff, G.; Lorimer, G.W. The quantitative analysis of thin specimens. *J. Microsc.* **1975**, *103*, 203–207. [CrossRef]

60. Manceau, A.; Calas, G. heterogneous distribution of nickel in hydrous silicates from New Caledonia ore deposits. *Am. Miner.* **1985**, *70*, 549–558.

61. Pelletier, B. Serpentines in Nickel Silicate Ore from New-Caledonia. In Proceedings of the Publications-Australasian Institute of Mining and Metallurgy, Kalgoolie, Australia, 27–29 November 1996; Volume 27–29, pp. 197–206.

62. Wartel, M.; Skirer, M.; Auger, T.; Boughriet, A. Interaction of manganese II with carbonates in sea water: Assessment of the solubility product of $MnCO_3$ and Mn distribution coefficient between the liquid phase and $CaCO_3$ particles. *Mar. Chem.* **1990**, *29*, 99–117. [CrossRef]

63. Jouon, A.; Douillet, P.; Ouillon, S.; Fraunié, P. Calculations of hydrodynamic time parameters in a semi-opened coastal zone using a 3D hydrodynamic model. *Cont. Shelf Res.* **2006**, *26*, 1395–1415. [CrossRef]

64. Clavier, J.; Chardy, P.; Chevillon, C. Sedimentation of particulate matter in the South-west lagoon of New Caledonia: Spatial and Temporal Patterns. *Est. Coast. Shelf Sci.* **1995**, *40*, 281–294. [CrossRef]

65. Baltzer, F. *Géodynamique de la Sédimentation et Diagénèse Précoce en Domaine Ultrabasique, Nouvelle-Calédonie*; ORSTOM: Paris, France, 1982.

66. Odin, G.S. *Green Marine Clays*; Elsevier: Amsterdam, The Netherlands, 1988.

67. Michalopoulos, P.; Aller, R.C. Rapid clay mineral formation in Amazon delta sediments: Reverse weathering and oceanic elemental cycles. *Science* **1995**, *270*, 614–617. [CrossRef]

MDPI

St. Alban-Anlage 66

4052 Basel

Switzerland

Tel. +41 61 683 77 34

Fax +41 61 302 89 18

www.mdpi.com

Water Editorial Office

E-mail: water@mdpi.com

www.mdpi.com/journal/water

www.ingramcontent.com/pod-product-compliance
Lightning Source LLC
Chambersburg PA
CBHW051721210326
41597CB00032B/5557